J.S. Knowlson

— 1981 —

PREFACE TO THE SECOND EDITION

Since publishing, in 1963, the first textbook on Project Planning and Control, "FUNDAMENTALS OF PROJECT PLANNING AND CONTROL" (Library of Congress Catalog Card No. 63-13195), the author has accumulated additional insight into, and practical information on the application of the technique of Network Planning (the most common generic name for planning projects by the CRITICAL PATH METHOD (CPM) and PERT (Program Evaluation and Review Technique) to a variety of projects as a practicing consultant in this field. This edition is an expansion of the first basic textbook, with the incorporation of new concepts (Invoiceless Cost Control and the Continuous Milestone Chart); observations (Current Abuses of the Technique, the Negative Float/Slack Fallacy, the limitations of Precedence Diagramming) and re-evaluations (The Time Scale Chart recommendations).

This book is an attempt to present to the line manager the fundamental methodology of a scientific information system. The advanced mathematical concepts of parametric linear programming are left to others. A bibliography is included for those who would be interested in the formal construction and derivation of the technique.

The updating of the material herein is the result of consulting work on over 150 projects of various types and magnitudes, ranging from a small local political campaign effort to the engineering and construction of a $200 million dollar nuclear power plant. Appearance as an expert witness in several arbitration and litigation cases accounts for the thoughts on the use of the technique in litigation and claims. Finally, the author subscribes fully to the Aristotlian theory that, "The Teacher learns more than the Pupil"; the fielding of questions from practical minded engineers and executives in over 100 seminars has forced him to defend, re-evaluate and clarify all aspects of the material herein, from the rudiments to the advanced concepts of multi-project scheduling.

The essence of any business "control" system is the timely, accurate and explicit flow of information. "Control" is a decision-making process, made by humans or groups of humans in concert. The process is contingent on judgements made from available information. The computing machine serves only as a tool to expedite information flow, and its basic manipulations. The methodology unfolded herein is primarily a manual one; allowing the manager to use it on projects of reasonable size. Information is included to determine size, time and cost boundaries, beyond which use of Electronic Data Processing equipment is recommended.

The style of presentation is informal. By making the material more readable, and less pedantic, it was felt that the reader would gain more confidence and understanding in a technique that is really relatively simple.

A. James Waldron
Haddonfield, N. J.
December, 1968

TABLE OF CONTENTS

CHAPTER 1

INTRODUCTION

The modern economic world, of which we are a part, has come to its present form because most individuals have made themselves specialists to some degree, and because large amounts of wealth and resources have been concentrated on the production of things that are demanded by the public. The former "Jack of all trades" has disappeared, and each individual, even in routine tasks, has become more efficient. In this way, each of us can produce far more in a given period of time than our forebears. Even though extremely large amounts of capital have been employed, it has been necessary to keep selling prices down to survive in competitive markets. This has created the situation in which the use of modern technologies must be efficient in order to create goods and services of high quality at maximum rates with minimum rejects. Modern technology has been extended into the management field — thus the rise of the term "Management Sciences".

The manager in this milieu, responsible for the planning and control of such operations, found himself in last decade with inadequate managerial tools and systems. Traditional methods started revealing defects in their nature and capabilities. In the growth of our modern economy, several factors came into focus as causes of severe burden to the manager or supervisor's ability to plan, execute and control the areas of his responsibility.

They are:

1. A "Project" Oriented Economy. More and more business and governmental ventures become "projects" in nature—that is, the undertaking has a specific, unique goal, to be accomplished at a specific target date. Even production plans in manufacturing have a definite goal, and a time boundary.

2. Increasing Complexity and Size of Projects. The advance of our technology has created more and more fields of narrower and narrower technical specialties. The manager is constrained to guide and direct more and more specialist, down the road towards his specific, unique goal. The magnitude of the project, in capital cost, amount of resources applied and expended, has grown almost exponentially. Since more and more specialists are required to contribute in their relatively narrower fields, ancillary problems of communication and supporting services has increased the complexity of administration. Collection, dissemination and evaluation

of information in order to obtain time and cost status, and trends of the project has become a problem of the first order to the manager. In addition, due to a wide diversity of tax and other governmental regulations, the reporting and control of expenditures, performance, acquisition of assets; all have become more involved, and affects the complexity of doing business.

3. Increasing Time Pressure. Specific target dates are often set in advance; usually without top management's knowledge of the extent or complexity of the project. This is due to economic market pressures or transiency, or a necessary defense or social posture on the part of the government. These milestone dates, by fiat, sometimes capricious, severely limit the manager in the use of a major resource — time. He is hard constrained to optimize his other resources of men, money and machines to effectively execute his project within fixed time boundaries.

4. Increasing Profit Pressure. With increasing tax regulation of corporate and company profits, the only large area to turn to in order to improve, or even maintain a reasonable return on assets employed, is the area of cost control. For every dollar reduction in direct costs on a project, up to an equivalent 67 dollars (depending on the nature of the company's business) in new sales must be brought in to show the same return. Time and costs are intimately related, as investigation of this new technique of project planning and control will reveal. Traditional methods can only give the manager a crude approximation of the relationships.

A traditional planning tool, long in use, has been the bar or Gantt Chart, Fig. 1.0. This represents, on a calendar time scale, estimated durations of activities on the project, their hoped for starting times and some gross form of relationship.

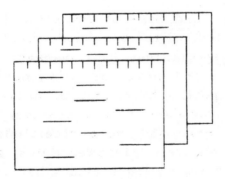

Fig. 1.0

The length of the bars indicate rough time estimates, and generally show a gross activity. For example, bar #1 may indicate an engineering activity on a project, and bar #2 a drafting or design effort. It shows a partial dependency; someone estimates that after a certain period of engineering effort, sufficient information will have been generated to allow the start of drafting activities. This type of planning is extended for every major operation on the project.

One shortcoming is the difficulty of seeing immediately and exactly if and how the overall project duration will be affected by changes in any particular activity. A slippage, a failure to obtain approval on time, contingencies, an extension in scope of an activity, while resulting in change in a bar length, cannot specifically inform the manager of consequential effects on other activities, or on the over all duration. He does not have a specific enough model to allow trade offs of men or machines; if there is an overtime requirement, he cannot predict exactly who should so be expedited and by how much.

The bar chart, while large in size or numbers, is inefficient. It does not allow a manager to generally plan every group's work in detail, but rather on a gross, general basis. His costs for each activity, no matter how fine, cannot be based on the bar chart. Economic conditions, such as latest possible time to shut down a profit producing plant, or machine, or facility in a plant; cannot be specifically planned ahead. The manager cannot adequately plan specific physical operations, such as the last possible time to pour concrete, or introduce catalyst, etc., from the bar chart. He can only guess at the affect of having limits on his resources of men, money or machines.

Thus, the manager has to think and act in generalities. Because of the limitations of the "old fashioned" methods, he cannot be aware in detail of the whole of an operation and all of its parts. He cannot visualize, at any given time, the exact state of progress. He is incapable of anticipating delays or other troubles soon enough to forestall them. He is not always aware of those activities which are critical and which, therefore, require prior prediction, special anticipation, and supervision. He does not always know precisely how a delay or failure in one activity will affect others following it. When following activities are affected, the ability to meet subsequent key milestone deadlines is affected, and hence, the success of the operation as a whole.

Because the critical activities or areas in any multi-activity project cannot always be readily ascertained, one common solution is to treat all activities on any project in the same manner, with the same expediting effort and same executive attention, resulting in lower efficiencies at every level of management.

Traditional methods of planning and (concurrently) scheduling do not show specifically which activities to control and expedite, and by how much. The anticipation of contingencies (delivery slippages, bad weather, situations of "Force Majeure", etc.) is usually handled by the addition of safety factors, by the "packing" of estimates towards a highly conservative number. Negotiations are started from this basis, and a final "intuitive" judgement as to realistic durations of activities becomes the ultimate criterion for acceptable contractual conditions.

With the occurrence of a contingency, such as a real or imagined slippage of the status of the project, compared against a realistic or theoretical datum (the widely used "S" or Gompertz Curve of accumulated hours and/or direct cost dollars), the only plausible decision of a manager is a massive, general acceleration of effort. The notorious "panic button" philosophy is applied. From consideration of traditional managerial techniques and systems, this is the only rationalized solution. A technique that is selective; that immediately picks out the chain of activities that directly affects the attainment of a desired or contractual target date; and then further selects the optimum of the least additional expenditure of capital and resources to make up lost time, would prove of considerable value to management. Managerial strategies will vary with the time and cost importance of the affected operations.

It became clear that a simple but logical Master Plan was first needed, one which would provide the executive with an accurate, timely, explicit and easily understood picture of his project at any and all times. This model plan should follow a uniform system which would be understood by all levels of supervisory personnel.

Traditional methods do not provide a quick and effective means by which an executive can see at a glance the over-all plan and the progress achieved to date. This is particularly true in a "turnover" during the progress of any particular project, or over-all corporate program. The loss of a key man, for whatever reason, causes the replacement a great deal of effort to learn the nature of the project, its exact status, and indicated future trouble spots. If they had such a model, management would quickly grasp the scope of the project and the exact function of each element. They could also quickly determine which activities are likely to be in difficulty and therefore to require special watching as the plan unfolds. Not only would such a model assist a supervisor in reaching decisions, it would also promote understanding and coordination among other management officials. It would enable executives to visualize operations in considerably more detail and with much more accuracy than is possible by other means. In addition, it would simplify the tasks of briefing and issuing orders.

In the negotiation of claims for "extras", intuitive judgment has been used to determine, and defend the additional costs involved, and the extension of time. A real need has developed for a graphic, unequivocal analysis of causes and effects – a method of analysis that will stand up in civil litigation of claims, if need be.

The technique of "network planning", described herein, is the solution to the managerial problems discussed above.

The history of "network planning" (PERT, Critical Path) starts about 1957-1958 for the first practical applications. It is difficult to specifically determine which system occured first in time, due to conflicting claims in the literature. The difference in time is so slight that it can be assumed that PERT and Critical Path were developed almost concurrently. Originally, the fundamentals of the "logic" network, and some of the basic calculation methods, were identical in both. However, the original objectives of PERT and Critical Path were poles apart. PERT started as an event oriented, probablistic system, attempting to predict the probability of attaining a specified target date. Critical Path started as an activity oriented, deterministic system, generating a schedule on a project for which there is sufficient historical and experience information. (Refer to the GLOSSARY OF TERMS AND SYMBOLS for the contextual meaning of "event" and "activity". There is a distinct difference in meaning in network techniques between these terms–daily usage quite often results in synonymous meanings).

The first large project on which PERT was employed, was by the U. S. Navy's Special Projects Office on The Polaris program. This extremely large program entailed the co-ordination and execution of some 60,000 time consuming operations, involving some 3800 major contractors. The Polaris program's goal, of making militarily operational a nuclear powered submarine capable of launching an intercontinental ballistic missile from below the surface of the ocean, was accomplished some 18 months ahead of schedule.

The first large project on which an integrated Critical Path system was employed involved the construction of a large synthetic fibre plant. A team from Remington Rand and E. I. duPont deNemours Co. devised and executed the system. The results were completion of a production unit well ahead of an estimated target date, generating a large flow of unanticipated income from the product sales.

With increasing usage of both techniques over the past three years, the systems have evolved into almost a single generic system. PERT is definitely trending towards the activity orientations of the original Critical Path (Refer to "DOD and NASA Guide, PERT COST SYSTEMS DESIGN," June 1962, issued by the Office of the Secretary of Defense).

This text shall consider just a single system of network planning, taking the best features from PERT and Critical Path, and combining them into an extremely powerful information system for the manager.

PROBLEM 1.1 Use whatever methods you currently employ in estimating the length of a project, and determine how long, in working days, it will take to complete the building of the house described below. Accept the activity descriptions below, do not break them down into finer or more detailed operations.

BUILDING A NEW HOME

This project involves the construction of a new home in a developed area. It is a two story Cape Cod design, with a field stone veneer to the roof. The incoming utility services (gas, water, sewer and electrical power) are to run across the property before the installation of the sidewalks and driveway. Windows and exterior doors are put in after the outside masonry work is finished. The house is to be closed in (roof on, windows and doors in) before the erection of the interior dry walls and painting. The basement walls are cinder block and the field stone covers the basement wall, starting 12" below grade elevation.

The activities involved are listed below, not necessarily in logical or chronological sequence. The duration of the activities are in working days (assume the contractor's proposal has provided for adequate crew sizes to accomplish the activity in the listed duration time).

Activity	Duration (Working Days)
1. Sign contract	0
2. Layout and excavate for building	3
3. Construct basement walls and backfill outside	5
4. Install outside water, gas, sewer lines and electrical power	4
5. Frame first floor to second floor	5
6. Stone veneer to first floor sill	4
.7. Stone veneer to eaves	7
8. Frame and shingle roof	8
9. Frame second floor to roof	6
10. Pour and cure basement floor slab	3
11. Install furnace, water heater, basement plumbing	10
12. Erect interior first floor walls, and rough flooring	2
13. Install first floor plumbing, ducts, wiring	7
14. Install windows, exterior doors	4
15. Install sidewalk, curbs, driveway	12

16. Erect interior second floor walls, and rough flooring   3
17. Install second floor plumbing, ducts, wiring   9
18. Put up inside drywalls and paint   10
19. Install kitchen equipment   7
20. Grade site, sod and landscape   10
21. Finish flooring   5
22. Install millwork, interior doors   4
23. Finish plumbing, electrical tie-ins   3
24. Punch list fix up and inspection   4

FUNDAMENTALS OF PLANNING

It was pointed out in chapter 1 that the traditional Gantt Chart is inadequate as a project planning and control tool. Time and costs are related on a project, and the bar chart at best will produce crude approximations. A second defect in the bar chart is the simultaneous planning and scheduling feature; the bars are listed in a rough serial and parallel sequence, and the length of the bar is a schedule. It indicates an approximate starting and finish date, since it is placed on a time scale.

The technique of "network planning" is based on a simple but logical thesis: planning must be separated from scheduling. The logical sequence and inter dependencies of all activities (time consuming operations, tasks, functions) must be simply and graphically shown. Once such a plan is created, it can be rapidly reviewed by all levels of management concerned, and ultimately revised to general acceptance of the logical sequence and dependencies of all activities.

The technique starts on a chart; each activity is represented by an arrow. There is no time scale. This is not a vector technique, the length of the arrow is meaningless.

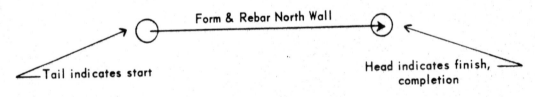

Fig. 2.0 Activity Arrow

The network is usually started from the beginning and the arrows are connected to show the logical flow of work, as each operation is added to the diagram, we examine it in relation to the other operations, and ask ourselves the following questions:

1. What other activity(s) must be completed before this activity can start? (Precedence)

2. What other activity(s) can be done <u>while</u> this activity is being done? (Concurrence)

3. What activity(s) cannot start until after this activity is done? (Subsequence)

Note that concurrence here is only a logical relationship, not one of time. Time durations are not assigned in this phase, only the logical execution of the individual activities. The fact that one operation may take 2 days, and a concurrent activity may take 6 weeks does not enter into the pictorial display. The key questions to be asked and answered, about a particular arrow (activity) are:

(1) Logically, what is the earliest (in terms of completion, or head of arrow, of precedent activities "arrows") point in the project that this particular operation or activity <u>could</u> start - the location in the diagram of the tail of this particular arrow.

(2) Logically, when must this particular operation or activity end (the location of the head of this particular arrow in the diagram).

Some examples of the development are as follows:

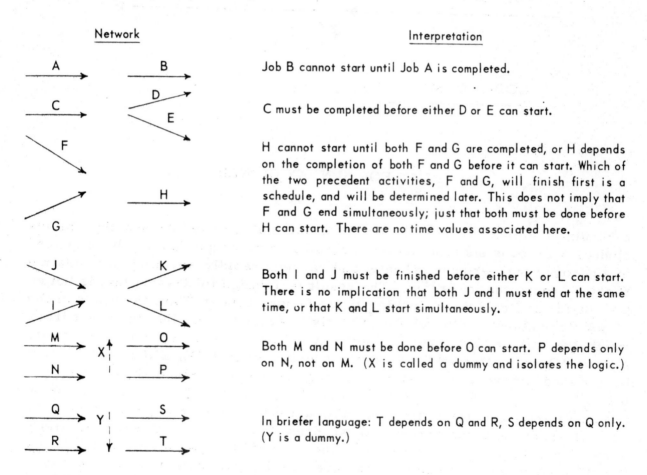

Network	Interpretation
	Job B cannot start until Job A is completed.
	C must be completed before either D or E can start.
	H cannot start until both F and G are completed, or H depends on the completion of both F and G before it can start. Which of the two precedent activities, F and G, will finish first is a schedule, and will be determined later. This does not imply that F and G end simultaneously; just that both must be done before H can start. There are no time values associated here.
	Both I and J must be finished before either K or L can start. There is no implication that both J and I must end at the same time, or that K and L start simultaneously.
	Both M and N must be done before O can start. P depends only on N, not on M. (X is called a dummy and isolates the logic.)
	In briefer language: T depends on Q and R, S depends on Q only. (Y is a dummy.)

The juncture points in the diagram where arrows come together, and leave, are called "EVENTS", and will become Time boundaries for the activities entering and leaving them.

9

To develop the method of this technique, the plan for a Sunday Dinner will be considered below. The main course will consist of a roast with vegetables. The roast and the vegetables must both be cooked before the dinner can be served and eaten as the plan indicates. Physically, the vegetables will be cooked on the top of the range, and the roast in the oven. Note that several options are available to the cook. (a) The roast may cook for three or four hours and be kept warm in the oven while the vegetables are being cooked, or (b) the vegetables may be started, cooked for their 45 minute duration, and kept warm while the roast finishes cooking. In the planning phase we are not concerned with when each operation starts or finishes--that is a schedule. All we are concerned with is the logical sequence of operations. At this point the only thing concerning us is that both the roast and the vegetables must be cooked before the dinner can be served and eaten. This is what the PLAN below indicates.

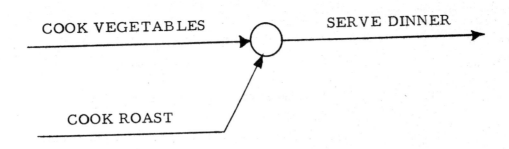

PLAN FOR SUNDAY DINNER

Extending the plan further, the exotic dessert of a Baked Alaska shall be included. This consists of an ice cream roll in a meringue layer. When placed in a hot oven the meringue is heated and hardens rapidly, forming an insulating barrier around the ice cream. The result is a delightful dessert that is hot on the outside and cold on the inside. In the Sunday Dinner Plan, the heads of the "Cook Vegetables" and "Cook Roast" activity arrows come into the event from which the tail of "Serve Dinner" activity arrow leaves. To add the activity arrow, "Prepare Dessert" we ask the question of PRECEDENCE--what must be finished before we can start "Prepare Dessert"? Since the oven is required, the "Cook Roast" activity must be finished. The vegetables were cooked on the top of the range, thus this activity has no relation to the "Prepare Dessert" activity arrow. If we draw the "Prepare Dessert" activity arrow from the point or event node at the head of "Cook Vegetables" and "Cook Roast" arrows as shown below, we have an illogical plan since now the start (Tail) of "Prepare Dessert" is dependent on both "Cook Vegetables" and "Cook Roast" activities being finished (Heads).

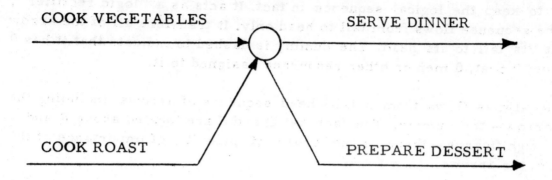

AN ILLOGICAL PLAN FOR SUNDAY DINNER

It is necessary to isolate the logic by the "Dummy" arrow. The Dummy is a logic transfer agent whose function is to transfer completed activities (or arrow heads) from its tail to its head. In the next plan below, the Dummy serves to establish the logical sequences by transferring the completion "Cook Roast" to the start of "Serve and Eat Dinner." The tail of the "Prepare Dessert" is only affected by what comes into it--it is not transferred by the Dummy. Note that the concurrency in the logic planning phase means "dependent on a common predecessor" not simultaneity. There are no time relationships in the planning phase, just logical sequences. When the logic question of CONCURRENCE is asked it means other operations can start in this plan at this point or event. Note also that heads affect subsequent tails. Tails do not affect other tails at the same event node; heads do not affect other heads at the same event. In the plan below, the activity "Serve Dinner" has been redefined into two activities, "Serve and Eat Dinner" and "Serve and Eat Dessert."

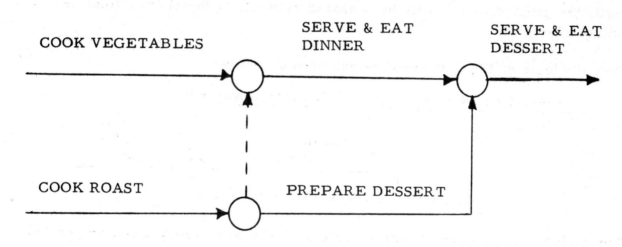

EXPANDED PLAN FOR SUNDAY DINNER

11

Note that we used a dashed arrow (called a dummy), which is not in itself an activity, to keep the logical sequence in tact. It acts as a "logic rectifier", in that the sequence flows from tail to head only. It transfers all of the arrow heads at its tail to its head. The dummy is dashed to indicate that it has 0 time value, 0 cost, 0 men or other resources assigned to it.

The logic always flows from a tail-head sequence of arrows, including the dashed arrow—the dummy. The fact that Q and S are located above R and T has no significance—there is no order of priority of importance of the location of the arrows.

The relationship of activities just follows a tail-head (start-finish) sequence. Most diagrams will read from left to right, but any individual arrow may go up, or down, from right to left. The logical relationship, or dependencies, only follows the path of arrows in the tail (start) to head (completion) sequence.

Generally, the development of an actual project plan arrow diagram will proceed on this basis: a chain of activities is drawn, in its tail-head continuity sequence, until the planner can go no further with this particular series of operations. The last arrow is left hanging open. Then a second parallel chain of operations is started from a logical starting point, and developed as far as possible; then a third, fourth, and so on. Ultimately, the individual paths, or chains of arrows, will interconnect. The development of any project arrow diagram will come together just as a jigsaw puzzle; slowly at first, with isolated pieces meshed (individual chains of arrows) partially. Finally, on an exponential curve of closure, the individual path will close, or interconnect, just as the last pieces of the puzzles go together.

Since this is a project planning technique (the reaching of a specific, unique goal), the final diagram will be a closed network, with only one final or objective event.

Thus the logic of Fig. 2.0 would be extended as below:

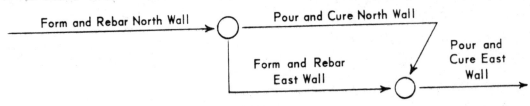

Fig. 2.1 Logic Development of Activity Sequence.

The nodes, or juncture points (circles) in the arrow diagrams, are called "Events". By definition in this technique, an event is just a point in time. The logic above reads as follows:

The forming and placing of reinforcing steel in the East Wall will not start (tail of arrow) until the forming and rebaring of the North Wall is finished (head of that arrow). The pouring of concrete for the East Wall will not start (tail of arrow) until both the forming and rebaring of the East Wall, and the pouring of the North Wall, are finished (both heads of the activity arrows come into the node or event, from which the pouring of the East Wall leaves).

There are conflicting statements in literature as how to develop a network plan. Particularly in PERT, the claim is made to build the plan backwards, from the last activity to the first. The argument proposed is that for an unknown type of project, such as a research and development project, the end objective is known, but it is indefinite as to what will transpire in between. From experience, the author finds that planning a project from the beginning results in a more logical and quicker drawn plan. There will be considerable "seesawing" in laying down sequential chains of operations. It must be noted that a chemist, embarking on a research project, does not rush into the lab, and indiscriminately mix chemicals a la a science fiction movie. He plans his steps as best he knows how, from a literature search, through preliminary testing to a certain point. The results obtained thereof will determine the next phase of his plan. In general, plan a project from the beginning, showing initial activities, and develop it from there.

Consider the following example in the development of an arrow diagram:

The XYZ Company announces the sale of a new type of reproduction machine. It utilizes a radically new method of reproduction: a source of nuclear radiation transfers the markings from a tracing to a permanant sheet. It is a rapid, high fidelity, low cost system — average cost of a print is one tenth of a cent. It makes 25 copies of a 30" x 40" drawing in one minute.

The unit consists of a two section unit; the reproduction unit, and the operating console. The reproduction unit contains shielding for personnel protection from radiation.

Your company decides to order one. Authorization is granted to procure. Since it is a more sophisticated piece of equipment, a higher level of intelligence on the part of the operator is required. Thus, the management authorization includes permission to hire a new operator. It is recommended that training of the operator be done after installation of the machine; so that the operator may sit at the console and make "dummy" unpowered runs with the operating manual at his side, for familiarization.

A health physics inspection of the reproduction unit for satisfaction of state codes as to radiation level is mandatory. In dealing with this agency before, considerable trouble was encountered in getting the inspector to come to the plant.

The activities involved in the project are:

> Authorization
> Issue P. O.
> Delivery Time of Equipment
> Installation of Equipment
> Hire Operator
> Train Operator
> Bring the Inspector
> Inspect the Installation

The steps in developing the plan (forget time units now, we are only planning. We will schedule later) are:

1. We'll develop the ordering and installation chain of Activities. (This is an arbitrary selection, any chain of operations could be selected as a start.)

It would look like this:

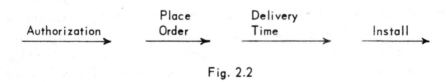

Fig. 2.2

2. Now we'll develop the hiring and training of the operator.

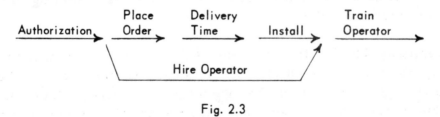

Fig. 2.3

Remember, it was recommended that operator training start after the installation.

14

3. Now we'll consider the health physics Inspector.

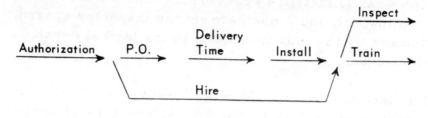

Fig. 2.4

Are there any logic flaws in this? There is one — the hiring of the operator restrains, or limits, the start of inspection (tail of arrow). This is false; inspection is only for installation compliance to codes. We must use a dummy to isolate logic relationships. Correct diagram development is shown in Fig. 2.5 below:

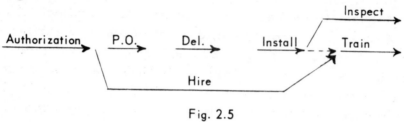

Fig. 2.5

Next, we bring the inspector aboard. We can start doing this right after authorization, and must finish before inspection.

Is there anything wrong with Fig. 2.6 below?

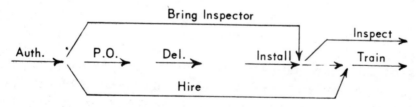

Fig. 2.6

The foregoing diagram is incorrect in that the bringing of the inspector aboard limits the start of the training of the operator. The training can only start after the completion of hiring, and completion of installation. The correct network plan is this:

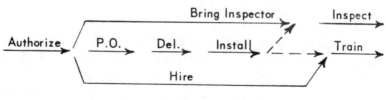

Fig. 2.7

15

Note that the plan is independent of time. It may take a week to get authorization; 3 days to issue P.O.: 6 weeks to deliver; one week to install; 4 weeks to hire; one week to train; one day to inspect; and 7 months to get the inspector aboard. At this phase we are only interested in the logical development of the plan.

Consider the use of the "dummy" as a tool to keep the logical relationships intact. The logical sequence of arrow diagram always flows from tail to head sequence. This applies to solid arrows as well as dashed "dummy" arrows.

Dependency and Fineness of Activity

There is no such thing as partial dependency with the logic network technique. The arrow represents an activity which is defined as any definable and time-consuming task, job or function. Therefore, in the consideration of the logical sequences and inter-relations between activities on a project, it is not permissible to start an arrow from the mid point of another arrow; each solid arrow represents that specifically defined and time-consuming job or activity. However, the method is extremely flexible, in that it allows you to consider the grossness of one activity, and to break it down in multiple arrows, representing the fineness to which it is desired. Remember, this is a tool for the manager, and the information that will flow from the network planning technique is ultimately based on his particular managerial needs.

Every solid arrow (activity) must have a definition placed over it on the plan. Thus, if the first diagram has an arrow, representing operation S, as below:

$$\xrightarrow{\hspace{2cm} S \hspace{2cm}}$$

and an analysis of the plan indicates that another operation, M, could start after 20% of S is complete, and as each additional 20% of S is finished, another proportionate part of the original M could start. It would be broken down thusly:

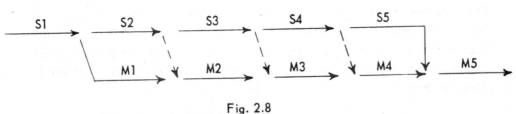

Fig. 2.8

Let us examine a project to point out the multiple arrow capability for fineness of control information. Consider the following: Company A has received an order to manufacture some 100 identical sections of a rocket engine. The steps involved are: (1) spin the section to a certain shape, (2) drill holes for certain attachments, (3) weld a ring around the periphery of the section, (4) pressure test the welded seam, and (5) affix the attachment. The gross diagram will be as shown below:

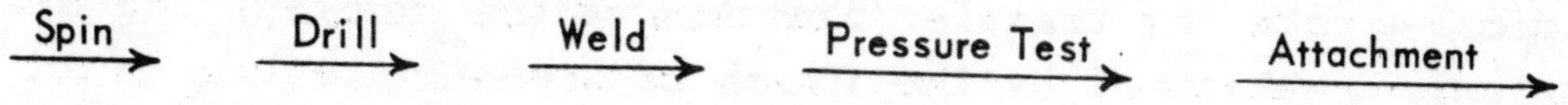

Spin → Drill → Weld → Pressure Test → Attachment →

Fig. 2.9

On first glance, this is the correct sequence of activities, and the plan looks acceptable. However, it is apparent that all 100 sections will _not_ be spun, _then_ 100 sections drilled, _then_ 100 sections welded, _then_ 100 sections pressure tested, and _then_ all attachments made. It is desirable to set up cycles, dependent on the available resources of men and machines, so that the duration of this project will be minimized, and maximum use of resources effected. Therefore, each activity arrow will be defined and broken down to a level to suit the particular situation and resources available. The breakdown could go to each individual unit, if this was desirable. Or it may be more reasonable to consider lot sizes of 5, 10, 25 units. Whatever the fineness of breakdown, the network technique is a tool to handle it.

The diagram below indicates a breakdown by lots; the project engineer or manager will decide what degree of fineness (this degree of fineness is called "level of indenture" in PERT terminology) is desirable.

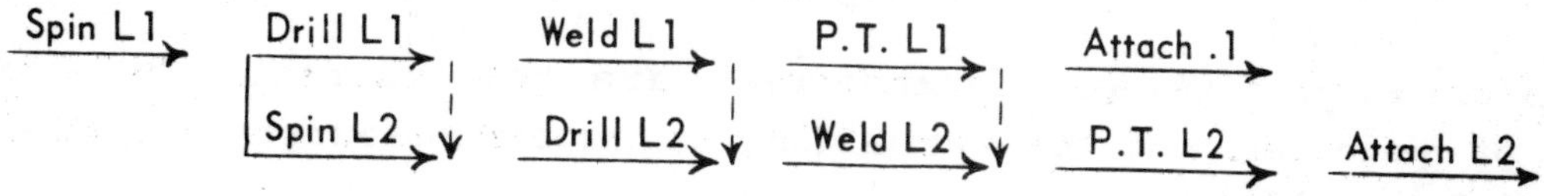

Spin L1 → Drill L1 → Weld L1 → P.T. L1 → Attach .1 →
Spin L2 → Drill L2 → Weld L2 → P.T. L2 → Attach L2 →

Fig. 2.10

As the logic sequence unfolds, it appears realistic, based on cycles of operations. We can start spinning lot #2 after completing lot #1. We do not start drilling lot #2 until the previous lot (lot #1) has been drilled, and until lot #2 has been spun. We do not start welding lot #2 until the previous lot has been welded, and also until lot #2 has been drilled, and so on with our logical development of the plan. Note again, time durations have not entered our picture. We shall plan first, completely, and only after the logic has been reviewed and accepted, shall we consider time estimates.

The plan is extended to show lot #3, in Fig. 2.11.

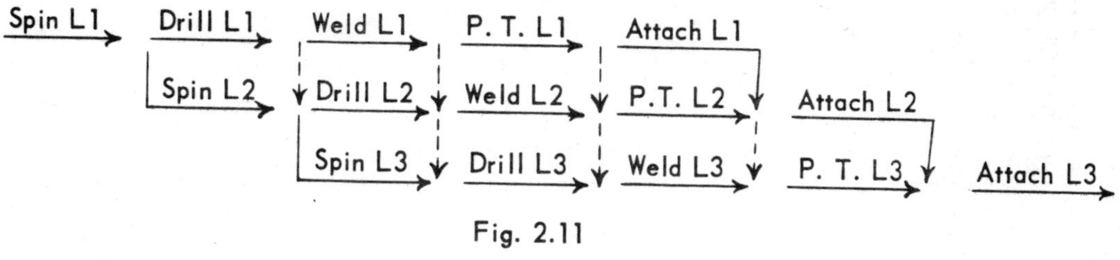

Spin L1 → Drill L1 → Weld L1 → P.T. L1 → Attach L1
Spin L2 → Drill L2 → Weld L2 → P.T. L2 → Attach L2
Spin L3 → Drill L3 → Weld L3 → P.T. L3 → Attach L3

Fig. 2.11

Examination of Fig. 2.11 shows some potent errors in our logic. As it is shown, the spinning of L3 cannot start until lot #2 is spun, and also the drilling of lot #1 is finished; the drilling of lot #3 is shown dependent on the completion of the drilling of lot #2 (logical) and on the finish of welding of lot #1 (illogical — note the transfer function of the dummy); the welding of lot #3 is falsely restrained by the finish of pressure testing of lot #1; the pressure testing of lot #3 is artifically limited by the finish of the affixing of attachments to lot #1.

It is necessary to return to the logic diagram, and "open" up the arrows to add dummies that will keep the logic relations and dependencies correct. This is shown below:

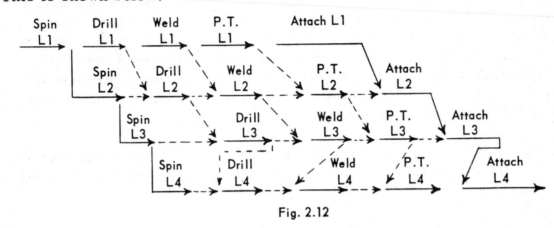

Fig. 2.12

The logical sequences and dependencies are now correct. The effect of delivery of material, and the availability of certain resources can be shown also, as in Fig. 2.13 below:

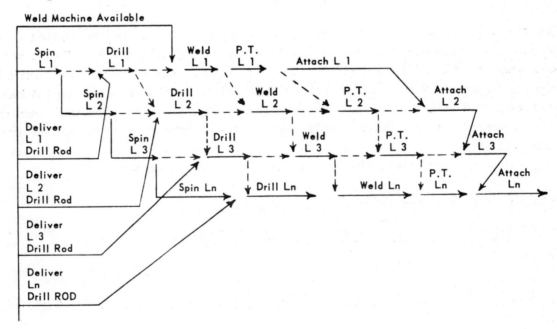

Fig. 2.13

Complete Project Plan

18

Examination of the complete Project Plan, Fig. 2.13 indicates several aspects of the technique. The last chain of activities (the final lot, Ln), does not need the logic dummies, so no further isolation is required. A dummy had to be added between "Spin L1" and "Drill L1" for the "Delivery of L1 Drill Rod", in order that this delivery arrow would not illogically limit the start of "Spin L2". A dummy was added between "Drill L1" and "Weld L1", so that the Welding Machine Availability arrow would not illogically restrain the start of "Drill L2".

The initial activities that are concurrent, can either emanate from one node, or event, the origin, as a "burst" array, or that event can be shown as a "base line", as indicated on Fig. 2.13. It makes no difference, since the master plan is to be read from a tail-head sequence of arrows. A "base line" is an extended event, and may occur anywhere in an arrow diagram.

The technique developed herein is only for a closed network; there can only be one Start event and only one Finish or Objective event. Many parallel operations may start from the initial event (which may be represented by a vertical event or base line) and there may be many parallel terminal activities all ending in the Objective event. This could be shown as a base line. The base line is primarily a drafting convenience.

At this point of planning do not be overly concerned with the Dummy arrow. It is not worth the effort to try to eliminate the Dummy. When in doubt, insert the Dummy. It will be noted that every arrow in any plan may be preceded and followed by a Dummy and the logic is in no way impaired or faulty. The schedule that will be obtained will be in no way erroneous.

Plan With Available or Reasonable Resources

There is a school of thought in some of the literature extant that advocates some of the initial plan be based on unlimited resources. The author has found this to be highly impractical: it assumes an infinite number of men, specialists, machines, etc., poised expectant, ready to execute an operation immediately. Very few proposals are based upon unlimited resources and even fewer projects are executed with unlimited resources. Unlimited resources is a metaphysical concept; only God has them. The best approach is to consider available resources in the initial planning. If a particular resource (craft, machine, facility, specialist, etc.,) is not available as standard to the planner's organization, then the planner is to start with what he considers reasonable assumed resource. Once the schedule is obtained the "reasonableness" of that availability assumption will rapidly be analyzed. If inadequate, the schedule will soon indicate this and point out when in time the inadequacy occurs. Then an alternate strategy can be quickly tested with the Network Analysis technique by changing the sequences of activity arrows, reflecting additional resources or methods. This will establish the adequacy for meeting contractual target dates and budgets. Plan on making two "passes" with this technique to obtain desired schedules; the first schedule will probably be inadequate.

Thus, a construction contractor can lay out his initial plan with the number of sets of forms he has available for concrete work, or the number of forms he feels reasonable as a starting point; or a manufacturer will start with the number of presses he has available, or a reasonable number he feels his supplier or sub-contractor should have available.

In the SUNDAY DINNER Plan developed previously in this chapter, there was an implicit assumed resource - one oven in the kitchen. Thus, the "Prepare Dessert" activity could not start until after "Cook Roast," in order to logically obtain use of that resource. If the SUNDAY DINNER Plan is to be established for a kitchen with two ovens (both operable, of course,) the initially assumed reasonable resource would establish the Plan below in Fig. 2.14. The logic question of PRECEDENCE for the "Prepare Dessert" activity produces this answer--there is no predecessor. Thus, "Prepare Dessert" becomes an initial activity.

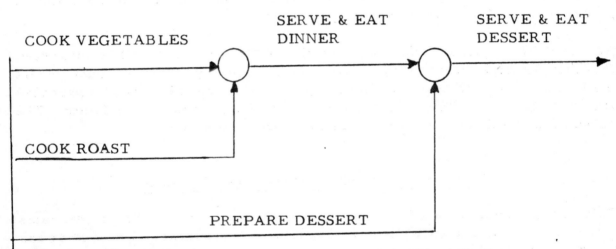

Fig. 2.14 Sunday Dinner Plan For Two Oven Kitchen

It must also be noted that a definite work package is represented by an arrow. The description of the work represented by the arrow should be clear and un-ambiguous. When an arrow diagram is received with a definition such as "Start Overhead Conduit First Floor ," "Continue Overhead Conduit First Floor," "Finish Overhead Conduit, First Floor," it indicates a nebulous amount of work, and the estimate of time for that arrow cannot be substantiated. If arrows are identified by "Lag" or "Lead" time, this shows an attempt to schedule while planning, and contravenes the purpose and power of this tech-nique. Since, with this technique of Network Planning, <u>the planner will never schedule again</u>. He just plans, and the methodology will produce a schedule for that plan with its particular estimates.

"Banding" A Diagram

Quite often it is desirable, in the final drafting of an initial Network (arrow) Diagram plan, to group or re-group identical operations, or work by the same contractor, for purposes of easy identification on the plan. The sequences of identical operations are shown in a horizontal section of a diagram. This is called "Banding" a diagram. The manufacturing plan of Fig. 2.13 is partially shown "banded" in Fig. 2.15 below. The logic of the plan is identical to that of Fig. 2.13, and the schedule that will be obtained for Fig. 2.15 will be identical to that of Fig. 2.13. In actual practice, Banding is a graphic display, and is done after the construction of the basic plan and the acceptance of the initial schedule.

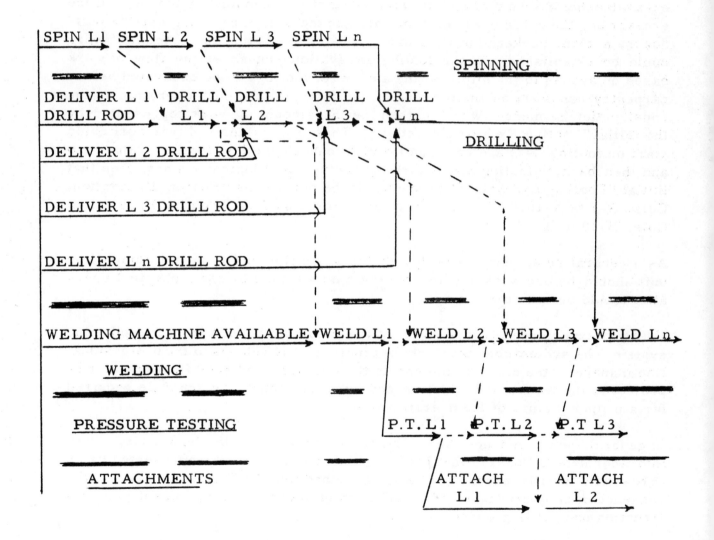

Fig. 2.15 A "Banded" Plan

21

Level of Indenture

The arrow may represent as fine a definition of work as desired, nail by nail in the construction of a house if so necessary; or as gross as possible, such as "Divert the River." It is difficult to advise the neophyte network planner how fine a definition of work (the Level of Indenture) should be. It is up to the planner to decide, based on his need for specific information. As a guide, the basic work time unit may be used to establish the amount of work to be represented by the activity "work package." Thus, if the basic time unit is going to be a work day, or work shift, the amount of work in a particular phase of the project that can be executed with the reasonable resource (crew, machines available, etc.,) in that period may determine the smallest work package represented by an activity arrow. For example, in a construction project, the planner in the excavation phases may decide that half the area could be rough excavated before the footing excavations start. Depending on the size of the excavation, the cubic yards of removal, and the resource being used (a bull-dozer, a clam bucket shovel, a crew of laborers, etc.) the amount of work could be established. In the footing excavation sequence, the flow of work based on access to the site, and the amount of footings to be excavated before carpentry can start on the forming of the footings, the next operation in the construction sequence, would determine the amount of work represented by the initial "Footing Excavation" activity. Thus, if it is desired that work could start on footing A-1, and continue through to A-10 (the last footing in that row) and then back to footing B-2 before carpentry could logically start, then that initial "Footing Excavation" arrow would be defined as "Footing Excavation, Cols. A-1 to A-10." The next activity arrow would be "Footing Excavation, Cols. B-10 to B-1."

As a general rule, using work day as the basic time unit, the minimum time unit should be one work day and the maximum time unit for a physical operation would be ten work days.

Note that experience is required to get the maximum benefit from this planning system. The system does not work for amateurs--it belongs to the professional line manager: the one who interprets the drawings and specifications of the project to the level of indenture needed by him to obtain the schedule required for a proper picture of the overall project.

In general, when in doubt, break the work represented by the activity arrow into finer detail. Observe that grossing the amount of work represented by an arrow will usually product an illogical schedule. In Fig. 2.16 below, the sequence of shop drawings and installation of mechanical equipment in a construction project is grossed.

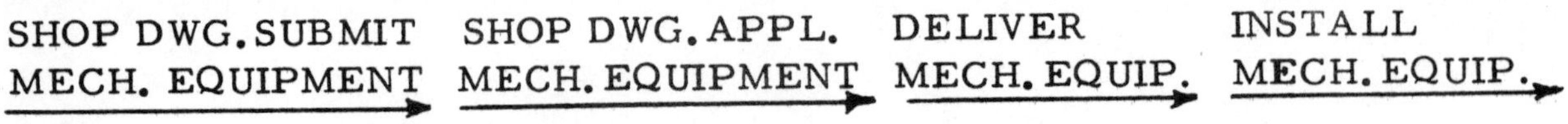

SHOP DWG. SUBMIT   SHOP DWG. APPL.   DELIVER   INSTALL
MECH. EQUIPMENT ⟶   MECH. EQUIPMENT ⟶ MECH. EQUIP. ⟶ MECH. EQUIP. ⟶

Fig. 2.16 Too Gross An Activity

This gross plan indicates that no mechanical equipment will start to be installed until all mechanical equipment is delivered. A mechanical contractor may be responsible for the installation of boilers, condensate system, air conditioners, incinerator, roof ventilators, louvers, etc. These are in different physical locations throughout the project, yet by Fig. 2.16, it is indicated that all of this equipment must be delivered to the site before installation starts, a patent absurdity.

The Equipment Plot Plan is a good source from which to determine the major pieces of equipment, and the arrow diagram will be set up with activity arrows for the major pieces. If the roof is fairly large, roof fan delivery activity arrows may be broken down to represent those fans in quadrants on the roof. This, in turn, would have been established from the level of indenture, or fineness of work detail from the roofing construction sequence.

Once the detailed "Master Plan" is constructed and an acceptable schedule obtained, it may be summarized into a sequence of major operations, with milestone dates obtained from the Master Plans schedule, and marked thereon. A summary diagram will be developed later on in this text.

RECAPITULATION

1. Each activity is represented by a solid arrow on a diagram, and the arrow tail indicates start of that activity; the head its finish. An Activity is defined as "Any definable and time consuming operation, task, function or time limited condition".

2. The Arrow Diagram is a logical sequence diagram, showing the necessary or assumed reasonable flow of work or conditions. Arrow heads affect subsequent tails.

3. The length, direction and position of the activity arrow has no significance; only heads and tails of arrows.

4. The nodes, or connection points in a diagram where arrow heads come in, and where arrow tails leave, are called EVENTS. These events become time boundaries for the activities.

5. The logic transfer agent, the DUMMY, is a dashed arrow. It is dashed because it has a zero value in time, men, money, machines and resources. Its function is to transfer all arrow heads (including Dummy arrow heads) from its tail to its head. When in doubt, put in a Dummy.

6. The "definable operation" represented by an activity arrow must be reasonable in definition. Too gross an amount of work assigned to the definition of an arrow will cause inadequate or falacious schedules. When in doubt,

break the activity arrow into finer definitions or categories of work. Experience will soon dictate the proper Level of Indenture, or fineness of detail.

7. Remember, an illogical Plan will produce an illogical Schedule. The technique is a Line Manager's technique; the plan must reflect experience or judgement by those who will execute the work, or be responsible for its estimates.

PROBLEMS

2.1 Draw the logic diagram for the following set of information. Letters represent the activities in this one plan.

    a. U is the first operation of the project
    b. S and L are concurrent and start after the completion of U
    c. P and T are in parallel, and can begin only after the finish of L
    d. M follows T, and precedes Z
    e. N and A can both start after S
    f. O, A, P, and Z must all be done before R, the last operation of the project, can begin
    g. O follows N

2.2 Draw the logic diagram for all the relationships below:

    a. L is the last task of the project
    b. R follows S
    c. Y and N are concurrent, and follow the initial project operation, C
    d. G precedes J
    e. D follows R
    f. N restrains the start of S, which also follows Y
    g. F and G are done concurrently, after the finish of N
    h. K is done after J
    i. R and F must both be finished before D starts
    j. F also restrains the start of K
    k. X can start after C is finished, and must end along with D and K before L can start.

2.3 Draw the plan for the following operations:

   a. P precedes M
   b. C, N, A are in parallel, and are the initial activities
   c. Operation E follows N
   d. C restrains the start of W, P, T.
   e. Job J can start after A is completed
   f. W follows J
   g. Function T occurs after E
   h. M, W, S must all be completed before X, the last operation, can be executed
   i. T has to be done before S

2.4 Construct the plan for the following operations. See if you can find a one dummy solution:

   a. B follows M, and is one of the parallel terminal activities
   b. S follows R, and is concurrent with L and H
   c. N is an initial job, in parallel with R
   d. J follows L, and is one of the last activities on the project
   e. L follows R
   f. T starts after the completion of S and H and N, and is a parallel terminal activity
   g. H must end along with N before M and T start. It (H) follows R.
   h. M precedes B
   i. R is an initial operation

2.5 Draw the arrow diagram for problem 1.1, page 6. From a study of your logic diagram, re-estimate the working day duration of the project. There may be more activities in parallel, or in series, than in your original estimate of problem #1.1. Compare the answer with your original estimated duration.

2.6 Draw this plan:

BUILDING A COIN OPERATED LAUNDRY

This project involves the construction of a Coin Operated Laundry in a local neighborhood. The building is a one story, cinder block walls, and a hot pour covered roof. The site has already been cleared for construction. The incoming services of water, gas and the sewer are to run across the property before the paving of the outside parking area. The roof is to be covered before installation of the washing machines, dryers, hot water heater, the heating-air conditioning unit, and before the installation of the plate glass windows. Assume the general contractor does all the ordering of items 15 thru 20 below.

The activities involved are listed below, not necessarily in logical or chronological sequence. The duration of the operations are in working days (assume the contractor's proposal has provided for adequate crew sizes to accomplish the activity in the listed duration time.)

Activity	Duration (Working Days)
1. Obtain Building Permit	3
2. Layout and Excavate	2
3. Award Plumbing Subcontract	12
4. Award Electrical Subcontract	10
5. Form, pour, cure footings	2
6. Electrical Rough In	2
7. Plumbing Rough In	4
8. Install outside Sewer, H_2O, Gas lines in trench	3
9. Form, pour, cure floor slab	2
10. Erect cinder block walls	5
11. Erect roof rafters and purlins	3
12. Waterproof exterior walls	5
13. Interior plumbing	3
14. Outside sewer, H_2O, Gas lines hook up	2
15. Order and deliver plate glass windows	19
16. Order and deliver Electrical material, lights	8
17. Order and deliver Hot H_2O Heater	17
18. Order and deliver automatic dryers	26
19. Order and deliver automatic washers	22
20. Order and deliver heating-air conditioning unit	20
21. Cover roof	6
22. Run interior conduit, pull wire	3
23. Hang Ceiling	2
24. Install counters, millwork, doors	3
25. Paint interior	2
26. Lay flooring	3
27. Install overhead lights	2
28. Install plate glass windows	1
29. Install automatic washers	8
30. Install automatic dryers	6
31. Install Hot H_2O Heater	2
32. Install heating-air conditioning unit	5
33. Final Plumbing, testing	4
34. Final Electrical tie ins, testing	2
35. Grade and pave outside parking area	3
36. Final inspection	1

CHAPTER 3

EVENT NUMBERING

Each arrow is always identified on the network diagram with a description. It is also necessary to establish an event numbering system. That is, each node or juncture point in the diagram will have a specific identifying number. This originated from the computer programs used to calculate schedules. Each arrow was assigned an "i" or "j" number. The letters "i" "j" have no significance, they were just selected arbitrarily from the alphabet. The "i" number identifies the tail (start) of the arrow, the "j" number the head (finish).

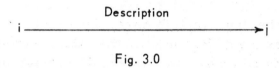

Fig. 3.0

Consider the network below, from Fig. 2.7

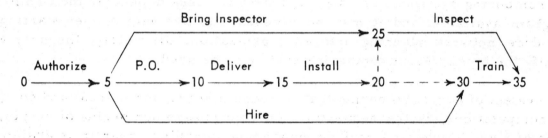

Fig. 3.1

Each arrow has its own distinctive "i - j" combination, including the dummies. The j of a precedent arrow becomes the i of a subsequent arrow, i.e., "10-15 Deliver", "15-20 Install". Thus a unique identification system for each arrow is established. We shall use this as a bookkeeping identification and a check, in the manual methods discussed herein. It is mandatory to use an i-j identification system when using digital computers to generate schedules, since the computer "sees" only numbers. The alphabetical descriptions never enter the computation section of the computer.

There are two methods of event numbering a network diagram.

1. Sequential numbering. The j number is always higher than the i. This is further categorized into two sub-categories.

   (A) Sequential consecutive. Here, every whole integer from 0 or 1 at the origin, is used (that is, 1, 2, 3, 4, 5, with no omissions) from the origin to the final event. This is a requirement of certain computer programs. It is a limiting restriction, in that any changes in a plan will force renumbering of the diagram. The rule for sequential event numbering is very simple. Before a node can be numbered, all of the arrows entering that event must have event numbers at their tails.

(B) Sequential non-consecutive. Here, every j is higher than the i, but there is no need for complete use of all numbers. Fig. 3.1 is sequentially non-consecutively numbered, in blocks of 5. If any revisions to the plan are required, the missing numbers may be be used, as long as the number at the head of the arrow (j) is higher than the number at the tail (i).

2. Random Numbering. In this case, the result of computer sophistication, it makes no difference whether the j number is higher or lower than the i. The computer will internally assign sequential consecutive event numbers.

The numbering system to be used, will in many cases depend on the Computer Program available, and it may be necessary to live with it when starting to introduce network planning into an organization. Ultimately, the user will specify the type of event numbering system to be used.

The system of event numbering that is recommended, for manual and computer computations, is the sequential non-consecutive numbering of events, in steps of 5 or 10, even if a random numbering computer program is available. It is fast, versatile and flexible. It also serves as a check on the logic. Consider Fig. 3.2 below, an initial development in a diagram. After the path of A, B, C, D was developed, a concurrent path was started from the head of A, and developed thru M, N, O, P. Then it was decided that the completion of C and M would restrain the start of D and P. Then the logic was further extended to show S, R, T. It was decided that T should end before N started.

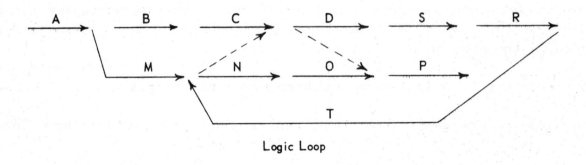

Logic Loop

Fig. 3.2

But the dummy from the head of M to the tail of D will create a logic loop, indicating that a chain of operations could finish before they start. In an actual diagram of several hundred arrows, loops are extremely difficult to visually detect. The random numbering of events (j higher or lower than i) will not detect loops, and it is necessary to have a sub-routine in the computer program to search the network for loops. When it becomes time to event number the diagram, the basic rule of sequential event numbering is followed, i.e.; the tails of all arrows coming into a node must be numbered before that node is numbered.

Generally, up to 3 digits can be used for event numbering. There are many programs for computers that allow more digits to be used for event numbering. Chapter 16, "Computer Orientation", will list some guide formulae for ascertaining the number of activities in a diagram from the last event number, and for determining the adequacy of the computer memory capacity for the size of network used.

Another reason for the sequential, non-consecutive event numbering system is that such a system produces a sequenced list of activities, in i major ascention, on an output schedule. When working on a real project, information will be sought from the plan to the schedule, and the required information is found much easier if the output is listed in a sequence.

The Numbering Dummy

In the network in Fig. 3.3, note the activities M and T. The diagram has been sequentially consecutively numbered, and M and T have the same i-j combination.

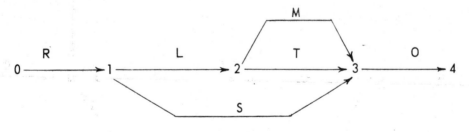

Fig. 3.3

With the exception of very few sophisticated computer programs, it is mandatory that each arrow have its own unique i-j combination. It will be necessary in the manual system to have a separate and unique i-j designator for each arrow, solid or dashed.

Thus, the dashed arrow, the dummy will serve a second function. We shall use it to create a unique i-j activity number for activities occurring in parallel between the same nodes. The "numbering" dummy may be added at

either the head or tail of a parallel arrow. It will not affect schedule information. In general, do not be afraid to add dummies. More time is wasted
trying to eliminate dummies than gaining valuable information.

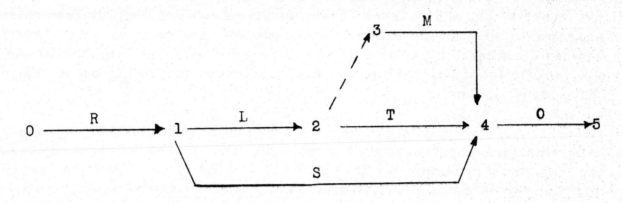

Fig. 3.4

Note that the consecutive event numbering system had to be changed to handle
the addition of a dummy arrow. If Fig. 3.3 had been sequentially, non-consecutively numbered in steps of 3, the new event at the head of the dummy
arrow could be numbered with a fill-in number from the available block. This
is shown in Fig. 3.5.

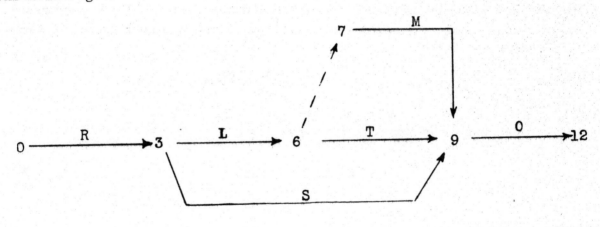

Fig. 3.5

It makes no difference that the pattern of event number skips numbers irregu-
larly; the only important thing is that each arrow have its own i-j combination,
with j higher than i.

The Transfer Event

There are no standard symbols established by industry for Network Plans. The planner is at liberty to organize his own symbolism. There are short hand techniques that speed the construction of the diagram, and reduce a confusing number of cross connecting number of arrows and dummies. One such short hand technique, involving the alphabetic coding of event nodes in the planning phase, is described below.

Consider the following example in the development of a section of an Arrow Diagram for the construction of a high rise (multi-story) office building. The basic structure consists of a structural steel framework erected on concrete footings. The forms for the floor slabs are metal pans which are welded to the structural steel frame, as shown schematically in Fig. 3.6 below. Before the concrete is poured, a bottom layer of reinforcing steel (rebar) is placed into position on the pan. Then the Mechanical and Electrical sub-contractor will place (rough in) their sleeves, and any pipes, ducts and conduit that are to be embedded in the concrete floor slab. Then a top steel mesh is placed into position, the concrete poured, and allowed to cure. The section of the plan developed starts after the structural steel erection, and with the first floor metal pan decks.

Operations

1. Set Metal Pan Deck
2. Place Bottom Rebar (Reinforcing Steel)
3. Mechanical Rough In
4. Electrical Rough In
5. Place Top Steel Mesh
6. Pour Slab
7. Cure Slab

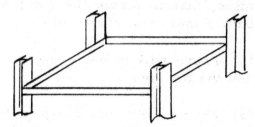

Safety Requirement:

While Steel is still being erected, there must be at least six (6) decks in place overhead before work may start on any slab.

Fig. 3.6 Typical Floor Slab, Multi-Story Building

In addition, there is a safety requirement to be incorporated into this plan. It is necessary to have at least six (6) Metal Decks in place overhead before any work is performed on a particular deck (if work is going on overhead). This is to eliminate the possibility of injury from tools or objects accidentally dropped from overhead. This constraint ends when the roof deck is installed, and the Steel Installation crew is no longer overhead.

The sequence of these operations in this segment of the plan will be developed for the first three floor slabs of the building. It must be remembered that the

31

precedent activities of site preparation, excavation, steel erection, etc., have been done previously. The plan here is a section of the overall plan.

Note several aspects in the development of the first floor sequence in Fig. 3.7 below:

(a) Activities (work packages) will be alternately grossed and refined in the definition of work, as the plan is developed. This is to furnish information as it is needed. Thus, the erection of the first seven metal floor pans will be represented by one arrow, to satisfy the initial safety requirement, Then each subsequent metal floor pan installation operation will be represented by separate arrows to allow the proper work flow.

(b) The plan will start with the assumption of available or reasonable resources Mechanical and Electrical Sub-contractors will either furnish their own arrow diagrams indicating the sequence of their work and what they feel must preceed each of their operations; or they will advise the General Contractor, in the development of the overall plan, the crew size and the number of work crews they will have available for their work. In the development of the plan below, the following initial assumption will be made:

   (1) There will be one crew each for the Mechanical and Electrical sub-contractors. Thus a cycle of work will be shown, using dummies, that indicates the completion of the work on one floor slab before that crew starts on the next floor slab.

   (2) There will be one crew for both the bottom reinforcing steel and the top mesh.

   (3) There will be one hoist or crane available for concrete pours, so that a pour of any floor slab depends on the completion of the pour of the slab below.

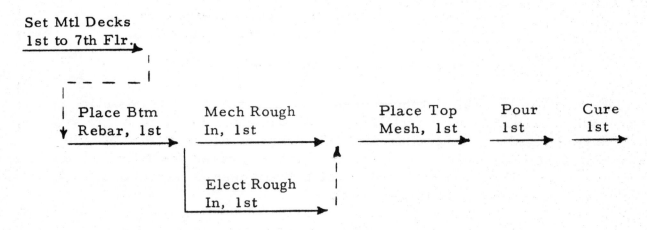

Fig. 3.7 First Floor Slab Operations

32

Note in Fig. 3.7 a numbering dummy has been inserted in anticipation of Event Numbering. The dummy from the head of "Set Mtl Decks, 1st to 7th Flr" appears superfluous now, but remember that this is just a section of a complete project plan. If the question of PRECEDENCE is asked about the activity "Place Btm Rebar, 1st," another answer will be "Delivery of Initial Rebar," whose head would also tie into the tail of "Place Btm Rebar, 1st."

Fig. 3.8 below now develops the next cycle of work, the 2nd Floor Slab sequence.

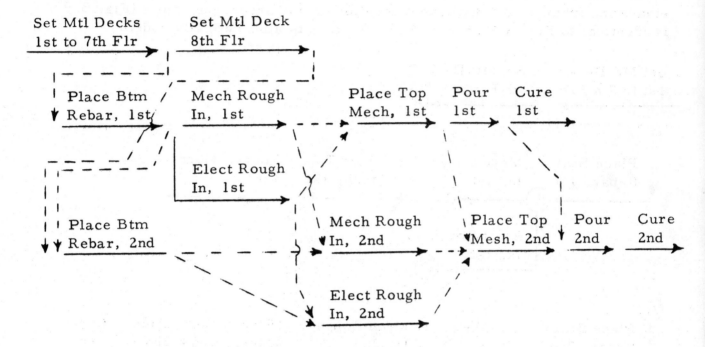

Fig. 3.8 First and Second Floor Slab Operations

When constructing your Network Plan, leave at least one inch between lines in order to have space for descriptions, and the possible future additions of new dummies and activity arrows.

Note that several "logic rectifier" dummies had to be added to maintain proper logic relationships. If the dummy from the head of "Mechanical Rough In, 1st Floor" to the tail of "Place Top Mesh, 1st Floor", had not been added, then the Electrical rough in operation on the first floor would have also been transferred to the tail of "Mechanical Rough In, 2nd Floor." Via the dummy from the head of "Mechanical Rough In, 1st Floor" to the tail of "Mechanical Rough In, 2nd Floor." This is illogical, since these are two separate, unrelated trades at this point in the plan. Once the cycle of relationships is set in the first phase of repetitive operations, it will be repeated for each successive phase The dummies also reflect the initially assumed reasonable resource; in this case one crew for each operation. Remember there is no guarantee that this will produce an acceptable schedule.

33

By now, the arrow diagram in Fig. 3.8 is becoming confusing to read, with the potential of increasing number of cross radiating dummies. In order to lessen the confusion and use a short hand notation, we shall start coding the events alphabetically (events are not numbered until the plan is complete, in order to have the sequential numbering feature check our logic for loops). These alphabetically coded events in our planning phase become Transfer Events. They will ultimately be numbered and a record will be kept of the particular event. Once the event is identified by a letter, it may be shown anywhere else in the diagram. A dummy from it to another event is the same as a dummy emanating from the original event location on the arrow diagram. (Fig. 3.9 is identical to Fig. 3.8 with the Transfer Events alphabetically coded.)

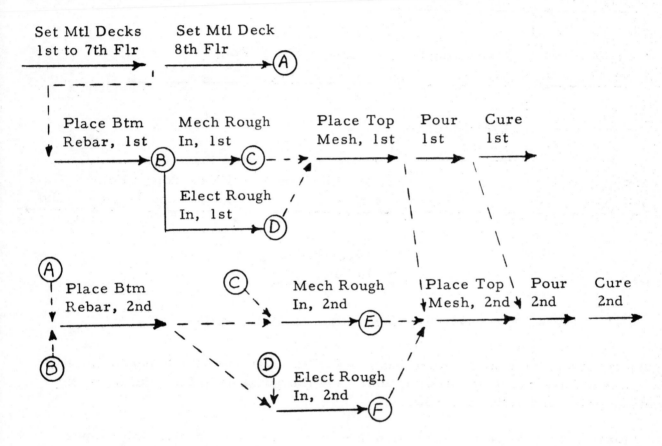

Fig. 3.9 First and Second Floor Slab Operations
with Transfer Events

In Fig. 3.9, the dummy from event C to the tail of "Mechanical Rough In, 2nd Floor" is the same as the dummy connecting the head of "Mechanical Rough In 1st Floor" to the tail of "Mechanical Rough In, 2nd Floor" in Fig. 3.8.

Fig. 3.10, using the coded Transfer Event method, continues the plan to the 3rd floor.

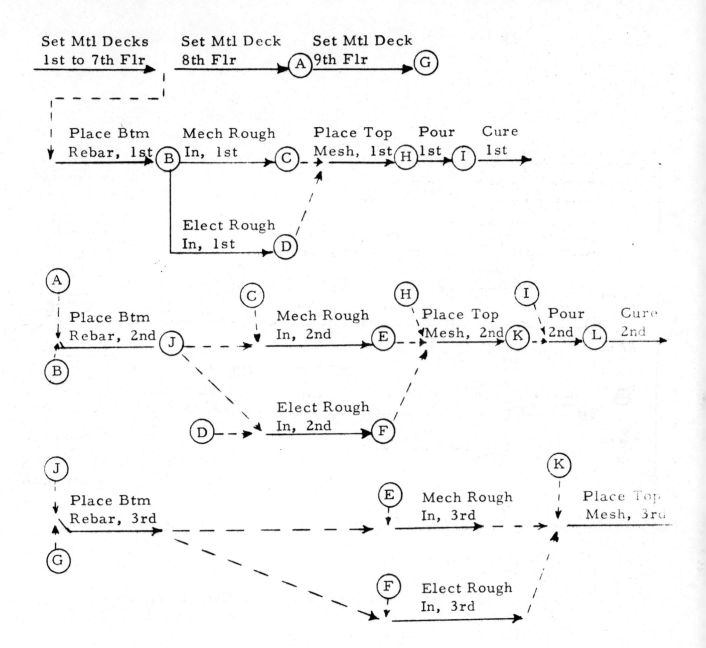

Fig. 3.10 First, Second and Third (Partial) Floor Slab Operations

The diagram is easier to read, and is still logically interconnected. This approach is continued to the end of the plan for the project. Once the letter "Z" in coding is reached, then double and triple letters (AA, AAA) are used. Of course, for purposes of example, an excess of transfer events is shown.

While coding the Transfer Events, a record must be kept to avoid the problems of duplicating event numbers. A tabular form is placed on the original plan drawing in order to maintain a record of letters assigned, then eventually the corresponding event number. One such table will be observed in Fig. 3.11.

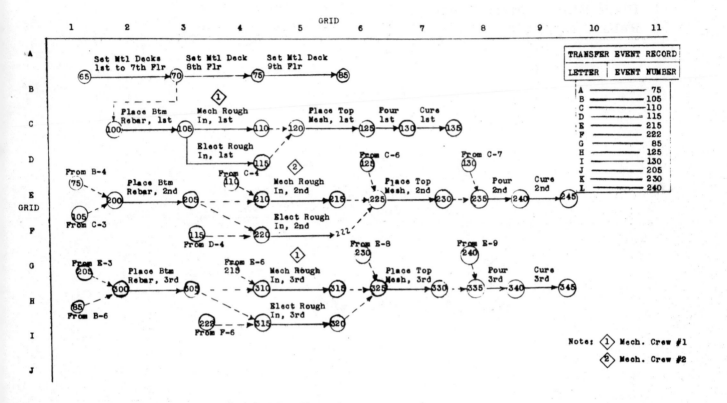

Fig. 3.11 Event Numbered Plan, Floor Slabs 1, 2, and 3

After the complete project plan is constructed with the alphabetical Transfer Events, the next step in the technique is identifying each arrow, activity and dummy by an Event Number. Following the previous recommendation, a sequential non-consecutive numbering system will be used to indicate this step in Fig. 3.11. Fig. 3.11 is a section of an overall plan. The first activity shown therein, "Set Metal Pan Decks, 1st to 7th Floors," has, of course, in the complete project plan been preceded by a whole series of operations from contract award through excavation and footing installation, plus the concurrent operations of shop or vendor drawings submitted and approved, and the delivery and erection of the structural steel. It shall be assumed that a sequential numbering series, in steps of 5, has been used, and in the Event Numbering of the complete project plan, the event at the tail of "Set Metal Pan Decks, 1st to 7th Floors", has been arrived at with Event Numbering. Fig. 3.11 indicates the Event Numbered section as developed from Fig. 3.10. A record is kept of each coded Transfer Event, to change it from a reference letter to an Event Number. Thus, every time A is observed on the plan, it will always be Event Number 65. Thus, the need for the Transfer Event Table. Once the plan is numbered, the reference letters disappear.

While Event Numbering is arbitrary, we can take advantage of it by reserving blocks of numbers for specific areas, buildings, systems, etc., on our plan. In Fig. 3.11, the 100 series has been used for the 1st Floor, 200 series for the 2nd floor, 300 series for the 3rd floor, and so on. This will speed information retrieval when a schedule is obtained and used.

Note that with the Transfer Event technique it is almost mandatory to use the sequential (j greater than i) numbering method. It is virtually impossible to visually detect any loops that may have been erroneously put into the plan.

Another aspect worth considering is the Gridding of the Network Plan chart like a Road Map, by putting an alphabetical reference (one letter every three inches) on the vertical margin, and a numerical reference (one number every three inches) on the top horizontal margin. This allows anyone reading the plan to quickly ascertain where the Transfer Event originated. For example, at the tail of "Place Bottom Rebar, 3rd," activity #300-305, a dummy is noted coming into Event #300, which is dummy #85-300. The Plan of Fig. 3.11 notes that Event #85 comes from location B-6 on the Grid. In essence, a dummy has been drawn from 85 to 300. The Transfer Event is a shorthand method.

The gridding of the plan is recommended in the initial planning stage, on the rough charts first constructed. When the Arrow Diagram is redrawn for presentation purposes, these formal plans are also gridded. This is particularly effective where the size of the formal plans are desired to fit Stick Files, such as 30" x 40" drawings. The grid reference can also refer from one sheet to another sheet where several sheets are required to contain the complete project plan.

Work Flow Portrayal

Another approach to arrow diagramming is to lay out the position of the activity arrows on the chart physically representing the work flow--the first floor chain in a high rise building would be at the bottom of the chart; the second floor above it; the third floor sequence of arrows above the second floor, etc. The diagram is built literally the same way the building is built, from the bottom up.

Within each floor vertical work such as "Install Vertical Conduit" is shown as a vertical arrow; horizontal work would be shown as horizontal arrows, with horizontal floor activities such as "Ceramic Tile Floors" shown as a horizontal arrow under a horizontal activity representing overhead work such as "Install Ceiling Grid." Fig. 3.12 is such an example.

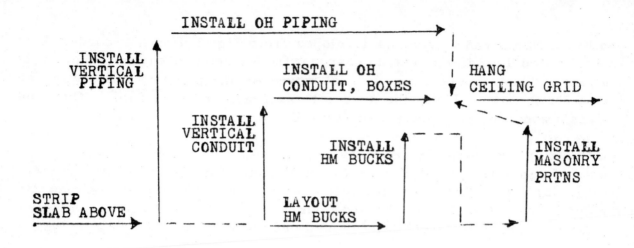

Fig. 3.12 Work Flow Portrayal

Common Mishaps

In the Event Numbering of Fig. 3.11, two common occurrences are shown. The Event Numbering proceeded to the end, but an event was overlooked. This is the event at the head of "Electrical Rough In, 2nd Floor," Events J, D and E of Fig. 3.10 had been numbered. When the oversight was discovered (either) in the manual calculation of the schedule--Chapter 8; or in the preparation of data sheets for a computer calculated schedule--Chapter 16, the network was examined to see what numbers were available from the non-consecutive numbering system. Inspection shows that the missing event number must be an unused number between 220 and 225. The number 222 is arbitrarily selected and written. Omitted Event Numbers will be discovered eventually, certainly by the time data is to be prepared for a computer calculated schedule.

Of even graver consequence is the duplication of an event number. This has occurred in our plan of Fig. 3.11. Event #315 has occurred twice in the diagram. It appears at the head of "Mechanical Rough In, 3rd Floor," and at the tail of "Electrical Rough In, 3rd Floor."

If an event number is duplicated on a diagram, it will produce an erroneous schedule in a computer output and quite possibly an error in a manual schedule particularly if a Transfer Event is involved. In Fig. 3.13 below, two unrelated chains of operations are shown, with an unfortunate duplication of an event number.

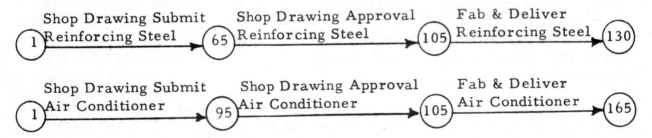

Fig. 3.13 A Duplicated Event Number in Original Plan

Since a computer only "sees" Event Numbers, the computer in calculating a schedule would see a Network Configuration, not the logic of Fig. 3.13, but the logic of Fig. 3.14 below. As far as the computer is concerned, there is only one Event Number 105.

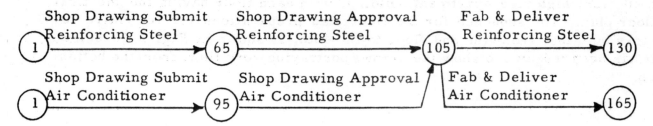

Fig. 3.14 The Plan as the Computer Sees It

One method of preventing the duplication of an event number is the use of a mechanical stamp that serves as a "memory." It consists of a series of cylinders in a row on a frame. Each cylinder has the digits 0 to 10. Every time the handle is pushed down, the cylinders imprint the digits on the paper. Upon release, the cylinders revolve through a ratchet mechanism to the next highest number in the series used. Thus, no duplication can result. This "numbering machine" is available in stationery stores, and is available with mechanisms that allow stamping in series of 2's, 3's, 5's or 10's.

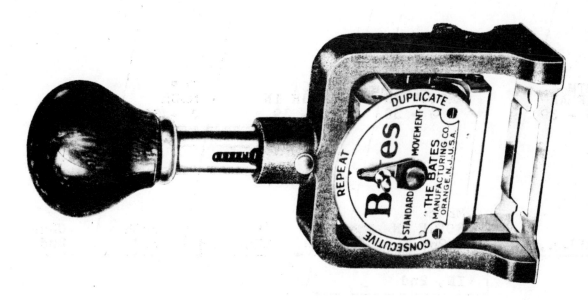

Fig. 3.15 An Event Numbering Machine
(Courtesy of Bates Mfg. Co., Orange, N. J.)

Repetitive Cycles of Work

When a project has a certain sequence of operations, or cycles of work, that are done repetitively, such as the floor slab preparation sequences for a multi-story building in Fig. 3.11, it is not mandatory, nor necessary to repeat each cycle in detail. Once the initial cycle of work is shown in detail on the arrow diagram, <u>identical</u> repetitive cycles can be grossed from the next cycle up to and including the next to last cycle of work. The last cycle of work is then shown in detail, as had been the initial cycle of work.

For example, in the diagram of Fig. 3.11, let it be assumed that this was for a 40-story high rise apartment building, with each floor having the identical floor plan. The sequence for the floor slab preparations for the entire 40-story building, from the 2nd floor to the 40th, is shown in Fig. 3.16 below. It has been redrawn to show the arrows portraying work flow, from the bottom up.

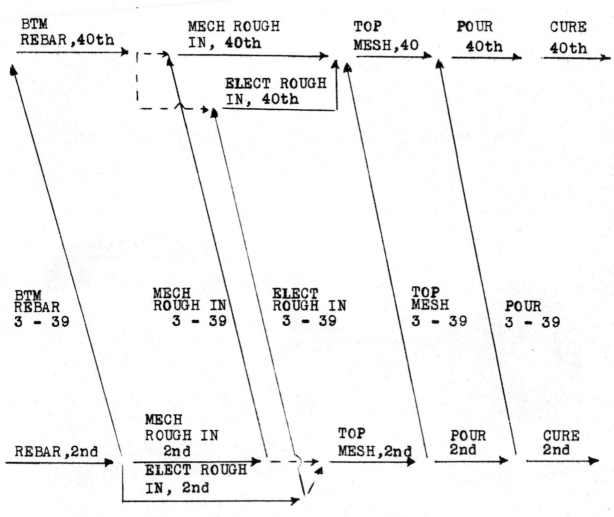

Fig. 3.16 Repetitive cycles of Work

This approach will produce the overall project schedule, the detailed schedules for the second and fortieth floors, and establish the "lead times" for each operation on any floor, so that a "typical" schedule for any floor, in detail, can be established manually. The typical floor schedule can best be shown by the Time Scale Chart described in Chapter 11.

Event Symbols

As previously mentioned, there are no standard symbols in the Network technique. The reader may feel free to establish his own. The author has found the following of help in the construction of Project Arrow Diagrams.

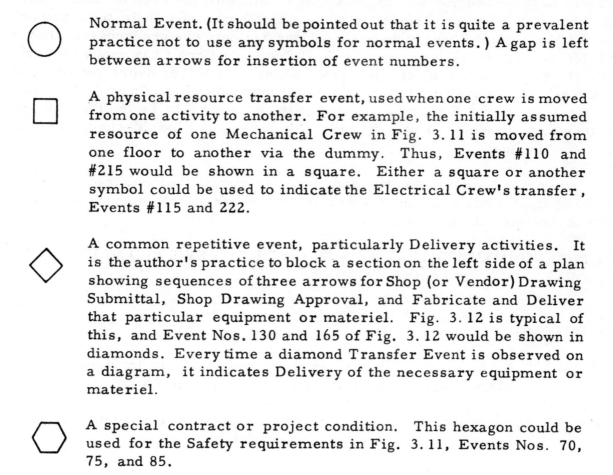

Normal Event. (It should be pointed out that it is quite a prevalent practice not to use any symbols for normal events.) A gap is left between arrows for insertion of event numbers.

A physical resource transfer event, used when one crew is moved from one activity to another. For example, the initially assumed resource of one Mechanical Crew in Fig. 3.11 is moved from one floor to another via the dummy. Thus, Events #110 and #215 would be shown in a square. Either a square or another symbol could be used to indicate the Electrical Crew's transfer, Events #115 and 222.

A common repetitive event, particularly Delivery activities. It is the author's practice to block a section on the left side of a plan showing sequences of three arrows for Shop (or Vendor) Drawing Submittal, Shop Drawing Approval, and Fabricate and Deliver that particular equipment or materiel. Fig. 3.12 is typical of this, and Event Nos. 130 and 165 of Fig. 3.12 would be shown in diamonds. Every time a diamond Transfer Event is observed on a diagram, it indicates Delivery of the necessary equipment or materiel.

A special contract or project condition. This hexagon could be used for the Safety requirements in Fig. 3.11, Events Nos. 70, 75, and 85.

Some Comments on the DUMMY

The most powerful aspect of the Network Planning Technique is the use of the dummy. Note for example the several functions of the dashed arrow in the Plan of Fig. 3.11.

Dummy 70-100 is a Safety Requirement function

Dummy 115-120 is a numbering Dummy

Dummy 110-210 is a Resource Transfer Dummy

Dummy 205-210 is a Logic Rectifier Dummy

This is the device used to determine the proper resources to apply to a plan in order to reach an acceptable schedule and budget. As a case in point, suppose that the schedule that is obtained from the plan of Fig. 3.11 is too long, and the Mechanical Contractor is the cause of the excessive length of time (he is Critical, which will become more meaningful when Chapters 4, 5, and 6 are studied). The originally assumed reasonable resource of one Mechanical Crew proves to be inadequate. It is decided to put on two Mechanical Crews; Crew #1 will work on odd number floors, Crew #2 will work on even number floors. By eliminating Dummy 110-210 in Fig. 3.11, the plan now indicates that work on the Mechanical Rough In on the 2nd Floor has only one predecessor, that of the Mechanical Rough In work on the first floor (now Crew #2 is ready to start). Dummy #215-310 is replaced by a new dummy, #110-310, indicating now that the Mechanical Rough In work on the 3rd Floor depends on Crew #1 finishing on the 1st floor (as well as the placing of the bottom rebar on the third floor), rather than previously requiring the work on the 2nd Floor Mechanical Rough In to be finished.

The dummy is the device the planner uses to determine the number of machines to be applied to his plan, the number of sets of forms for concrete work that the contractor will use, etc.

IT IS RECOMMENDED THAT THIS CHAPTER BE REVIEWED AGAIN AFTER CHAPTERS 4, 5, and 6 ARE STUDIED.

PROBLEMS

3.1 Event number sequentially, non-consecutively in steps of 5, the arrow diagram drawn for problem 2.5, page **25**

3.2 Event number the diagram below sequentially, consecutively, starting with the origin event numbered 1.

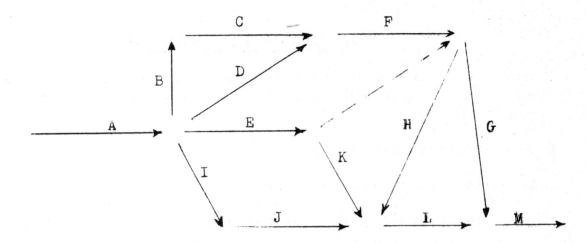

CHAPTER 4

COLLECTING TIME ESTIMATES

Once the project arrow diagram plan has been constructed, and event numbers assigned, the next step is the collection of time estimates, in consistent units of duration, or elapsed time. Estimating time for an arrow diagram will be done on a different basis. If an organization uses unit time or costs, or man-hours, this information will suffice to start. It will be reworked into units of time duration. The rules of estimating the time for the activities on arrow diagrams are:

1. Estimates are obtained from the best possible source. There is no change in the basic requirement from past procedures; the person, group, section or sub-contractor responsible for the execution of the activity is the source of the estimate. If a central estimating group is used, they will be the source of time duration estimates. Delivery of equipment will come from purchasing personnel. Estimating must obviously be done by experienced personnel.

2. Estimates for each activity (arrow) will be on a "disassociated" basis. Since we have separated planning from scheduling, we shall continue to do so in the estimating phase. "Disassociation" means that each arrow is considered separately, one at a time, and the other arrows are ignored. It is as though figuratively, the arrow is physically removed from the diagram, and held in the estimators hand. The estimate for that particular arrow is made completely independent of every other arrow in the diagram.

3. In the estimate for the individually disassociated arrows, the elapsed time duration is based by the assignment of the usual practical resources available. "Usual" means the regular, or desired crew size, number of machines, the normal working day, etc. If any doubt arises as to what the regular crew size should be for the particular activity, consider a feasible size, and use that for the initial estimate. Re-estimating will occur after the technique generates the schedule, identifies the Critical Path, and establishes calendar dates. "Practical" means that physical conditions, safety requirements, labor regulations must be considered. Thus if a company has a normal complement of 12 laborers, and an activity on the diagram is "Dig Section 31-33 of Trench", the estimate must be based on the actual number of laborers that can physically get into the trench for digging purposes. For example, it was originally estimated that section 31-33 of the trench would take 120 man-hours. But only 4 of the 12 laborers

can physically dig that section. Thus the duration to be assigned to the arrow diagram will be $\underline{\text{120 man-hours}}$, or 30 working hours.
$$\frac{\text{120 man-hours}}{\text{4 men}}$$

This is perhaps the most difficult aspect of network planning to grasp. The tendency is still to simultaneously plan and schedule. The disassociation of the arrow (estimating that particular arrow while ignoring the others) will not make much sense in the beginning. It will be only after the technique of scheduling, and analyzing the schedule, is mastered, that this requirement becomes logical.

Three usual objections can be shown from Fig. 4.0.

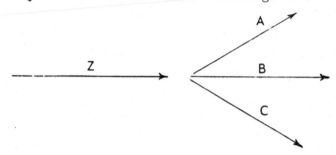

Fig. 4.0

Operation A, B, and C are shown concurrent, all emanating from the same event. Suppose, in Case I, that the same craft is to be assigned to all three activities. When we "dis-associate" A, we ignore B and C, and assign the usual practical resource of the number of men in that craft to A, and obtain the duration of A. Then we disassociate B, and assign the same number of men of that craft; then to C. The usual objection is, if the men are working on A, how can they be working on B and C? We do not know that now – the schedule that will be separately generated will tell us whether we have enough, inadequate or surplus resources (men) to execute the plan as shown.

A case II objection is, from the same logic configuration in Fig. 4.0, that A entails the craft of electricians, of which our normal or usual crew is 8; B uses pipefitters, of which the available crew is 10; and C uses welders of which the available number is 6. However, operations A, B, C from the diagram in Fig. 4.0 are all in the same small room. Thus if we get 8 electricians in there, we will not be able to get more than 2 pipefitters and one welder. How will we handle this case II situation? The answer is the same as case I. We do not know the actual schedule. We must first obtain the schedule, then see if all crews are scheduled to be in that room at the same time. If so, we'll adjust our estimate, but only after we analyze the initial schedule. A Case III situation for the same logic in Fig. 4.0 – if the welder in operation C is in that small room, safety requirements

44

demand that no other workers be present. Again, we shall have to wait to see the initial schedule before we decide the actual conditions. For the plan, the initial estimate is based on the disassociated estimate, and an assignment of the usual practical resources to that disassociated or separated activity arrow is made.

4. The estimates use consistent time units throughout (working days, working hours, weeks, etc.;) for the particular project. Decimals may be used where a basic time unit, say working days, is too large for a particular activity arrow. Working hours can then be espressed as decimal part (4 hours equals 0.5 days).

5. In the repetitive cycle of work situation, as described in Chapter 3, the detailed activities are estimated first. Then those gross activities between the initial and final detailed activities are estimated by a straight multiplication of the number of cycles times the individual activity's duration estimate.

PROBLEMS:

4.1 A sub contractor indicates it will take 160 man-hours to set lighting fixtures on the second floor. He plans to use a crew of 4 men. The estimating of this project uses working days as the basic consistent time unit. There are 8 hours in the working day. What is the estimated duration to be assigned to the arrow "Install Light Fixtures, 2nd Floor"?

4.2 Piping in a mechanical equipment room will take 60 man-hours. The normal crew of pipefitters is 12, but the room is of such a size that only 5 of the craftsmen can work in there at a time. What duration, in working days (8 working hours per working day), will be assigned to that activity on the arrow diagram?

CHAPTER 5

ESTABLISHING THE PLAN'S TIME BOUNDARIES

Once the network plan has been drawn, event numbered, and estimated, the next phase is the use of the estimated elapsed time durations to establish time boundaries in the plan. The time boundaries will be used as the basis of establishing the project schedule. The methodology in network planning techniques comprises a simple arithmetic operation. In essence, the planner will add up all the time paths of activities into each event, and establish the earliest time at which any particular event can be reached. Then the same summation method is used to work backwards and set the latest time that any particular event can be reached. Event times, therefore, are time boundaries.

The Forward Pass to The Earliest Event Time

The Earliest Event Time shall be designated by T_E. Each arrow, including dummies, shall have an Early Event Time (T_E) at its tail and head, or TE_i and a TE_j. The manual method will insert the T_E in a square block at the event.

Consider Fig. 5.0, which has its activities identified, estimated and event numbered.

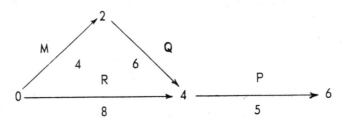

Fig. 5.0

The estimated durations are entered under the individual arrow. From the logic of the diagram in Fig. 5.0, we note that we cannot start operation P until both Q and R are completed. We shall start at the origin Event (#O) and start at 0 time. 0 time is the morning of the first day. We note it is estimated that activity M will take 4 time units. Thus, the earliest we can hope to reach Event #2 is time unit 4. We thus shall add up all the durations of the paths into each event in the forward pass. It is a summation process, each event from the origin is first summed, the highest number entered into the T_E block at the event, and from there the path to the next event is summed. Before the Early Event Time for any event can be established, all of the immediate predecessor events [tails of the arrows coming into the node (event) we are evaluating] must have been calculated.

Thus we note that before we reach Event #4 (the tail of Activity P), operations M, then Q must be executed, in concurrence with R. All must be performed, but it takes longer to execute M then Q (4 + 6 = 10) than it does R. Thus the earliest we can reach Event #4 is time unit 10. All paths into each event must be summed up and the highest total entered into the square at that event, for the T_E. Dummies are paths, even though they have 0 time value, and they must be summed up along with all the other paths into an event.

Fig. 5.1 shows the T_E's for Fig. 5.0.

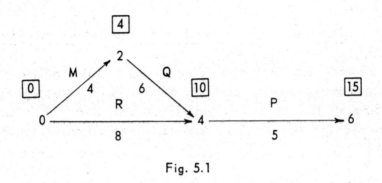

Fig. 5.1

To continue the method, Fig. 5.0 has been expanded to show additional paths. A "bookkeeping" notation has been shown, to indicate the sums of the various paths into each event.

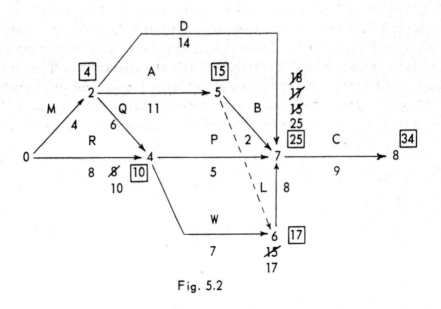

Fig. 5.2

47

Note that the dummy from event 5 to event 6 must also be evaluated. Also, there is a $\boxed{T_E}_i$ and a $\boxed{T_E}$ j for each arrow in the diagram. The forward pass establishes the duration of completion date of the project plan. Examination of the simple network in Fig. 5.2 easily reveals the longest path through the network. An actual project plan of several hundred arrows or more does not reveal the longest path so readily.

The Early Event Time, TE, indicates the earliest time, in project time units (working days, hours, weeks, etc.) that an activity which leaves that event (tail of its arrow) can start.

The Backward Pass to The Latest Event Time

Once the forward pass has been completed to the final, or objective event of the project arrow diagram, and all of the TE's established, the next step is the calculating of the second event time boundary, The Latest Event Time. This will be designated as T_L, and is inserted in a circle at the particular event.

In the forward pass, we proceed on a summation process, in a tail - head sequence from the origin. In the backward pass, a similar approach is used, except that it is a negative summation (subtractive). We have added all of the paths into the Final Event in the forward pass; finding the sum of one long chain of elements in a closed network. Thus the starting point for the backward pass will be the Final Event. The backward pass shall be a subtractive effort, subtracting the activity estimated durations from the Final Event, and entering the smallest remainder in the circle. The smallest remainder, because we are looking for the longest path back from a fixed number, the Event Time of the Final Event. Subtracting the largest number of several numbers from a fixed number will give the smallest remainder. The method goes back, now in a head - tail sequence, an event at a time, and subtracts all paths back into that event from the successor event. Before the Latest Event Time (T_L) can be entered for any event, all of the successor events into that node must have been established.

The T_L's for each event in Fig. 5.2 are shown in Fig. 5.3 a similar bookkeeping system is noted.

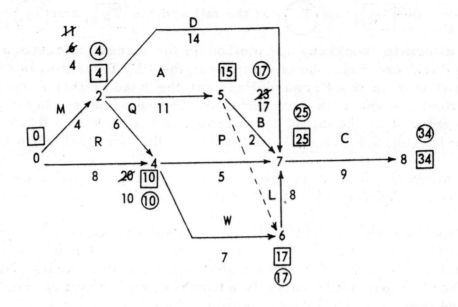

Fig. 5.3

Note that there are two paths back into Event #5, one from Event #7 (through B) and one back from Event #6 (through the dummy). The smaller remainder is 17 (17-0 compared to 25-2 or 23). This can be logically explained by studying the sequence. If 23 were the T_L at Event #5, this would state that Event #5 could be reached as late as time unit 23. But the logic sequence of the diagram shows that L cannot start until both W and A (through the dummy 5-6) are finished. Thus if A finished on time unit 23, then the duration of the project would be 23 plus L (8) plus C (9) or 40. The integrity of the closed network would be violated.

Thus too, at Event #2, there are three paths back into it. If any time later than time unit 4 is accepted as the T_L of Event #2, two of the three paths emanating from Event 2 might not affect the final event time, but the third would give a higher project duration erroneously, if a new Forward Pass were made from that event and starting with the higher T_L.

As a final check, one of the many paths back into the Origin Event must end with a zero time for T_L at the origin.

The Latest Event Time, T_L, indicates the latest time, in project time units, that an activity which comes into that event can finish.

Event times (T_E and T_L) are time boundaries, not a schedule. While much information can be obtained from Event times, much more will be obtained from the schedule times. Activity Times are schedules, Event Times are boundaries.

49

The forward and backward passes establishes four time boundaries for each arrow, the $\boxed{T_E}_i$ and $\left(T_L\right)_i$ at the tail and the $\boxed{T_E}_j$ and $\left(T_L\right)_j$ at the head.

An alternate "bookkeeping" method in the manual calculation of the Forward and Backward Pass, quite common in the PERT system, is to enter the individual sum in the Forward Pass, at the head of that arrow as it enters a particular event, with a bar under it. The remainders, in the Backward Pass, are entered at the tail of that arrow coming back into that event, with a bar over it. Fig. 5.4 shows this methodology, that was derived in Figs. 5.2 and 5.3.

On manually calculating the event times for an actual project, it is recommended that different colored pencils be used for each Pass.

As will be seen in Chapter 16, the recommendation will be to use manual calculations for networks up to 100 activities. For higher activity networks, if updating a schedule is contemplated, the computer should be used. While the author has calculated manually networks of 800 activities, he does not recommend this. For a single calculation, the manual approach should not be considered for over 200 activities.

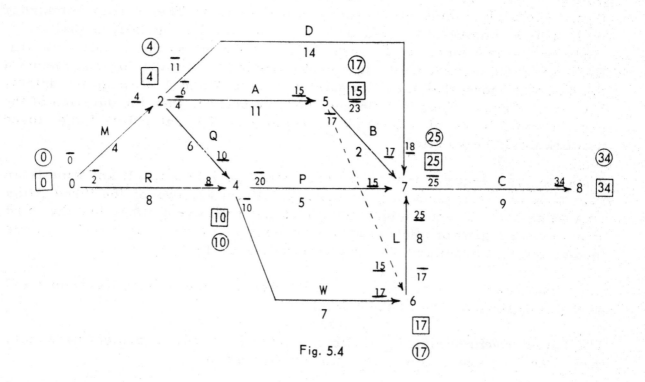

Fig. 5.4

SUMMARY OF MANUAL METHODOLOGY CALCULATIONS

FORWARD PASS	BACKWARD PASS
Start with zero at origin	Start with result of Forward Pass at End or Last Event

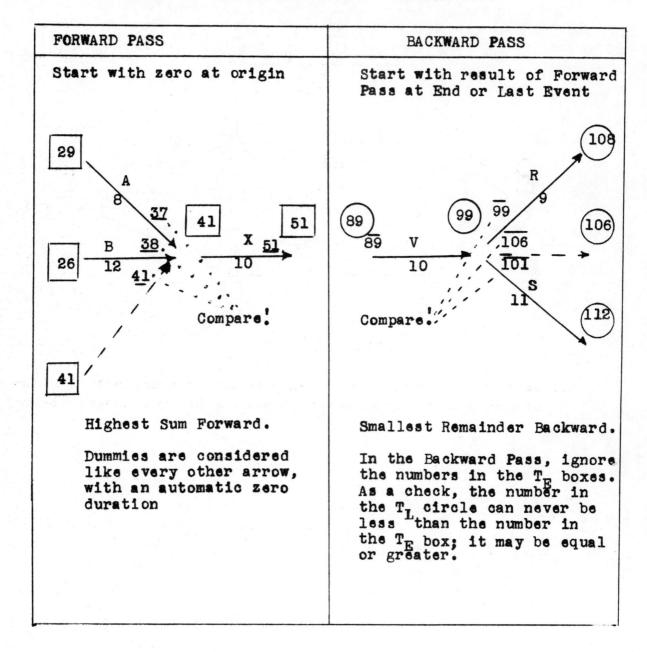

Highest Sum Forward.	Smallest Remainder Backward.
Dummies are considered like every other arrow, with an automatic zero duration	In the Backward Pass, ignore the numbers in the T_E boxes. As a check, the number in the T_L circle can never be less than the number in the T_E box; it may be equal or greater.

PROBLEMS

5.1 Calculate each TE and TL for the following network.

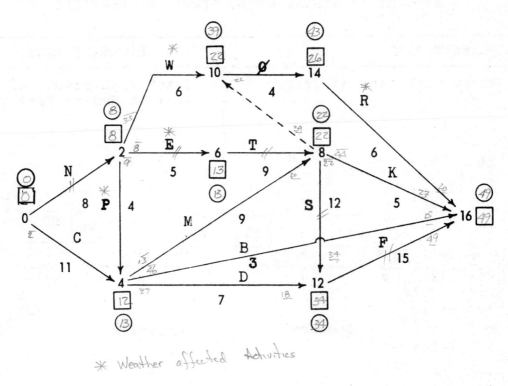

✱ Weather affected Activities

5.2 Sequentially consecutively event number (starting with 3 at the origin), and calculate the event times for the arrow diagram below.

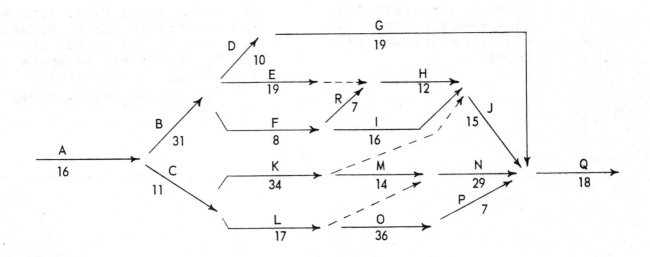

5.3 Draw the arrow diagram plan for the 41 operations listed below, find the completion date, and the Latest Event Times (T_L) for each activity.

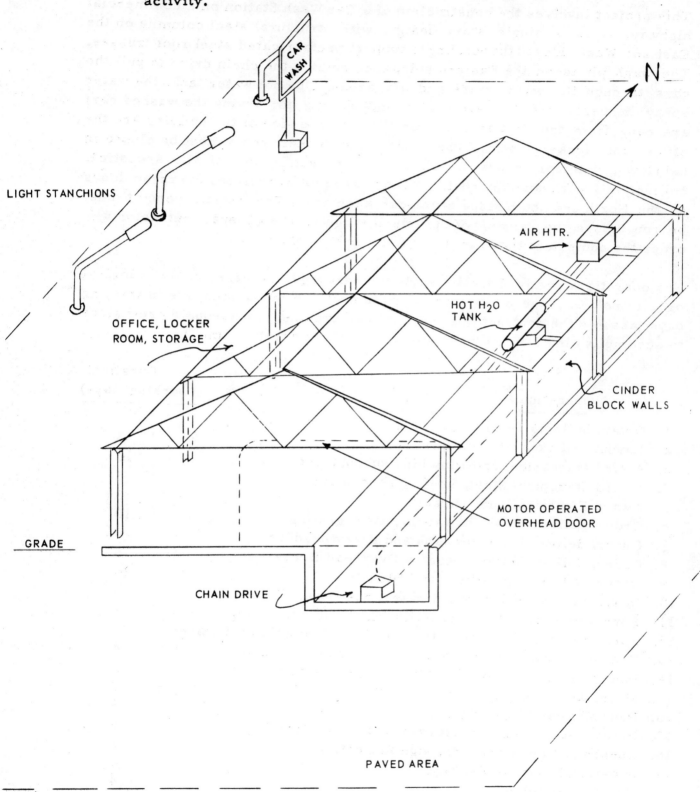

CAR WASH

N

LIGHT STANCHIONS

AIR HTR.

HOT H$_2$O TANK

OFFICE, LOCKER ROOM, STORAGE

CINDER BLOCK WALLS

MOTOR OPERATED OVERHEAD DOOR

GRADE

CHAIN DRIVE

PAVED AREA

BUILDING A CAR WASH STATION

This project involves the construction of a Car Wash Station on a commercial highway. It is a single story design, with structural steel columns on the East and West sides of the building to support prefabricated steel roof trusses. The wash pit is on the Eastern side, and contains the chain drive to pull the cars through the wash, rinse and dry areas. The hot water tank, the water spray headers, and the air heater and blower (for drying the washed car) are hung from the roof trusses. On the Western side of the building are the office and locker room. The office and locker room must be closed in (windows, interior doors installed) before installation of the acoustical ceiling, millwork and floor tiling. There are motor operated overhead doors on the North and South sides of the car wash area. The interior conduit must be run before the hanging of the interior lights. The general contractor will order the equipment listed in items 6 through 9 below.

The operations involved are listed on the next page, not necessarily in logical or chronological sequence. The duration of the operations are in working days (assume the contractor's proposal has provided for adequate crew sizes to accomplish the specific operation in the listed duration time.)

Operation	Duration (working days)
1. Obtain building permit	5
2. Layout and excavate building	3
3. Award Mechanical and Plumbing subcontract	10
4. Award Structural Steel, Roofing subcontract	5
5. Award Electrical subcontract	10
6. Order, deliver Wash Pit railing, steel grating	10
7. Order, deliver hot water tank and spray headers	10
8. Order, deliver Motor Operated Overhead Doors	15
9. Order, deliver Air Heater and Blower	20
10. Install outside underground water, gas, sewer lines	5
11. Form and rebar Wash Pit Walls, East column footings	2
12. Form office-locker room section slab, West column footings	1
13. Pour and cure wash pit walls, East column footings	4
14. Pour and cure office slab, West column footings	2
15. Mechanical rough in (underground)	3
16. Pour and cure Wash Pit Slab	3
17. Deliver steel columns, trusses, galvanized roofing	15
18. Construct light standards, sign foundations	2
19. Install light standards, sign	4
20. Erect Steel columns	4
21. Install chain drive in pit	3

22.	Erect roof trusses, ridge pole, and plumb steel	2
23.	Cover roof	6
24.	Erect cinder block walls	4
25.	Install windows, doors	2
26.	Install plumbing in Wash Pit	3
27.	Install Hot H-2° tank, spray headers, plumbing fixtures	6
28.	Install railing, steel grating over Wash Pit	2
29.	Install motor operated overhead doors	2
30.	Install air heater-blower	2
31.	Hang acoustical ceiling, office and locker room	4
32.	Hang interior ceiling light fixtures	2
33.	Install millwork, office and locker room	2
34.	Final plumbing connections, and test	3
35.	Grade, compact and pave outside area	6
36.	Final electrical connections, and test	3
37.	Paint interior	3
38.	Install floor tile	2
39.	Final Inspection	1
40.	Install interior partitions	5
41.	Run interior conduit	5

5.4 If it is the responsibility of the Mechanical sub-contractor to order the items in #6 through 9 above, will this affect the completion date? By how much? Which items affect the completion date?

CHAPTER 6

THE CRITICAL PATH

In any closed network there is one continuous long path through the network that establishes the duration of that project; and, from actual experience it generally represents 15% or less of all the activities in a real project. The larger the project, in terms of the number of arrows, the smaller the percentage of activities that controls the duration of that project.

In some projects there may be several parallel critical-paths; they are identified in the same way. The criteria for identifying the critical activities which make up the Critical-Path are as follows:

1. The Earliest and Latest Event Times at the tail of an arrow for a given activity are identical, or

$$\boxed{T_E}_i \quad = \quad \left(T_L\right)_i$$

2. The Earliest and Latest Event Times at the head of an arrow for a given activity are the same, or

$$\boxed{T_E}_j \quad = \quad \left(T_L\right)_j$$

3. The difference between the equal Event Times at the head of the arrow and the tail of the arrow is the same as the expected time duration of that arrow.

Note that all three criteria must be satisfied in order for an activity to be critical.

It is only along the Critical Path that savings in time can be effected. Once the manager has that "main chain" identified, then he can exert judgement, and make decisions as how to reduce, or make up time on his project. Decisions to increase the number of men, go on overtime, subcontract the work, change the logical sequence (an alternate strategy), etc. can be made on an optimum selective basis. Note however, the Critical Path can only be reduced so far; the next shorter path will then take over as the new largest path in the project.

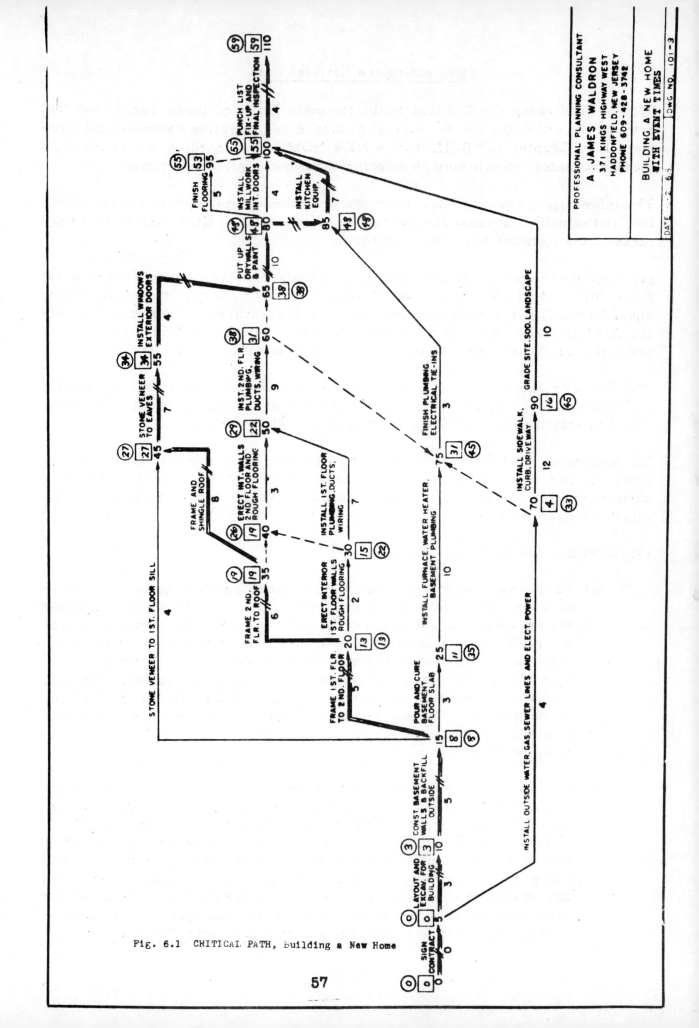

Fig. 6.1 CRITICAL PATH, Building a New Home

57

Examining the Critical Path

Fig. 6.1 indicates the Critical Path, the main chain of those operations that control the end date, for the logical sequence and duration estimates of Problem 1.1, Chapter 1, "BUILDING A NEW HOME." According to this plan and these estimates, it will take 59 working days to complete this project.

The information now available from the time bounded plan proves most selective and valuable. Several situations will be examined to illustrate salient features of the Network Planning technique.

(1) Note that there are two operations in the plan. 15-45, "Stone Veneer to 1st Floor Sill" and 80-100, "Install Millwork and Interior Doors," that while having equal Event Times at both head and tail, are not Critical. They do not satisfy the third criteria: the difference between their equal Event Times at the head and tail must be the same as the expected time duration of that arrow.

The Event oriented PERT system (Chapter 10) originally identified an activity as Critical if it lay between "Critical Neighbors" or critical events. This criteria would falsely identify activities 15-45 and 80-100 as critical.

(2) A dummy, 80-85, lies upon the Critical Path, since it meets all three criteria. These Critical dummies are important because they indicate a change in criticality from one agency to another, such as between sub-contractors, suppliers, departments, etc.

(3) Consider the following situations:

(a) On day #8, there is a labor jurisdictional squabble between the stone masons and the carpenters as to who will erect the wooden scaffold for the veneering. An initial settlement by the Project Manager awards the scaffolding to the carpenters. As a result, on day #8 the stone masons walk off the project and are gone for 5 days. This means that operation 15-45 "Stone Veneer to the 1st Floor Sill" is delayed 5 days or its equivalent duration goes up to 9 days. What effect does this have on the project? The answer is: no effect! There are 19 days available (from day #8 at the head of the arrow to day #27 at the tail of the arrow) to perform the 4 day task.

(b) On day #13, the stone masons return to work in response to a change in the award--they are to erect the remaining scaffolding. Now the carpenters are upset and they leave the project. Operation 20-35, "Frame 2nd Floor to Roof" is delayed two days to the end of day #15 before a compromise is arbitrated. What effect does this two day delay have on the project? The answer: a direct effect! For every day this Critical activity is delayed or extended the end date of the project is correspondingly extended. In this example, our completion date would be day #61, according to this plan with these estimates, rather than the original day #59. Remember, just numerical information is being examined to explain a methodology. Whether a project can stand an extension of two time units (here in working days-depends on the actual circumstances of that real project.

(c) There was a fog and a cold mist encountered on day #8. While the carpenters could proceed with their work on operation 15-20, "Frame 1st Floor to 2nd Floor," it was decided not to pour the basement concrete floor slab, operation 15-25. What effect would this situation have if it is delayed 4 days in the start of the pour? The answer is: no effect! There are 27 days available to execute the original 3 day estimated duration of this activity.

(d) The furnace and hot water heater are purchased from a local distributor who promises delivery by day #11. However, the delivery of this equipment is delayed until day #31. What effect does this late delivery date have on the overall completion date? What is the very latest delivery date that can be tolerated? Delivery by day #31 will have no effect on the project's end date. Note activity #25-27 "Install Furnace, Hot Water Heater and Basement Plumbing." It can start as late as the 35th day (the T_L at Event #25). The equipment to be installed therefore must be delivered by the 35th day. Any time after this day will extend the project's end date.

(e) Examine activity #70-90 "Install Sidewalk, Curbs, Driveway." Note its large estimate of 12 days. On a usual development plot, even up to an acre in size, this is a very conservative estimate, that does not appear realistic. What would we gain if we cut this sub-contractor's estimated duration in half. The answer is: nothing. This operation does not affect critical Event #90. As you will see later, this activity has 29 days of float or slack time. Since this conservative estimate does not affect our end date, it becomes that sub-contractor's problem (he will not make as much profit as he should if he takes that amount of time).

The point here is that all estimates are originally accepted as equally valid; the optimistic, the conservative, the honest and the false estimates are all used on the first Forward and Backward Pass run. Once we find the Critical Path we will then re-evaluate those critical estimates, which will be only 10% or less of all the estimates of activity durations, in light of our time and budget goals.

Let us evaluate another possible situation that could occur with activity #70-90. We accept the conservative estimate. On day #4 this activity was started. By day #10 it was one half complete. At this point the owner decides upon a change and issues a Change Order. It seems the owner learned of a new, exotic, multi-colored asphaltic material and now wants his driveway so installed. The Change Order authorizes removal of the installed work, and for the replacement by the new

59

asphaltic material. The contractor submits a new price, to account for the demolition work (removal of installed originally specified material. He also estimates that as of day #10 it will take him an additional 15 working days to complete this activity. He asks for an extension of time on the contract (the end date) of 15 days. Is he entitled to both the additional costs and an extension of time? The answer is: just the additional costs. If a new forward pass is made from Event #70 at day #10, with a remaining duration of 15 days, it will be noted that there is no change in the end date, the Critical Path remains the same. You will note that time and costs are not proportionately related.

The situations in this section are discussed to emphasize an important part in planning. In the laying out of the Arrow Diagram Plan, there are no contingencies built into the plan. There are two reasons for this: (1) For every contingency thought of by the planner, the chances are that another was forgotten. There are so many possible mishaps and emergencies that they all could not be put into a plan. (2) As evident from the discussion of 3 (a), (b) and (c) above, a different managerial or supervisory decision will be made when a contingency arises, depending on whether a critical or non critical activity is affected. If a non-critical activity is affected, the pressure for immediate action is much less than if a critical activity is affected.

Planning does not contain contingencies. The work is planned as if nothing would go wrong, and it is to be performed in ideal weather. The effect of inclement or limited weather periods which can be predicted by meteorological methods will be introduced when the project schedule is converted to a time scale schedule (Chapter 13). If an outside activity is found to be scheduled in an inclement weather period (anticipated by meteorological history and reports) its equivalent duration will be extended or it will be delayed. Again, if this particular outside activity is not critical, the extended duration may have no effect on the project whatsoever. Weather affects a schedule, not a plan.

(4) Upon examination of the initial plan and the 59 day end date, the owner of the house decides that the project duration is too long. He requests the builder to study the plan, and reduce the overall time. The builder offers the following proposition from the Critical Path diagram.

> --The Subcontractor who can most easily be expedited is the one installing the Kitchen Equipment (operation 85-100) The builder offers to reduce this Critical Operation from 7 days to 4 days if the owner will pay $100 a day premium for each day of reduction, or a total of $300 to reduce the Critical Operation, 85-100 "Install Kitchen Equipment," three days. Is this a good proposition? The answer is: it is only two-thirds of a good proposition! Examination of the Arrow Diagram of Fig. 6. 1

reveals that between Events 80 and 100 there are three parallel paths: 80-95, 95-100; 80-100; and 80-95, 85-100. If 85-100 is reduced three days, only two days will be dropped from the end date. In parallel with the 7 day activity 85-100 is activity 80-95, "Finish Flooring" of 5 day duration. 85-100 can only be reduced two days to a 5 day duration. Then there will be a parallel Critical Path from event #80; 80-95, 85-100 and 80-95, 95-100. Further reduction of the duration of 85-100, without reducing 80-95, will not improve the end date since 80-95 will take over as a Critical Activity. From a duration of 5 days down to 4 days, both 85-100 and 80-95 must be reduced.

A segment of the Critical Path can only be reduced in time so far, until it is joined between time boundaries by a parallel equal duration segment, then both parallel segments must be reduced if further improvement of the end date of the project is expected.

Once the Critical Path is ascertained, the end date obtained may then be compared to the contractual completion date, or other established target dates. If the end date is discovered to exceed the required final date, then examination of the durations along the Critical Path and the methods and resources assumed in the initial plan are to be re-evaluated. Quite often, different methods radically changing sequences are called for by this analysis.

Another part of the testing of alternate resources is the increasing of resources in the plan. Where before there may have been a series of arrows, now they may be placed in parallel if additional resources are to be extended. The "Dummy" will be used to increase or manipulate resources to improve a schedule. This will be demonstrated in a later chapter.

Remember, there is no guarantee that the Critical Path Method will produce an acceptable Completion Date and Schedule on the first analysis. In general (80% or more of actual case histories), the initial schedule must be analyzed, decisions made, and a second run made to obtain desired goals.

Near Critical Activities

It should be obvious that while the Critical Path is the longest chain of operations in the plan, the other parallel and shorter paths are not going to be ignored or relegated to a secondary level of consideration. Once the Critical Path is found, attention must be then focused on those paths that have the least amount of flow time. As a guide, it is recommended that those activities where "float" time (see Chapter 7) is 10% or less of the project duration be considered "Near Critical," and that they deserve almost as much managerial attention as the Critical activities. If a project has an overall duration of two years, those activities that have four weeks of float time or less, deserve surveillance. It is easy to lose 10 working days in a six month period. This technique will develop this information early in the project--it should be acted upon then.

PROBLEMS

6.1 Is operation D in **Fig.** 5.3, page 48 a critical activity?

6.2 Determine the Critical Path in Problem 5.1, page 50

6.3 Determine the Critical Path in Problem 5.2, page 50

6.4 Determine the Critical Path of the Coin Operated Laundry Plan, Problem 2.6, page 25

6.5 What is the Critical Path in the Car Wash Station, Problem 5.3?

CHAPTER 7

SLACK OR FLOAT

Those chains of activities that do not lie on the Critical Path have a certain time flexibility, or freedom, in that there is a difference between the earliest and latest event times for a particular event on that path. This latitude is called "slack" in the PERT nomenclature, and "Total Float" in the Critical Path literature.

Consider the network in Fig. 7.0.

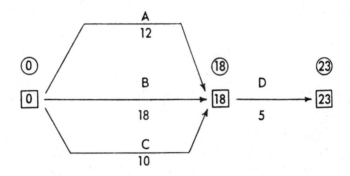

Fig. 7.0

It can be seen that operation A is 6 time units (days) shorter than the controlling, or critical activity B, operation C is 8 time units (days) shorter. This is "float", the latitude that can be tolerated in the execution of these non-critical activities. Slack or total float for an activity is the difference between the widest time boundaries established by its events, and duration of that activity between those boundaries.

$$\text{Activity Slack} \atop \text{or} \atop \text{Total Float} = \left(T_L\right)_j - \boxed{T_E}_i - \text{Duration} \quad (7.1)$$

Total Float indicates the number of days delay, or extension of an activity, that can be tolerated without affecting the project scheduled completion date.

This formula, mathematically valid, does present misleading information to the uninitiated. This will be seen by the expansion of Fig. 7.0 into more segments in the paths, shown in Fig. 7.1, with same path total durations.

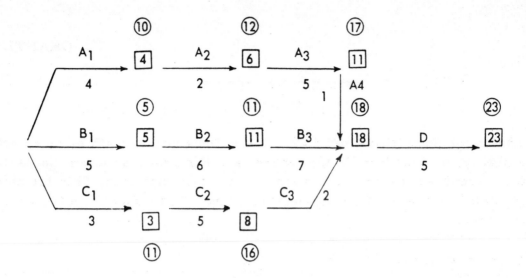

Fig. 7.1

By using the formula (7.1) for each of the arrows or path A, we find that

$$
\begin{aligned}
\text{Total Float} \quad A_1 &= 10 - 0 - 4 = 6 \\
\text{Total Float} \quad A_2 &= 12 - 4 - 2 = 6 \\
\text{Total Float} \quad A_3 &= 17 - 6 - 5 = 6 \\
\text{Total Float} \quad A_4 &= 18 - 11 - 1 = 6
\end{aligned}
$$

In other words, the methodology assigns total float to each individual arrow on a path, but total float is only that latitude of a non-critical, shorter path between critical events. Thus, if A_2 should slip, and use up 4 days of total float, then every arrow on that path loses 4 days of float. This may be verified by changing the duration of A_2 from 2 to 6, making new forward and backward passes to establish new T_E's and T_L's, and then apply formula 7.1 to each arrow.

There is a rationale for this seeming inconsistency that will be noted in the next chapter on scheduling.

The deterministic system of Critical Path also has several other categories of float: Free Float, Independent Float and Interfering Float. Again, these are mathematically valid numbers, but of almost no use to a manager in his decision-making position.

Free Float occurs only at the last arrow on a path that joins a longer path. The formula for Free Float is:

$$
\text{Free Float} = \boxed{T_E}_j - \boxed{T_E}_i - \text{Duration} \qquad (7.2)
$$

Its rationale is the assumption that if every activity on that particular path starts at its Earliest Event Time, and is completed within its estimated duration, then the last operation on that path of activities will have a latitude that will have no effect on the Earliest Starting Time ($\boxed{T_E}_i$) on the subsequent activity leaving the last event on that path. In a real project, such paths may comprise many major operations in series, and it is rarely the actual case where every activity will start and finish early. Life and people just do not seem to function that ideally.

Independent Float is given by the formula:

$$\text{Independent Float} = \boxed{T_E}_j - \left(T_L\right)_i - \text{Duration} \qquad (7.3)$$

There is no known practical use for this number.

Distributed Float

A method does exist, described in Chapter 12, that allows the manager to distribute all of the float in a network to each arrow in that network separately and proportionately to a priority numbering system. In essence, the manager may assign value judgment number to activities, which indicates which activities are more important, or more uncontrollable, and should receive a greater share of the path float for its own. The higher the value judgment number (called "activity weight") assigned, the more of that path's total float is given to that activity. The range of the activity weights may be any one selected; 0-9, 0-100, 0-1000, etc.

This will be called Distributed Float, and allows a manager to apply a safety margin of time to any particular activity he feels ought to have it. For example, in Fig. 7.1, if C_1 was an "Order Elect. Equipment" activity, C_2 was "Deliver Elect. Equipment" and C_3 was "Install Elect. Equipment", and the manager felt that C_2, the delivery activity was the most uncertain or uncontrollable, he could elect to distribute the major part of that path float to activity C_2, Activities C_1 and C_3 would get proportionately less, in accordance with the value judgment priority numbers assigned.

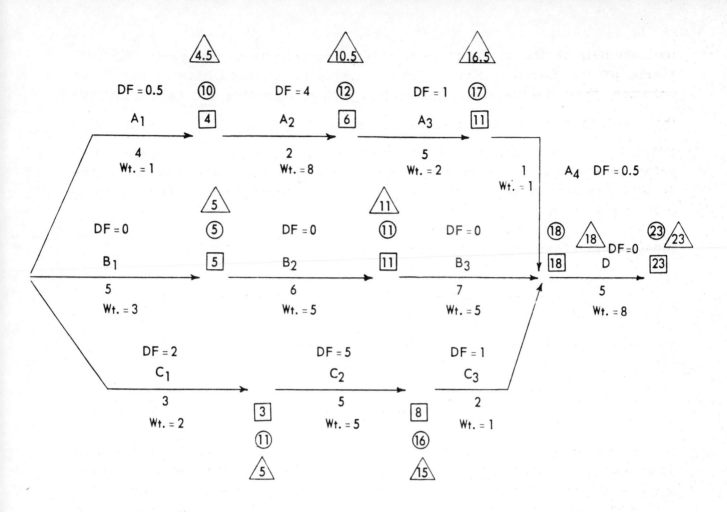

Fig. 7.2

Fig. 7.2 indicates arbitrarily assigned activity weights, based on a 0 to 9 range of uncertainty, with 9 the most uncertain, to the network of Fig. 7.1. Note every activity has been given a value judgment number. On a real project, the critical path will not be known, so the assignment of weights is made along with the disassociated estimate. Distributed Float is calculated by adding up the weights on a particular path, and proportioning the total float of that path to each activity on that path. Dummy arrows are given a value of 0 weight. The results may be shown as a Distributed Event Time (T_D) at the head of the specific arrow. The Earliest Event Time is the same as before. For purposes of example, the Distributed Event Times are shown in triangles on Fig. 7.2. Chapter 12 describes the formal methodology of calculating Distributed Float, and Distributed Event Times.

There is one computer program today that will automatically generate a form of Distributed Float (it is called "Scheduled Float"), based on the assignment of activity value weight numbers from 1 to 9. However, in the computer program, a time bias has been inserted, that tends to assign a portion of the "scheduled float" to the end of the project. This cannot be overriden or bypassed, so unfortunately, whether the manager wants it or

66

not, the distribution of float is controlled by the computer. The manager should have the perogative of controlling the distribution, based on his value judgment assignments. The fact that this computer program will re-assign the available float, based on a formula, is another case that finds the manager working for the system, rather than the system working for the manager.

PROBLEMS

7.1 What total float will a critical activity have? Free Float?

7.2 Calculate the Total Float for each activity on the network in problem 5.1, Chapter 5.

7.3 Calculate the Total Float for each activity on the network in problem 5.2, Chapter 5.

7.4 Calculate the Total Float for each activity on the network plan of problem 5.3, Chapter 5.

CHAPTER 8

NETWORK SCHEDULES (ACTIVITY TIMES)

Once the time boundaries of Earliest and Latest Event Times have been established, we can begin to schedule the project. Referring to figure 8.0 below, we will investigate the earliest we can start each activity. It is obvious that the Earliest Event Time is the limiting factor. We cannot plan to start any activity sooner. Thus, the Earliest Event time at the tail of the arrow becomes our Early Activity Start Time.

$$ES = \boxed{T_E}\ i \tag{8.1}$$

If we start an operation as early as possible, we can then expect to have it finished at the completion of its expected time duration; therefore, Early Activity Finish Time equals Early Activity Start Time, plus the expected time duration or:

$$EF = ES + D \quad \text{or} \quad \boxed{T_E}_i + D \tag{8.2}$$

We know that the latest any activity can finish is the other time boundary, or the latest Event Time at the head of the arrow. Thus, the Latest Activity Finish Time equals $\left(T_L\right) j$

$$LF = \left(T_L\right) j \tag{8.3}$$

Lastly, if we are to finish as late as possible, and still remain within the integrity of our time-bounded plan, then our latest possible activity start time is the Latest Activity Finish Time less the expected time duration.

Latest Activity Start Time LS equals $\left(T_L\right) j$ less D

$$LS = \left(T_L\right) j - D \tag{8.4}$$

Note that the activity times (schedules) will be those numbers entered <u>on</u> the heads and tails of the arrows, shown in the "bookkeeping" method of Fig. 5.4, page 31.

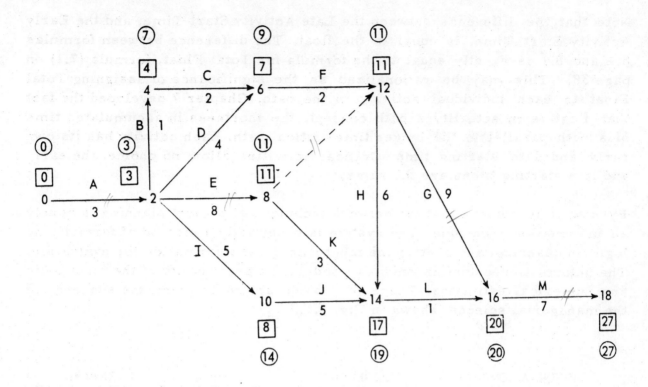

Fig. 8.0

The schedule for the above plan in working time units, is as follows:

Activity		Duration	ES	EF	LS	LF	TF
0-2	(A)	3	0	3	0	3	0
2-4	(B)	1	3	4	6	7	3
2-6	(D)	4	3	7	5	9	2
2-8	(E)	8	3	11	3	11	0
2-10	(I)	5	3	8	9	14	6
4-6	(C)	2	4	6	7	9	3
6-12	(F)	2	7	9	9	11	2
8-12	DUMMY	0	11	11	11	11	0
8-14	(K)	3	11	14	16	19	5
10-14	(J)	5	8	13	14	19	6
12-14	(H)	6	11	17	13	19	2
12-16	(G)	9	11	20	11	20	0
14-16	(L)	1	17	18	19	20	2
16-18	(M)	7	20	27	20	27	0

Table 8.0

69

Note that the difference between the Late Activity Start Time, and the Early Activity Start Time, is equal to the float. The difference between formulas 8.4 and 8.1 is exactly equal to the formula for Total Float, formula (7.1) on page 38. This may be rationalized as the significance of assigning Total Float to each individual activity in the path. Chapter 7 developed the fact that float is in actuality a path concept, the shortness in accumulated time of a path parallel to the longer time critical path. Each activity has its own early and late starting time. Critical activities allow no choice, the early and late starting times are the same.

By now it is obvious that the network technique of project planning is purely an information technique. The system is essentially a method of formalizing logic in planning, and sorting the input numerical information for evaluation. The information output is only as good as the practicality of the input logic and numerical estimates. There is a basic arrow diagram, the sequence of the managerial science, shown in Fig. 8.1.

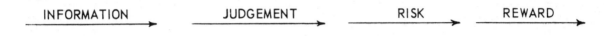

INFORMATION → JUDGEMENT → RISK → REWARD →

Fig. 8.1

This text only concerns the first arrow in Fig. 8.1; this is only an information system. Properly used, it will generate complete, timely, explicit, and accurate information. The philosophy of good management is fairly basic. Decisions based on complete, up to date, or predictive and explicit information by a capable manager should lower the business risk, and increase the business reward. The technique does not solve problems, humans or groups of humans acting in concert do that. In particular, this technique does not solve "people problems". If a particular manager or executive has a personal philosophy of waiting long enough, so that a worrisome problem will go away, he will not be greatly helped by this technique. It cannot be stressed enough, human judgement must be applied to the schedule, cost, and resource allocation information generated. It will be seen that some of the numbers generated in a schedule will be mathematically valid — but highly impractical or impossible for use.

Calendar Dating

Up to this point, we have been using consistent working time units (working hours, working days, working weeks, etc.), and have generated the project schedule composed of such time units. Now the project schedule shall be calendar dated, by constructing a project calendar.

For the schedule in Table 8.0, we shall construct that project's calendar reference. For purposes of example, it will be assumed that the project shall start on Monday morning (time 0), January 28th, 1963. The project shall work a 5 day week, and observe the holidays February 12th and 22nd. The calendar is simply constructed by listing in columnar form the project days, from 1 to 27 and listing the appropriate calendar dates next to the particular project day.

Project Day	Calendar Day	Project Day	Calendar Day	Project Day	Calendar Day
1	1/28/63	10	2/8/63	19	2/25/63
2	1/29/63	11	2/11/63	20	2/26/63
3	1/30/63	12	2/13/63	21	2/27/63
4	1/31/63	13	2/14/63	22	2/28/63
5	2/1/63	14	2/15/63	23	3/1/63
6	2/4/63	15	2/18/63	24	3/4/63
7	2/5/63	16	2/19/63	25	3/5/63
8	2/6/63	17	2/20/63	26	3/6/63
9	2/7/63	18	2/21/63	27	3/7/63

Table 8.1

Thus the project of Fig. 8.0, if started on January 28th, 1963 should finish on March 7, 1963, on a 5 day work week, 2 holiday calendar schedule.

Note several things, that now refine our initial estimating. Based on what has been learned so far, the original rules for estimating a project, given in Chapter 3, may be further modified, as follows:

1. When inserting estimates in the initial plan, they are based on the consistent working time units. Thus a delivery of equipment, which is quoted as being 60 calendar days, would be inserted as a ratio of the working days to the calendar days. (5/7 of the calendar days is a good approximation), or 42.5 working days.

2. In the initial estimate, every activity should be first considered, as being performed under ideal conditions. Outside activities should be initially thought of as being performed in a balmy, late spring period. After the project schedule is made, and calendar dated, the position of the activity in the calendar time schedule should be evaluated, and re-estimated if it occurs in a bad weather condition. For example, assume a 2 calendar year project starts in the spring, and it is a high rise building. The initial outside activities, based on spring time conditions, have been estimated accordingly. But the duration of the chains, or paths of physically outside activities extends into winter. Consider the erection of structual steel. From the initial schedule,

steel erection starts in July, and extends into January. Picking an arbitrary date for the start of bad or cold weather, at the site location. the initial schedule is examined to see what outside activities are scheduled to continue from this climotological reference point. Extending the example, let us assume we find the structural steel erection on the 12th floor of an eighteen floor building will occur about November 15th, and November 15th is the selected beginning of bad weather. Up to this point a three (3) working day period had been estimated for each floor of steel erection. But due to the possibility of bad weather after November 15th it is felt the erection crews will become more inefficient by about 75%. Thus, those steel erection activities that start after November 15th will be re-estimated at four (4) working days each.

The reason for the re-estimate is apparent. The critical path has been found, and the effect of a re-estimate on a chain of activities will differ in importance, according to whether that chain is critical, or on a float path.

The effects of bad weather can be major in any construction project, or any project involving operations affected by climate (the delivery of a control console for a portable nuclear power plant in the Antarctic was affected by the icing of McMurdo Sound). The specific day on which it rains or snows is difficult to predict in advance. One company actually uses the Farmer's Almanac in their initial evaluation of the first schedule to adjust the schedule. While most contractors know in their particular area that they may count on 22 working days in August, 20 in September, 14 in October, 10 in November, etc., they can't forecast which date will limit work due to bad weather. One common approach is to arbitrarily state that the later days of a week are working days, so that by November, they state they will probably work Thursdays and Fridays only. Remember, the concept is to get information, and this will produce conservative calendar completion dates.

Some programs use a "rain day" in the calendar date conversion. Where the early days of a period, such as a work week or a work month, are adjudged lost due to inclement weather. This is a fallacious approach since it assumes all activities are weather affected in this time period, and adjusts all of the activities schedules accordingly. If a delivery of an item, or a design activity, or an inside activity that is not weather affected, fall into the "rain day" period, their duration estimates are artifically extended. Only weather affected activities are to be adjusted for anticipated inclement weather. Weather affected activities are adjusted in Chapter 11.

3. The effect of natural time activities should be evaluated against the calendar schedule. Natural time activities are those which have their

durations established by nature, not the efforts of man. The curing of concrete, the drying of plaster, the settling of catalyst, heat treating a machined piece, are examples of natural time activities. Once started, these activities will consume their natural duration regardless of the calendar date. As a case in part, let us assume that activity G (12-16) of Fig. 8.0 is "form, pour, and cure concrete slab X." From the schedule we note that this 9 working day duration activity is scheduled to start on a Monday, February 11, 1963.

Let us assume that the forming takes 3 days (a day off for the holiday indicates forming finished 2/14/63). Allow one day for the pour, on Friday 2/15/63. Curing starts immediately and goes on over the weekend. Thus, while 9 working days were initially assigned, 2 of those 9 days occur in a natural time activity when no working day is scheduled. Under these circumstances, activity G could be re-estimated to seven working days. Since G was on the critical path, the project completion date will be reduced by 2 days.

Analyzing the Schedule

Judgement must be used by a human to establish the practicality of the information. From actual case histories, the following situations occurred.

1. In a pre-case concrete structure fabrication project, the critical path went through the design and installation of specially made forms. The pouring of the concrete mixture was not critical. Thus, the delivery of the concrete activity arrow had an Early Activity Start Time and a Late Activity Start Time. But the use of the Early Activity Start Time was highly impractical; it would be in essence saying the concrete truck could pull up outside the plant and wait 18 hours (the Late Activity Start Time) before the pour started. The ES time was mathematically valid, but highly impractical.

2. A maintenance project in a paper mill found the installation of a new conveyor motor critical. The shutdown of the paper machine had 3 days of float. The Early Start Time of the "Shut Down" arrow was extremely impractical since the machine produced a profit making product. Only the Latest Activity Start Time was of value to the superintendent. This told him the latest possible shut down time for the minimum loss of production.

3. In the construction of an 18-story office building, the site had a water table fairly close to the surface. The water table was only four feet below the site elevation of the sub-basement slab. De-watering was necessary, and the de-watering arrow started after the first set of caissons were sunk at the site. Here the Latest Activity Start Time was of no use, otherwise the contractor would find himself attempting to pour the sub-basement slab under several inches of ground water.

Blank Estimates

Quite often an activity, or several activities, are un-estimated when the scheduling step is reached. This may be due to a lack of understanding of a sub-contractor as to how to furnish a network estimate, or a lack of co-operation, or it is an unknowable time duration activity (how long does it take to lower by 12 inches the water table of Denver, Colorado? The de-watering example in the section above). There is still information available to the manager, to allow decisions to be made. For the blank activity he assigns a zero value.

In Fig. 8.2, operation J is a blank estimate.

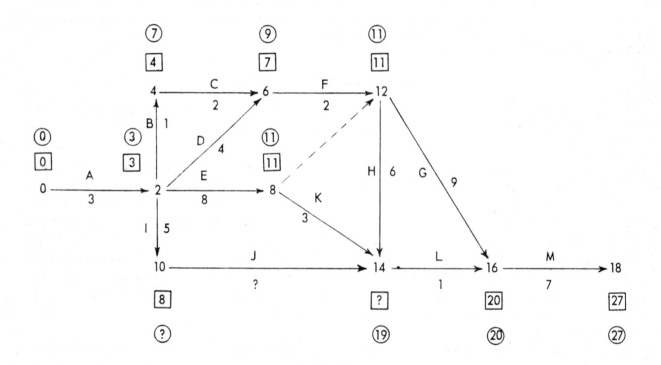

Fig. 8.2

However, if the manager desires, he can assume that the other paths of activities are valid, and that all time boundaries (event times) except those at event #10 and #14 are real and acceptable. Thus, he is able to have determined the outer time boundaries for operation J. He knows that J must occur between day 8 and day 19. He can use this information to advantage; by <u>directing</u> a duration equal to, or less than these time boundaries; or by going on record by saying that for every day after day 19 that J drags on, the project will be extended. This type of information can be used to make decisions or influence actions on uncontrollable parties or agents involved in the project.

Resource Assignments

It will be remembered that in Chapter 4, Time Estimating, that one of the criteria for estimating an activity was the assignment of the "usual, practical" resources available, or desirable. The other criteria was the "disassociation" of the arrow; that is the complete segregation of an arrow from the rest of the diagram while estimating its duration. Then a situation was given, where three concurrent activities, all emanating from the same event, had the same resource assigned to each one, in its disassociated estimate. Let us now re-evaluate that recommendation, in the light of what has been learned from the method of establishing time boundaries, and calculating the schedule.

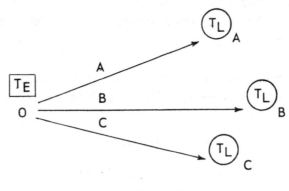

Fig. 8.3

Each arrow has the same Early Activity Start Time, but they have an individual Latest Activity Finish Time. From the schedule, we may obtain the following information:

1. If any one of the three arrows is on the critical path, then the same resources that were assigned to the other two arrows in the disassociated estimate, are obviously inadequate. Decisions must be made as to increasing the number of personnel, working overtime, assigning the work outside the "house", etc.

2. If none of the three arrows are on the Critical Path, then the time boundaries of each can be evaluated as below:

    Case 1. In Fig. 8.4, one of the three arrows is scheduled to start at its Early Start Time. Then the activity with the highest Latest Activity Finish Time is noted, in order to use that time boundary as a second one for evaluation. Then the third arrow is fitted between the Early Activity Finish Time of the First, and the Latest Activity Start time of the third.

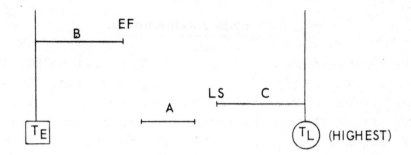

Fig. 8.4

The sequence has been selected at random. Note now this is essentially scheduling the three activities in series, even though they are shown in parallel on the arrow diagram. If this case occurs, there is no worry about craft interference in a small room, or the safety requirement of the welder only being in that room.

Any overlaps, as shown in Fig. 8.5, gives information to the manager for decision making as to overtime, obtaining additional help on a temporary basis, etc.

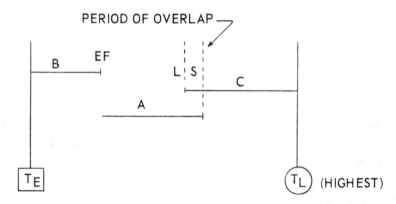

Fig. 8.5

The re-analysis, and re-estimating should only be done once. Continually testing new estimates, or new configurations of logic, rapidly arrives at a point of diminishing returns in valuable information. After the first re-evaluation, allow the plan and schedule to stand, until the project has sufficiently progressed to allow the first up dating of the progress of the project.

Contingencies

It should be apparent by now, that the initial plan does not have any provision for contingencies inserted into it. For every anticipated contingency (delivery slippage, extreme bad weather, jurisdictional strikes, etc.) considered there will be many alternate possibilities forgotten. We noted the re-evaluation on the calendar schedule for the start of limiting inclement weather. The best use of network planning is to start with the basic operations laid out in sequence. Only after a contingency occurs, or is definitely anticipated, should its effect be inserted into the logic, or activity duration estimates. The reason for not putting in "safety factors" in the beginning, should be obvious. The decisions the manager will make will differ, according to the path on which the contingency or upset occurs. If an emergency or slippage arises on a critical or near critical activity, action that will be taken will differ greatly than that taken on a floater activity that suffers a setback or an extension in duration.

Planning With Available Resources

The logic used in making the initial plan, will be based on the resources available, or felt to be initially desirable. The plan for the metal piece fabrication, shown in Fig. 2.9 on page 15, has an implicit resource behind it. The sequence of spinning a particular lot, then drilling, welding, etc. implicitly assumes just one facility to execute those activities. If the manufacturer had two facilities so that lots #1 and #2 could be executed concurrently, the initial plan would look like Fig. 8.6.

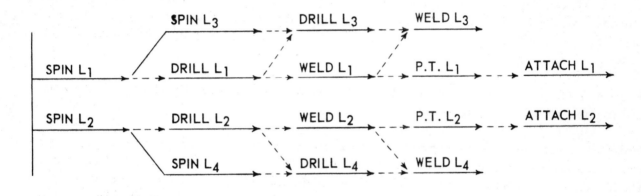

Fig. 8.6

From this it can be seen that another versatile aspect of the managerial tool, is the rapidity with which alternate strategies can be composed and evaluated. Thus is the case of Fig. 8.6, the management can explore the advisability of installing additional facilities, and compare the time and cost schedules, from network techniques, against additional rewards of higher profit, or contractual bonuses. This is quite a common approach in construction; the testing of the minimum number of forms to be employed in concrete work.

It has been noted that this intriguing aspect of network planning can create a form of a "people problem" — some planners can get so fascinated and involved with the alternate strategy aspects, that they tend to spend an inordinate amount of working time playing "Business Monopoly".

This concept of network planning, then scheduling, is an essence of the "Management by Exception" philosophy, wherein the manager learns the management and control of those critical items, and delegates the non-critical operations to lower echelon personnel. Thus, reasonable self-discipline or control is to be exerted by or over the project planner.

The actual network plan may be composed by a lower echelon person, either a staff or line assistant. It is mandatory that the planner have adequate working knowledge of the method of execution of the project. The manager has the ultimate right of review and acceptance, since it is his responsibility for the attainment of the specific goal of the project.

Fig. 8.8 indicates a typical Critical Path Output from a computer.
Fig. 10.7 indicates a typical Critical Path Output calendar dated.

"Loading" a Plan With a Contract Completion Date

If a specific completion date is known, or committed, it may be added to the network as one arrow from the origin to the objective or final event. Its duration is the elapsed working time units for the official start of the work to the contractual end date. If this arrow has float time, then the Critical Path of the plan will take longer than the contractually allowed time, and a management review and decisions are called for. If the "Contract Limit" arrow is Critical then the plan and its estimates indicate completion ahead of the required end date

Fig. 8.7 has added a "Contract Limit" arrow to the diagram of Fig. 8.0, with a duration of 30 working time units.

Note that since the new arrow 0-18 of 30 time units in the longest path, or critical, it in essence gives 3 time units of Float Time to the original Critical Path The Early Start Times (ES) are unaffected, since they were made by the Forward Pass. The Late Finish Event Times (LF) are unaffected, since the Backward Pass starts with the highest end date at the Objective Event.

In general, the loading of a network plan with a contract end date is not recommended The purpose of the technique is to identify the real Critical Path for managerial surveillance and decision. It may be difficult to find the real Critical Path in a manually calculated schedule if the plan's Critical Path is longer than the target date; the addition of the "Contract Limit" arrow serves no purpose other than a quick comparison of the two end dates. In a computer generated schedule, sorting the activities by total float time will indicate the real Project Critical Path.

It is in the area of dissemination of information that the greatest problem arises. If a project can be completed ahead of a contract end date, it may not necessarily be to the advantage of the contractor to do so and to reveal this information initially. When the aspects of resource allocation are discussed further on in the text, it will be seen that it may be economically desirable and ethically and logically allowable to use the additional time to further minimize the resources assigned to the particular project. In actuality, the ideal case is to make as many activities as critical as possible to minimize the resources. This, of course, is a maximum risk position. But in between, there is some point where the float time should be available to the contractor for the manipulation of his resources.

Another aspect is the Parkinson's Law effect--the work will fill up the time allowed, if it seems that there is more than adequate time available for the completion of the project, human nature being what it is, there will be a considerable slacking of effort with valuable float time being wasted in the beginning of the project.

Another approach, that of inserting target dates into a network, is discussed in the next chapter.

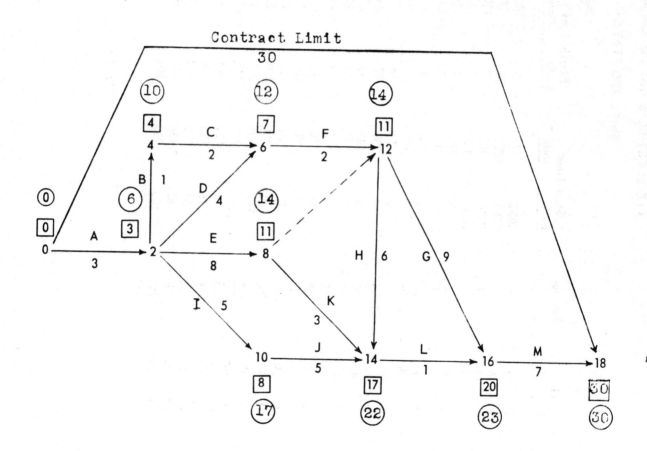

Fig. 8.7

XYZ ELECTRIC & GAS CO. C.P.M. COURSE

SWITCHING STATION

Critical Path	Job # I	Job # J	Duration	Start Time Early	Start Time Late	Finish Time Early	Finish Time Late	Float Total	Float Free	
	2	3	6	0000	174	6	180	174	0000	START ONE LINE
*	2	6	30	0000	0000	30	30	0000	0000	TRANS SPEC
	2	10	3	0000	107	3	110	107	0000	BRKR SPEC
	2	14	1	0000	131	1	132	131	0000	DISC SW SPEC
	3	4	3	6	182	9	185	176	0000	EST CONT REQ
	3	5	1	6	184	7	185	178	2	CONTINUE ONE LINE
	3	18	30	6	180	36	210	174	174	START ELEC GEN ARRGT
	3	5	0000	9	185	9	185	176	0000	DUMMY
	4	57	7	9	264	16	271	255	3	START CABLD DIAG
	4	62	8	9	239	17	247	230	2	START CARRIER SPEC
	4	66	30	9	241	39	271	232	171	CONT SCH
	4	67	3	9	267	12	270	258	30	START CONT RM ARRGT
	5	52	10	9	185	19	195	176	0000	FINISH APPROVE ONE LINE
*	6	7	60	30	30	90	90	0000	0000	TRANS BIDS
*	7	8	30	90	90	120	120	0000	0000	TRANS ORDER
*	8	9	90	120	120	120	210	0000	0000	TRANS SHOP DWGS
*	8	90	180	120	394	300	574	274	274	TRANS DELIVERY
	9	18	0000	210	210	210	210	0000	0000	DUMMY
	9	66	0000	210	271	210	271	61	0000	DUMMY
	10	11	45	3	110	48	155	107	0000	BRKR BIDS
	11	12	20	48	155	68	175	107	0000	BRKR ORDER
	12	13	35	68	175	103	210	107	0000	BRKR SHOP DWGS
	12	90	120	68	454	188	574	386	386	BRKR DELIVERY
	13	18	0000	103	210	103	210	107	107	DUMMY

Fig. 8.8

80

PROBLEMS

8.1 Calculate the schedule for problem 5.2, page 50

8.2 What would the acceptable schedule limits be for operation H in the network below, if all the other times are valid, and the basis of the duration of this project.

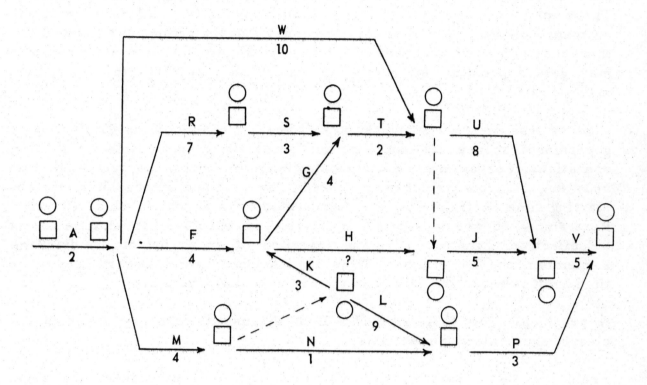

8.3 Calculate the schedule for the "Building a House" problem, problem #2.5 on page 25. Use the schedule forms found in the appendix.

8.4 Calculate the schedule for the plan of Building a Coin Operated Laundry, problem 2.6, page 25.

8.5 Calculate the schedule for the plan of Building a Car Wash Station, problem 5.3, page 52.

8.6 Show the change in schedule times due to the change in assignment of the Order and Delivery activities in problem 5.4, page 53.

CHAPTER 9

INSERTING A TARGET DATE

There is an increasing trend today for contractual requirements for completing a project, or a specific phase of a project such as beneficial occupancy of a building, at a pre-selected calendar date. Quite often this target date, or milestone, is arbitrarily selected. The basis of selection may be an intuitive guess as to a practical date, but it becomes rigid, contractual time boundary. There may be a liquidated damages clause, which is a unilateral penalty of so many dollars a day after this target day that the project is extended. Or, there may be a bi-lateral penalty — bonus clause, so that the project executor may gain a premium for every day before that target date that the project is finished.

Another type of a forced target date is found in road building. Each state or government agency has specification limits of calendar time for specific operations. For example, earth moving may only occur between April 15th and Nov. 1st., concrete paving between June 1st and October 30th etc. The starting time restraint may be represented by an arrow, labeled "Calendar Restraint — 66 working Days". However, the mandatory finish time cannot be effectively controlled by the insertion of an arrow into the diagram. Remember, the Latest Event Time T_L, is found by a Backward Pass. This time limit is set by the path from the arrow head to the end.

A graphical time scale method will be discussed in Chapter 11 which will clearly show calendar restraints.

From the PERT system, we shall explore a method of inserting a target date any where into the network, in order to learn if we can attain the desired target date. If not, we shall learn just what activities affect the attainment of that target date, and by how much.

The insertion of the target date, no matter how capricously it was established, comes after the normal forward and backward pass. The steps are:

1. Ignore the demanded or desired target in the planning of the project. The plan is laid out under the basic philosophy of separating planning from scheduling. The best logic is applied to the arrow diagram.

2. The usual estimating procedure of disassociated estimates, applying the usual practical resources to each arrow separately, is followed.

3. The normal forward and backward passes are made to establish the Earliest and Latest Event Time boundaries.

4. The Critical Path is established.

After this normal planning phase is accomplished, <u>then</u> we insert the outside target date. This new event time shall be called the Scheduled Event Time, and designated T_S. The technique consists of replacing the (T_L) at the particular target or milestone event with the T_S. To keep the symbolism intact, we shall insert the T_S in a hexagon $\langle T_S \rangle$.

A second backward pass is made from that target event, back to the origin. The (T_L) 's on that backward path are replaced by $\langle T_S \rangle$'s, by the same backward pass method of subtracting durations from the successor $\langle T_S \rangle$, negatively summing up all paths back into each predecessor event, and entering the new smallest remainder at the particular event. Normally, negative event times will appear, since the T_L represented the time boundary established in the normal backward pass. The T_S generally is less than the T_L at the target event.

If the target event is the final event of the project plan, and the inserted T_S is less than the final normal event time T_E, the original critical path stays the same, but now a negative total float or slack appears. The formula for Scheduled Float is:

$$\text{Scheduled Float} = \langle T_S \rangle \, j - \boxed{T_E} \, i - \text{Duration}$$

(9.1)

Consider the network in Fig. 9.0

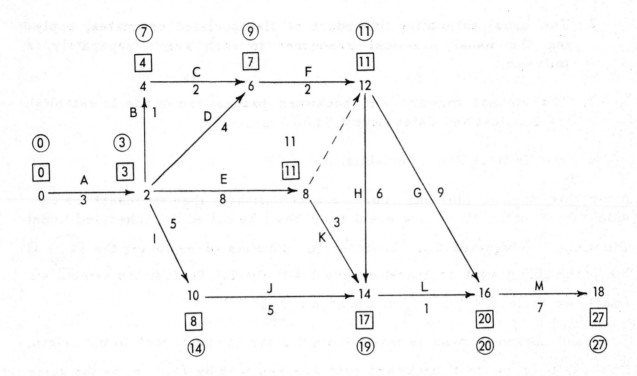

Fig. 9.0

Case I. Assume the time units are now in working weeks. The plan in Fig. 9.0 indicates the best logic, the best estimates and a completion in 27 weeks. However, there is a liquidated damages target date for completion in 23 weeks. If the $\langle T_S \rangle$ of 23 is inserted at the final event, #18, nothing will change. The same critical path, activities A, E, dummy, G and M still control the duration of the project. It is here that decisions will have to be made in order to make up the desired time. The original float of 0 or the critical path is now -4

From Scheduled Float Formula, (9.1)

Scheduled Float of M = 23 - 20 - 7 = -4
                    G = 16 - 11 - 9 = -4
                    E = 7 - 3 - 8 = -4
                    A = -1 - 0 - 3 = -4

(verify the T_S's of all the events by making a second backward pass from event #18, using a $\langle T_S \rangle$ 18 of 23.)

Actually, the information above can be seen by inspection. When a T_S is inserted at the final event, and it is less than the T_E of the final event, it is known that the Critical Path will stay the same, but now have a negative float or slack.

84

The valuable information for the decision making manager comes from the insertion of a $\langle T_S \rangle$ at an event that is not on the normal critical path, but on a normal float path. Then the information will be generated as to which paths affect the reaching of that original non-vital event by the Scheduled Event Time inserted there.

Case II. In Fig. 9.0, a $\langle T_S \rangle$ of 13 is desired at event number 14, Fig. 9.1 shows the results of the second, T_S backward pass.

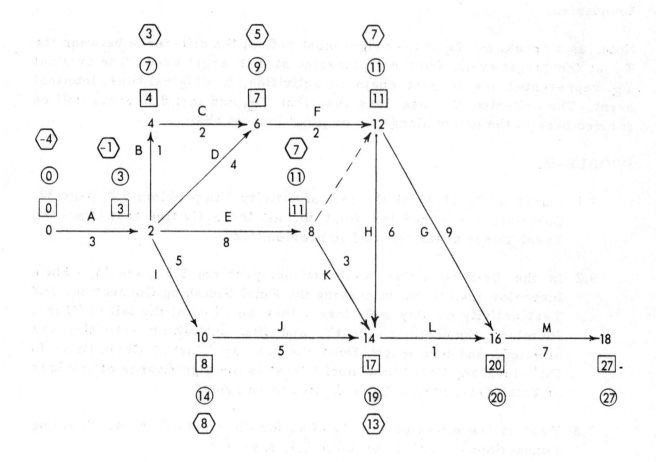

Fig. 9.1

The second pass, establishing the T_S for each event in Case II, now indicates different orders of criticality. Each one can serve to trip up the manager in attaining his target goal of week 13 for attain-event #14. There is a (-4) float path: A, E, H, which is the worst. However, concentrating on this chain of activities to the exclusion of others will result in a project slippage past the desired target date. There is a -2 float path; D and F. There are two (-1) float paths; B and C, and K. There is a 0 float path: I, J.

Remember, this is only an information system. The purpose of the use of the Scheduled Event Time is to point out the chains of activities that affect the reaching of the desired T_S. Once decisions are made for increased crew size, overtime, sub-contract some activities, and new estimates of durations attained, a new plan has been established. Then a second forward and backward pass must be made to establish the new duration, and new event time boundaries.

Note, as a check, the T_S at the origin must reflect the difference between the T_E at the target event, and the T_S inserted at that target event. The original T_E represented the longest chain of activities, in original time, into that event. The inserted T_S was less than that T_E, and that difference will be carried back to the origin along that original longest chain.

PROBLEMS

9.1 Insert a T_S of 75 at the tail of activity J in problem 5.2, page 32. Calculate the scheduled float for all the activities that affect the Event Times at the tail of J in problem 5.2.

9.2 In the Building a Car Wash Station, problem 5.3, page 33, a State Inspector insists on witnessing the Final Plumbing Connections and Test activity on Day #40. Insert this T_S of 40 at the tail of "Final Plumbing Connections – Test", and find out which activities are affected, and how much. Does the activity "Install Chain Drive In Pit" lose any float? How much? What is the significance of any loss of Total Float when a T_S is applied to an event?

9.3 What is the effect of this T_S of 40 for the start of "Final Plumbing Connections – Test" in problem 5.4, page 35?

86

CHAPTER 10

PROGRAM EVALUATION AND REVIEW TECHNIQUE (PERT)

PERT was originally devised as a tool to aid program managers in predicting probability of reaching specific milestones, and forecasting uncertain events in Research and Development programs. It is primarily a reporting and monitoring technique, and was designed for the user to report status, and predict the probability of attaining a target or milestone date.

Originally, PERT was event oriented — the points in time when specific activities, or operations would be finished. The project plan described in Chapter 2 for the installation of the new reproduction machine, would look like the diagram in Fig. 10.0.

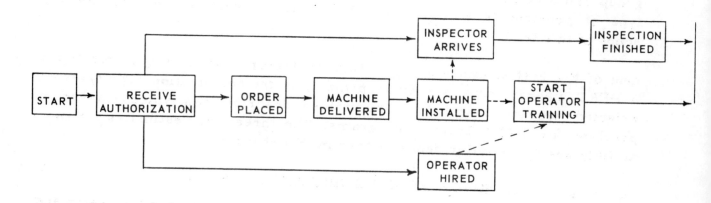

The arrows in Fig. 10.0 still represent time consuming and resource utilizing activities.

A source of confusion in the past has been the situation that at least ten different PERT "systems", living or now deceased, have been promulgated by various government agencies. These systems are all variations of the same fundamentals. However the variations are significant enough to cause confusion to private firms who are dealing with several agencies concurrently.

The variations are primarily between an event oriented and an activity oriented system. One major problem in the event oriented system was the definitions to be assigned to an event. Fig. 10.1 points this out.

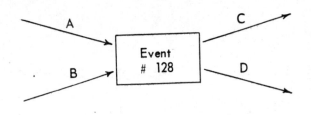

Fig. 10.1

In Fig. 10.1, Event #128 represents the completion of operations A and B, and the start of C and D. Whick of the four will be used in the definition? Lexicons have been issued, listing terms that may and may not be used on the PERT diagram. However, the method of logic development, that of Precendence, Concurrence, and Subsequence, still is the same as that described in Chapter 2.

Whether events are shown in circles or squares is trivial (however, don't under estimate "people problems" in this regard. It's amazing how rigid a group can become, and this may become a problem of major magnitude. This type of problem is solved by psychological approaches, and is beyond the scope of this text).

One of the best solutions to conflicting, confusing or vague requirements for a PERT diagram on a particular project, is the combination activity-event oriented diagram shown below. The problem described is a typical classroom problem for PERT training programs. The three time estimates for each activity are discussed in detail further in this chapter.

SPACE VEHICLE GUIDANCE SYSTEM

Inertial Instruments, Inc. (III), has received a contract for the design and development of an Inertial Guidance System for a prototype space vehicle. III's contract includes three main components:

1. A vertical vibrating pendulum type of sensor, mounted on a stable platform
2. A regulated power supply
3. A solid state Guidance Computer

Assembly of the guidance system must be in an extremely clean "White Room" environment. III must design and construct this assembly facility. An Environmental Test Facility, capable of simulating the temperature and vacuum extremes of outer space must also be designed and constructed by III.

While III's Engineering Department already has the necessary security clearance, the technicians who will make a "mock up" (for dimensional compatibility in the vehicle), and who will fabricate and assemble the system, do not have the requisite clearances. Provision must be made for obtaining these clearances.

SPACE VEHICLE GUIDANCE SYSTEM

ACTIVITY LIST

Description	a	m	b	t_e
System Engineering	4	7	11	
*Preliminary Design, Turbo-generator	5	8	12	
*Preliminary Design, Guidance Motor Cont. Unit	5	8	12	
Investigate areas for "White Room", Environmental Test Facility	1	2	2	
*Preliminary Design, Vehicle Frame Dimensions	3	5	8	
Preliminary Design, Sensor - Stable Plat.	3	5	10	
Security Clearances, Technicians	6	8	12	
Preliminary Design, Reg. Power Supply	3	4	7	
Preliminary Design, Guidance Computer	4	10	11	
*Final Design, Turbo-generator	7	10	15	
*Final Design, Guidance Motor Control Unit	6	10	18	
Design "White Room"	2	3	5	
Design Environmental Test Facility	1	2	3	
Fabricate, test Sensor-Stab. Plat. Proto.	2	3	6	
*Client App'l. Sensor-Stable Platform	1	2	3	
Final Design, Sensor-Stable Platform	3	4	6	
*Final Vehicle Frame Dimensions	4	6	9	
Fab. Reg. Pwr. Supply Breadboard	1	1	1	
Fab. Guidance Computer Breadboard	1	2	2	
*Client App'l. Reg. Pwr. Supply Breadboard	1	2	3	
*Client App'l. Guidance Computer Breadboard	1	2	3	
Final Design, Reg. Pwr. Supply	1	2	4	
Final Design, Guidance Computer	2	3	7	
Construct White Room	5	7	9	
*Client App'l., Environ. Test Fac. Design	1	2	3	
Construct Environmental Test Facility	2	3	6	
Acceptance Tests, White Room	2	2	4	
Acceptance, Tests, Environmental Test Facility	2	3	4	
Fabricate System Mock Up	1	3	4	
Del. Mock Up to Vehicle Mfr. Receive App'l.	2	2	2	
Build Sensor, Stable Platform	3	4	7	
Fabricate Reg. Power Supply	1	2	2	
Fabricate Guidance Computer	2	4	6	
Assemble Guidance System	2	3	4	
Preliminary System Tests	1	2	2	
Environmental Testing and Acceptance	4	5	10	
Deliver to Vehicle Mfr.	1	2	2	
*Construct Vehicle	9	10	16	

Time Units are in working weeks

*Activities by companies other than III

A gas driven turbo-generator will supply the raw power for III's regulated Power Supply. This turbo-generator is to be developed by another company. The Guidance Motors are gimbaled mounted small rocket engines, which will be positioned in response to command signals from III's Guidance Computer. The Guidance Motor Control Unit is to be developed by another company, and its input signal requirements will affect III's final design.

The system must have a dimensional "mock up" made. This cannot be made until after the final design of the sensor-stable platform, the regulated power supply, and the guidance computer.

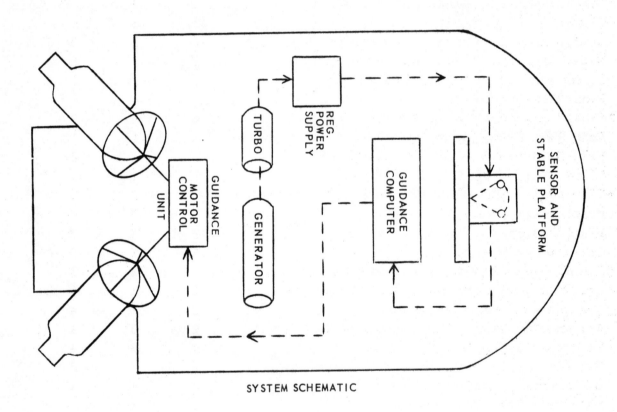

SYSTEM SCHEMATIC

Fig. 10.2

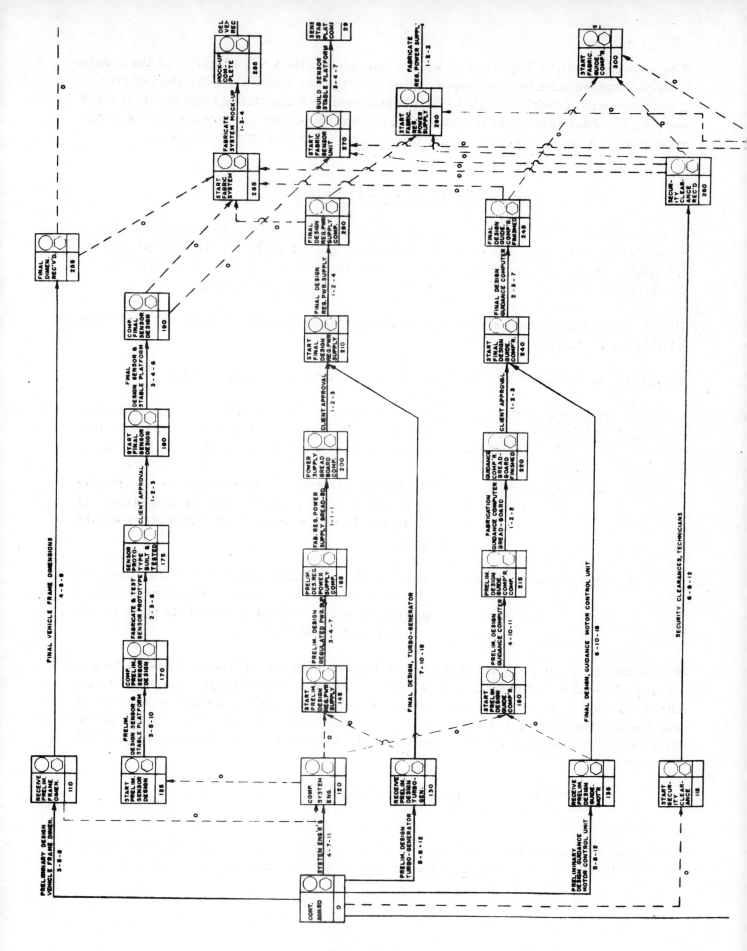

Fig. 10.3

91

The network plan in Fig. 10.3 is a problem sheet, thus the addition of the circle, hexagon and square for the event times of T_L, T_S, and T_E. Note that each activity arrow is described, as well as each event. Admittedly redundant, it does resolve the conflicting problems encountered. Whether the events are in large circles or squares depends on the particular project requirements.

PERT Time Estimates

In the beginning, PERT systems, due to the probabilistic return, utilized three time estimates. The three time estimates (a) optimistic, (m) most likely, (b) pessimistic, are used to derive a distribution curve of durations, and the activity expected time (t_e) is calculated as a statistical average.

PERT literature published by the military generally defines the various time estimates as follows:

From the same datum of usual practical resources available, then

1.	Optimistic Time	Time estimate if everything went better than expected. In this case there is probably little chance of completing the activity in less time.
2.	Most-Likely Time	This time estimate is one which the estimator believes will be required for the operation. If he had only one estimate to make, this would be it.
3.	Pessimistic Time	This is the estimate which would be required if everything went wrong. It represents an operation which had an unusually large amount of trouble or bad luck.

From these, a single time estimate is statistically calculated. This is based on a philosophy that you have a 50-50 chance of accomplishing the activity in the averaged time. The expected time is used to calculate the event times via the Forward and Backward Pass described in Chapter 5. Given these three estimates, the expected value (mean) is calculated using the following statistical average:

$$t_e = \frac{a + 4m + b}{6} \qquad \text{(expected time)} \qquad (10.1)$$

Where $\underline{a}$ is the optimistic estimate of time interval, $\underline{m}$ is the most likely estimate of time interval, and $\underline{b}$ is the pessimistic estimate of time interval.

In order to understand the background of the derivation of the expected time formula, t_e in equation number 10.1, a brief review of the basis of statistical distribution and probability is in order. Consider the payroll of mythical company XYZ in Table 10.0 below.

Company XYZ PAYROLL

POSITION	ANNUAL SALARY		
President	$ 30,000		
Exec. V. P.	23,000		
1st V. P.	14,000	Average Salary:	
Dept. Head A	13,000		
Dept. Head B	12,000	$\dfrac{\$150,000}{15} = \$10,000$	
Sr. Eng.	9,000		
Eng. A	8,300	Mean Salary:	
Eng. B	7,700		
Sr. Draftsman	7,500	$\dfrac{\$30,000 + \$3,000}{2} = \$16,500$	
Draftsman	6,500		
Sec. A	5,000		
Sec. B	5,000	Median Salary:	$ 7,700
Clerk A	3,000		
Clerk B	3,000		
Clerk C	3,000		
	————		
Total Payroll	$150,000		

Table 10.0

Average salary is the total divided by the number of employees: the mean (arithmetic) salary is the sum of the extremes of the range of salary (this assumed payroll ranges from $3,000 to $30,000 annually) divided by two; the median salary represents that point on a salary distribution curve where as many people on the payroll earn more as earn less. There are seven people earning more than $7700 per annum, and seven people earning less.

From the information in the payroll table of Table 10.0, a distribution curve can be constructed by first arranging a bar distribution chart, showing the frequency (the number of employees earning the same salary) on the ordinate, and the salary range on the abscissa. A distribution curve is made by connecting a smooth curve to the tops of the bars, as indicated in Fig. 10.4. The peak, or highest point, is called the "Mode" of the distribution curve.

Frequency

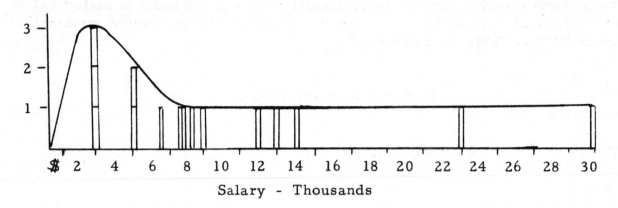

Salary - Thousands

Fig. 10.4   DISTRIBUTION CURVE

There are many types of distribution curves defined in statistics, depending on their shapes and other factors. In the three time estimates of activity durations in PERT, it is apparent that the optimistic and pessimistic times should occur the least often, and the most likely, or realistic time, most often. It was assumed that the peak or modal value would be represented by the most likely time in a probability distribution, and that this peak could be moved between the two extremes of the optimistic and pessimistic estimates. This characteristic could be best expressed by a particular statistical distribution curve called the "Beta" distribution. This curve may be skewed left or right between the extremes. From the preparation of the Beta Distribution Curve, it was decided that the expected time, t_e, should be interpreted as the weighed mean of m (most likely) and M (mid range average or $\frac{a+b}{2}$) with weights of 2 and 1 respectively. In other words, t_e can be located one third of the way from the mode to the mid range point. This represents the 50% probability point of the distribution, i.e., it divides the area under the distribution curve into two equal portions.

Fig. 10.5 indicates two possible conditions in which t_e divides the area under the distribution curve into equal areas.

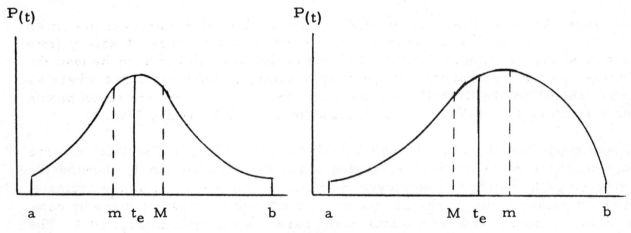

Fig. 10.5 BETA DISTRIBUTION CURVES

Note that the whole thesis of the three time estimates is based on a 50% probability that you are correct.

Another problem encountered in the event oriented PERT system, is the usefulness of Event Slack (Slack in PERT is the same as Total Float in CPM--Critical Path Method). Event Slack (Float) is

$$\boxed{T_L} \ - \ \boxed{T_E} \ = \ \text{Event Slack} \qquad\qquad (10.2)$$

The event oriented system would identify "Critical Neighbor" events, and thus state that an activity occurring between Critical Neighbor events would be on the Critical Path. An examination of Fig. 5.3 in Chapter 5 will show that operations D, P and R would be considered "Critical". This is not true. The line manager has enough problems already without getting additional falacious information. In general, the PERT systems are decidedly going activity oriented.

Consider the estimates, for the same activity, submitted by three different personnel, below:

Estimate	a	m	b	t_e	
I	1	7	7	6	(10.3)
II	4	6	8	6	
III	2	4	18	6	

The estimate I would indicate the person estimating has a fairly good grasp of the nature of the activity, the pessimistic and realistic estimate are the same. There is a certainty indicated.

Estimate III shows uncertainty, the range from optimistic to pessimistic is large.

The three elapsed time estimates in Estimate II are called a "balanced estimate"; in some cases this is really one estimate, since it is in the form of $\frac{M-x+4(m)+m+x}{6}$. If a group consistently gives balanced estimates, they are either lazy, or do not understand the rationale behind PERT.

In addition, the variance is also calculated from the following formula:

$$\sigma_{t_e}^2 \ = \ \left[(b-a)/6\right]^2 \qquad\qquad (10.4)$$

Variance is a statistical measure of uncertainty.

The mathematical basis is structured upon the Central Limit Theorem of Probability. Just as the more times you flip a coin the more the probability goes to one (certainty) that a head will turn up one time out of two, the probability is 0.5 that a head (or a tail) will show up. A calculus of Probability was used, based on a BETA distribution curve for the three time estimates of optimistic (a), realistic (m), and pessimistic(b) (See Appendix), which allows a probability on a Standard Normal Distribution function of one out of two that the $\boxed{T_E}$ OE of the final or objective event will be the same as the $\left(T_L\right)$OE, or the Latest Event Time for the final or objective event.

The expected time; t_e, is a statistical average, based on a 1 sigma (σ) deviation. The deviation formula is:

$$\sigma = \frac{b-a}{6} \qquad\qquad (10.5)$$

(b-a) is called the range of the deviation.

To find the probability of reaching a Scheduled Event Time T_S, we must find the deviation along the longest path to that event. If the event in question is the final or objective event (oe), then the deviation along the Critical Path would be found.

Case II of Chapter 9, finding the affects of inserting a T_S of 13 at event #14 in Fig. 9.0 will be evaluated on a probabilistic basis. Fig. 10.6 below is the network of Fig. 9.0, but three time estimates (a, m, b) have been assigned to each activity. The expected time, t_e, will be the same as in Fig. 9.0.

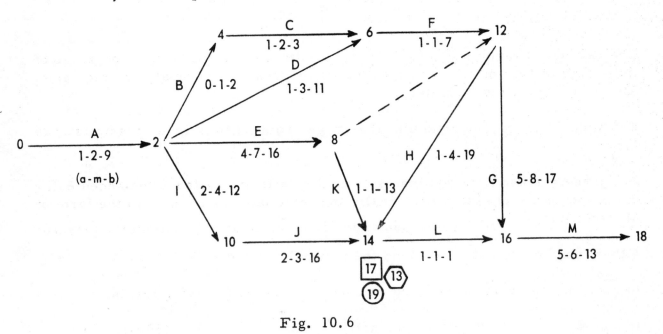

Fig. 10.6

What is the probability of reaching Event #14 by week #13. The longest path into event #14 is A, E, H. The deviation along this path must be found. Since the deviation is the square root of the sum of the squares (or individual variances of the longest chain of activities), we shall find the variance along the longest path. To simplify the calculation, the (ranges)^2 of each activity along that path shall be added, then the sum divided by 36, and then the square root of that path variance shall be found. From the particular calculus of probability, we shall find Z, a function from the Standard Normal Distribution Function.

$$Z \text{ is equal to } \frac{T_{S_{oe}} - T_{E_{oe}}}{\sigma_{\text{longest path}}}$$

$$Z = \frac{T_{S_{(Event\ 14)}} - T_{E_{(Event\ 14)}}}{\sigma_{\text{path to event 14}}} \qquad (10.6)$$

1. Longest path Variance

$$A = (9 - 1)^2 = 64$$
$$E = (16 - 4)^2 = 144$$
$$H = (19 - 1)^2 = \underline{324}$$
$$532$$

$$\sigma^2 \text{ (variance of longest path)} = \frac{532}{36} = 14.8$$

$$\sigma \text{ (deviation of longest path)} = \sqrt{14.8} = 3.85$$

$$Z \frac{13 - 17}{3.85} = -1.04$$

From the table of values of Z, the Standard Normal Distribution Function (see Appendix), Z = .1492, or there are 15 changes out of a hundred that event #14 can be reached by week #13.

What does this mean; that there is a 15% probability of reaching Event #14 by week #13? No specific answer can be given to this question. It can be interpreted that, based on the usual or accepted performance of being 50% right in estimates, the plan so analyzed should be further studied. However, this answer was previously obtained when the Forward Pass using the t_e average estimates produced an Early Finish date of week #17, instead of the desired T_S of week #14 at this particular event. There is more management information in the Forward and Backward Pass than there is in the probability and variance values.

The use of Probability in PERT has diminished to minor importance since more valuable information concerning the establishing of activities has been realized with use. It is a guide that is sometimes used as an indicator of uncertainty. One school of thought states that if the Probability for the Final or Objective Event is between 0.25 and 0.65, this is an acceptable status for a Cost-Plus-Fixed-Fee contract. If the Probability is more than 0.65 on a CPFF contract, then the agency has a right to suspect that too many resources are being applied to that contract (overloading). If the Probability falls below 0.25, then they may feel that the contract is understaffed, or insufficient resources are being applied.

One area of usage of the probabilistic aspects of PERT, is in the evaluation of alternate strategies (different logic configurations or diagrams) for a particular project. This will indicate the method of execution for the best probability of attaining a desired target date.

In general, the use of Probability is rather minor.

The use of statistics in any managerial decision is highly sophisticated, and the line manager must be aware of all the assumptions and rationales involved. Never forget, those children that use that toothpaste additive and have 21% fewer cavaties, may also have 25% fewer teeth! An excellent article for reference of statistics in business is Mr. D. Seligman's, "We're Drowning In A Sea Of Phony Statistics" that appeared in the November, 1961, issue of Fortune Magazine.

There is a noticeable trend to the use of one time estimate, or realistic time in PERT.

Another discouraging aspect of the three time estimates is a Parkinson Law reaction. Consider activity E of Fig. 10.4, which has the optimistic estimate of 4 weeks, realistic estimate of 7 weeks, and a pessimistic estimate of 16 weeks. The statistical average of 8 weeks was calculated. In practice, this was then given as a "directed" scheduled elapsed time. However, the estimator stated that "realistically", the activity E should only take 7 weeks. If the activity was given 8 weeks (the t_e, or average), it was found that in general 8 weeks would be consumed. Parkinson's Law states succinctly, "The work will fill up the time allowed available."

In addition to the use of one time estimate, PERT is becoming activity oriented (Refer to the "DOD-NASA Guide, PERT COST Systems Design" Manual).

Fig. 10.7 shows a typical PERT output.

PERT COMPUTER RUN ON RCA 501
REPORT DATE 10/1/62

SORT EXPECTED DATE

Start Event	End Event	Activity	Actual Date	Expected Date	Schedule Date	Latest Date	Slack	Dev.	Prob.
125	130	ANALOG SYSTEM SIMULATION		12/01/62		10/23/62	- 5.5	1.5	
210	260	CLIENT APPL TEST FACILITY DESIGN		12/10/62		03/14/63	-13.4	.9	
120	310	FINAL VEHICLE DIMENSIONS		12/12/62		02/15/63	9.2	1.0	
130	170	PRELIM DESIGN STABLE PLAT		12/15/62		11/21/62	- 3.4	1.5	
150	190	PRELIM DESIGN REG PWR SUPPLY		12/31/62		01/10/63	1.4	1.6	
130	180	PRELIM DESIGN SENSOR		01/02/63		01/09/63	1.0	1.9	
260	270	CONSTRUCT ENVIRON TEST FACILITY		01/02/63		04/06/63	13.4	1.0	
170	220	VENDOR SELECT STABLE PLATFORM		01/05/63		12/12/62	- 3.4	1.5	
190	230	FAB PWR SUPPLY BREADBOARD		01/07/63		01/17/63	1.4	1.6	
180	225	CLIENT APPL SENSOR DESIGN		01/16/63		01/23/63	1.0	1.9	
270	420	ACCEPT TEST OF ENVIRON TEST FACILITY		01/16/63		04/20/63	13.4	1.2	
230	235	CLIENT APPL REG PWR SUPPLY		01/21/63		01/31/63	1.4	1.6	
145	200	PRELIM DESIGN GUID COMPUTER		02/03/63		12/26/62	- 5.5	1.9	
135	235	FINAL DESIGN TURBO-GEN		02/07/63		01/31/63	- 1.0	1.8	
225	290	FINAL DESIGN SENSOR UNIT		02/08/63		02/15/63	1.0	2.1	
140	250	FINAL DESIGN GUID MOTOR CONTR UNIT		02/09/63		01/30/63	- 1.4	2.3	
200	240	FAB GUID COMPUTER BREADBOARD		02/16/63		01/08/63	- 5.5	1.9	
220	300	FINAL DESIGN STABLE PLATFORM		02/18/63		01/25/63	- 3.4	1.8	
235	320	FINAL DES REG PWR SUPPLY		02/22/63		02/15/63	- 1.0	1.8	
300	340	SGI REVIEW STABLE PLAT DESIGN		03/03/63		02/07/63	- 3.4	1.8	
290	380	FAB SENSOR UNIT		03/05/63		03/22/62	- 2.4	2.3	
320	380	FAB PWR SUPPLY		03/08/63		03/22/63	- 2.0	1.9	
240	250	CLIENT APPL GUIDANCE COMPUTER		03/10/63		01/30/63	- 5.5	1.9	
250	280	FINAL DES GUIDANCE COMPUTER		03/26/63		02/15/63	- 5.5	2.1	
340	360	FAB STABLE PLATFORM		04/02/63		03/09/63	- 3.4	1.9	
330	350	FAB SYSTEM MOCKUP		04/15/63		03/07/63	- 5.5	2.1	
360	380	DELIVER STABLE PLATFORM		04/15/63		03/22/63	- 3.4	1.9	
280	380	FAB GUIDANCE COMPUTER		04/23/63		03/22/63	- 4.5	2.2	
350	370	DELIVER MOCKUP		04/29/63		03/21/63	- 5.5	2.1	
380	390	ASSEMBLE GUIDANCE SYSTEM		05/20/63		04/12/63	- 5.4	2.2	
390	420	PRELIM TEST GUIDANCE SYSTEM		05/28/63		04/20/63	- 5.4	2.2	
410	440	TRAIN FIELD PERSONNEL		06/11/63		06/05/63	- .8	2.2	
420	430	ENVIRON TEST AND PLANT ACCEPTANCE		07/06/63		05/29/63	- 5.4	2.3	
430	440	SHIP SYSTEM TO VEHICLE MFR		07/13/63		06/05/63	- 5.4	2.3	
400	440	MFR VEHICLE		07/14/63		06/05/63	- 5.5	2.5	
440	450	INSTALL SYSTEM IN VEHICLE AND TEST		08/08/63	06/30/63	06/30/63	- 5.5	2.5	01

Fig. 10.7

The fact that PERT is more a reporting system, than a managerial control system, can be seen in the output of Fig. 10.7. Only two of the four activity times are in the output. The Early Activity Finish Time is the "Expected Date" in the output of this particular computer. This EF activity time is sometimes known as the "Predicted Time". The "Latest Date" is the same as the "Latest Activity Finish Time". Starting times for activities are not part of the PERT output. This information is in the program, but not given out. The methodology of the Forward and Backward Pass is the same as described herein. The "Schedule Date" is the T_S described in Chapter 9. T_S times are not available in CPM (Critical Path Method) programs. Note the calandar dating (part of the program), and the negative Slack (Total Float) caused by the insertion of 6/30/63 as a T_S at the end event #450.

PROBLEMS

10.1 What is the probability of reaching event #18, in Fig. 10.4 by week 23, instead of the normal duration of the project of 27 weeks?

10.2 Construct the PERT Diagram for the Inertial Guidance System described on pages 63 and 64. Calculate the schedule. What is the PERT probability of Starting Vehicle Construction by week #25?

THE TIME SCALE CHART AND PROJECT CONTROL

The typical Network Plan and Schedule format for a project was shown to the reader with Fig. 8.0 and Table 8.0 in Chapter 8, and the PERT variation of the output schedule in Fig. 10.7. These are the "digital" formats, with each activity listed separately in a row, with all the pertinent schedule, cost and code information. This format is the most commonly accepted and used one. Yet to the uninitiated, this could be confusing and frightening, particularly when unexplained arrow diagrams are accompanied by reams of computer outputs. There is a simple next step that generates a schedule that is readable and understandable in form at any level of supervision. That is the conversion of the schedule into a Time Scale Chart, a logically interconnected bar chart. This chart will become our PROJECT CONTROL CHART.

Time Scale Presentation

The arrow diagram shown in Fig. 11.0 may be shown as an interconnected bar chart, i.e., the arrow diagram redrawn on a time scale. The time scale drawing can only be made after the schedule for that diagram has been generated by the methodology thus far described. It is necessary to have had the Event Times established before the construction of a Time Scale Network.

The network of Fig. 11.0 will be shown in two cases on a time scale drawing, Fig. 11.1. Case I will indicate all activities starting at their Early Activity Start Time. Case II will show one alternate, wherein the overlap of activities is minimized. Case II is an attempt to minimize the total number of activities being worked on any one project day. The approach shown in Fig. 11.1 is that of an interconnected bar chart. The value of this display is that the inter-relationships of all activities is shown. It is easy to forget all the dependencies if separate bars, not interconnected, are shown. However, for information received, the activities may be shown as separate bars, with brackets indicating Early Start Times and Latest Finish Times. Fig. 11.2 shows the alternate presentation of Case II for Fig. 11.0.

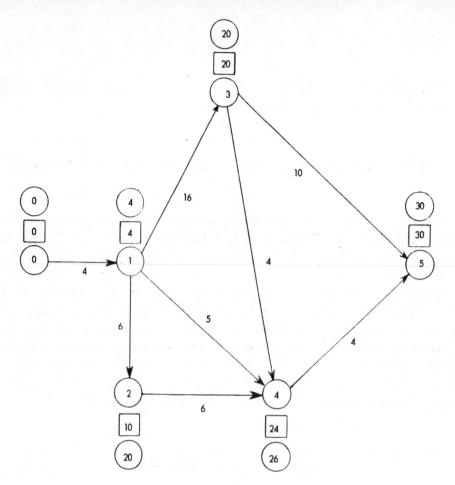

Fig. 11.0

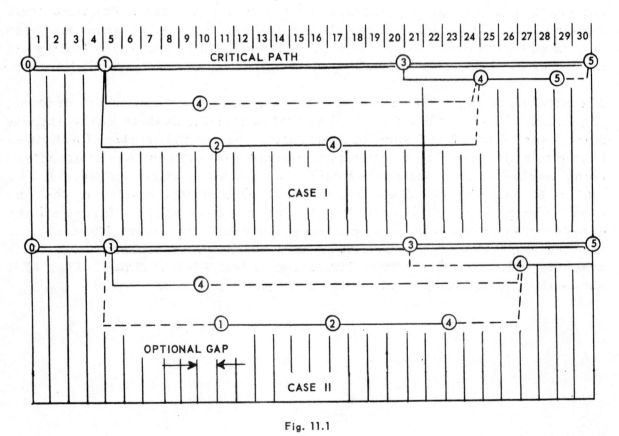

Fig. 11.1

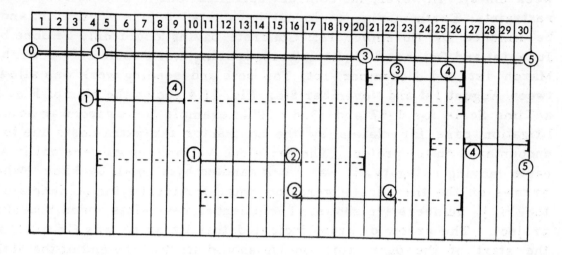

Fig. 11.2

Note that a Time Scale Drawing is shown on Project Working days, which may also represent calendar dates. Also note that time O is the morning of the first day; that Event Times indicate the finish of that day. This time boundary also represents the start of the next day. Thus, Event #2 has an Early Event Time, T_E, of day 10. This indicates that activity 1 - 2 may finish as early as the end of the 10th day, and that activity 2 - 4, which also has an Early Event time of day #10 at its tail, may start as early as the end of the 10th day, which is the same as the morning of the 11th day.

Another use of the Time Scale drawing of a network plan, is to show the effect of a Schedule Event Time, T_S. A calendar restraint on the start of an activity may be shown as an arrow on the network diagram, and if it is the longest time path into an Event, it will control the starting time of that activity. However, where a calendar restraint for a finish time of an activity is required, an arrow into the head of that activity arrow may not have any effect, since the T_L of the activity was set by the Backward Pass of the remaining paths to the end of the project. On a Time Scale drawing, calendar restraints may be graphically shown. Fig. 11.3 is the network plan for a turnpike construction project. The usual procedure of Forward and Backward Passes indicates one Critical Path, and a completion time of 91 weeks (Durations shown are in work week units). However, the contract specifications had the following calendar restraints: Earth moving could not be started before April 15th, and had to be finished by November 1st; Concrete paving could only be done between June 1st and October 30th; Deck (bridge paving) work could only occur between March 1st and November 1st; Top soil and seeding work was allowed between August 1st and November 8th. Fig. 11.4 shows the plan of Fig. 11.3 on a Time Scale for 1957 and 1958. This example comes from an actual civil litigation case for claims by the contractor for extra costs due to delays encountered on the project. The early excavations shown were not in a major earth moving category, it was work handled by a small backhoe. Where the arrows on the time scale were too small for identifying nomenclature, just the "i, j" numbers are shown. The calendar restraints added 8 weeks to the project. The arrow diagram indicated that Ramp 1A was Critical, and that the start of the deck work on 1A should start at the end of the 56th week. However, the specification restraint of March 1st forced the delay of starting this operation until the 63rd week. The eighth week of delay was caused by the specification restraint of concrete paving not starting until June 1st. Thus activity 65-70 and activity 190-195 could not start until week 76. This then made parallel Critical Paths from the start of paving operations on both the Ramp 1A section and the River Road section. Note another interesting facet involving the scheduling of resources. From the arrow diagram, activity 65-70, "Pave Approaches, Ramp 1A" is Critical, and has Early and Late Start Times of week 68. It is shown on the arrow diagram as having to be finished by week 72. On the River Road work, activity 190-195, "Pave Approaches, River Road" is the same type of operation as 65-70. 190-195 is shown as having to occur between weeks 64 (T_E) and week 72 (T_L), with 4 weeks of float. The normal use of this information would be to schedule the paving equipment during weeks 64 to 68 for the River Road work on 190-195, then move that equipment over to the Ramp 1A work on activity 65-70 for weeks 69 to 72 inclusive. However, the calendar restraints forced these activities to be scheduled simultaneously in weeks 76 to 80, inclusively. This forced the contractor to rent an additional rig during that period of time.

Inclement or Limiting Weather Periods

Expected bad or limiting weather periods, obtainable from local meteorlogical records, may be shown as a vertical section on the Time Scale Chart. On the First Forward and Backward Pass, those activities so affected would be observed to see when their Early Start Time would fall in this period. They are either delayed to after the Limited Period (by the addition of a Winter Restraint arrow on the diagram, whose duration is the time from the Start Baseline to the last Day of Winter); or if inefficiency in production is considered for this period of Time, their duration is extended by that amount of inefficiency. A second Forward - Back ward Pass is then made to place these activities in a reasonable Calendar period of Time on the schedule. Fig. 11.4 indicates the approach of Winter Periods on a Schedule.

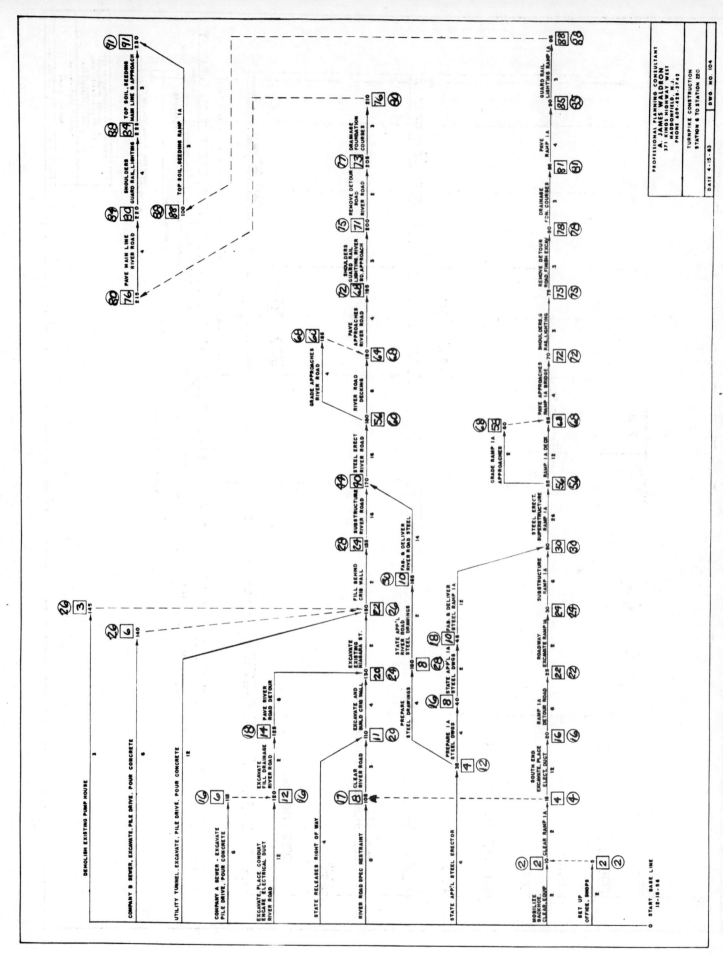

Fig. 11.3

105

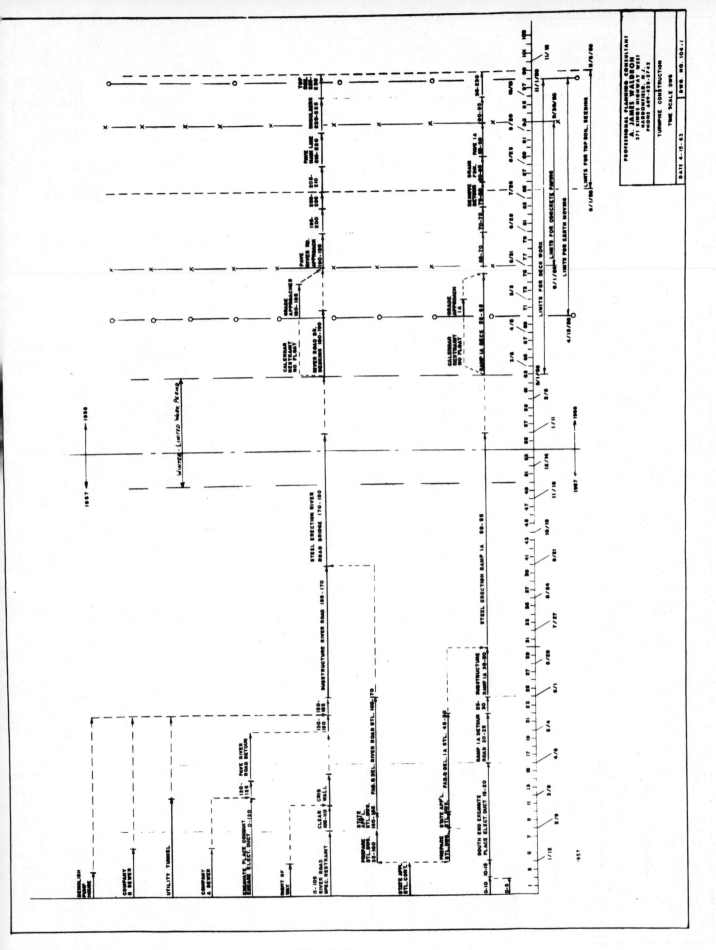

Fig. 11.4

This approach of Time Scale presentation is used in machine loading for manufacturing operations. Fig. 11.5 shows machines assigned to a production plan, the network of Fig. 11.0.

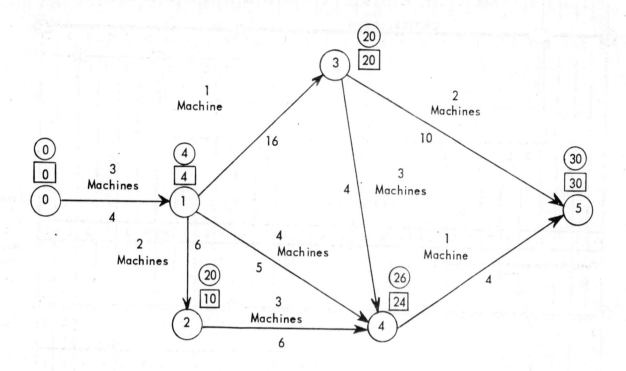

Fig. 11.5

Machines Assigned To Production Plan

Case I of Fig. 11.6 indicates machine loading, from the assignments on Fig. 11.0, based on all operations starting at their Early Start Time. A peak load of 7 machines at time units 5, 6, 7, 8 and 9 is shown. Case II indicates a better distribution, by the delaying of slack operations. One obvious use of this technique is in factory or manufacturing planning, to optimize the investment in capital equipment, based on sales forecasts and projections. For the already established plants, the technique will help to generate an optimum schedule for the fixed number of machines available. If it is possible to re-assign the machines, originally scheduled for the slack operations, to the critical operations, the overall time of the plan may be reduced. If it can be assumed that every additional machine that can be assigned to critical activity 1-3 will reduce that activity's duration proportionately, then the machines originally assigned to operations 1-4, 1-2 and 2-4 should be re-evaluated. These last operations can be extended, to the extent of their path slack, without extending the completion time of the plan of execution. Thus, if 1 of the 4 machines originally assigned to operation 1-4, is assigned to operation 1-3, the time of 1-3 might be cut in half, from 16 time units to 8 time units (with

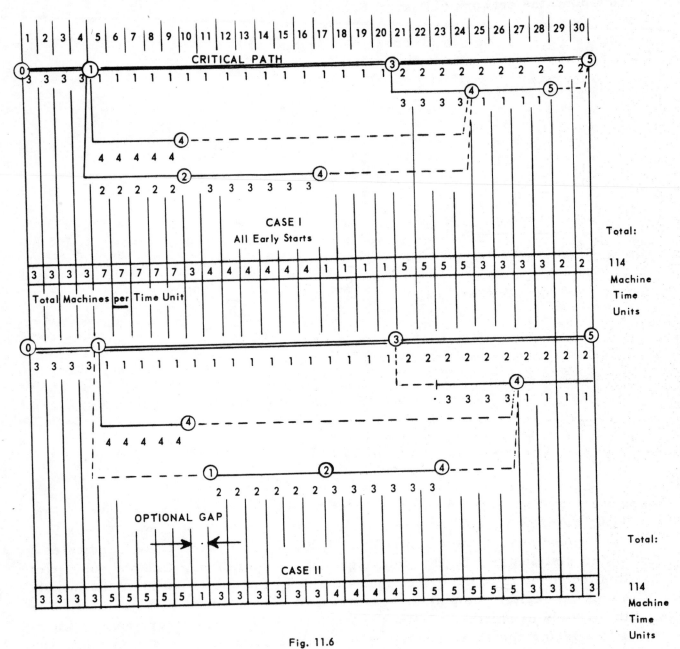

Fig. 11.6

Using Float Activities To Minimize Machine Loads

2 machines), while 1-4 is extended to 6 2/3 time units. The completion time of this plan, that is, the event time of the Final Event #5, might then be cut to time unit 22.

The example above is over-simplified, of course. There is no guarantee that machines can be swapped indiscriminately. However, the technique is used to handle each of many specific types of machines or resources, and for more than one project that occurs simultaneously in the house. Computer programs are available that will test every possible combination of schedules for each resource, and will generate the optimum machine load schedule. Don't overlook the fact that money is a resource; this technique will help determine economic optimum.

A simple method of scheduling more than one project at a time, be they simultaneous or overlapping in time, can be handled by the time scale approach. A large gridded chart may be hung on the office wall, with the activities from each network plan shown in time scale. As a new project comes into the company it is added to the overall company time scale chart. The large chart may encompass just the fiscal year of the company, or a several year span, depending on the length of the projects. The flow of people, machines, equipment, etc., can be rapidly established from the manipulation of float on the time scale, so that schedules of such things as concrete forms, welding machines, foremen, engineers, etc., can be graphically portrayed.

Preparing The Time Scale Chart (Schedule)

The steps to be followed in the preparation of a Time Scale Chart (Schedule) are:

1. From the Forward and Backward Pass of the Arrow Diagram, determine the Critical Path. This becomes a chain of continuous activities on the Time Scale, from start to finish of the chart. It is usually emphasized by heavy or wavy lines. Parallel sections of the Critical Path are also shown emphasized. There will be "steps" or "jogs" in the chain, due to dummies on the Critical Path.

2. Next, show all the Float activity chains at their Early Start (ES) Time. Event numbers should be shown on the Time Scale chart. The activities are to be shown in the same logical sequence as on the Arrow Diagram Plan.

3. One format, taken from an existing computer program, is to place the Critical Path at the bottom of the chart, and the Float chains of activities, in successive order of Float (TF), upward from the Critical Path. Thus the chain at the top of the chart would have the most Total Float. Another common format is the "Banding" (See Chapter 2) of a Time Scale by physical or geographical areas, or by trades, sub-contractors, etc.

4. Since a dummy has zero (0) duration, it becomes a vertical dashed line on the Time Scale.

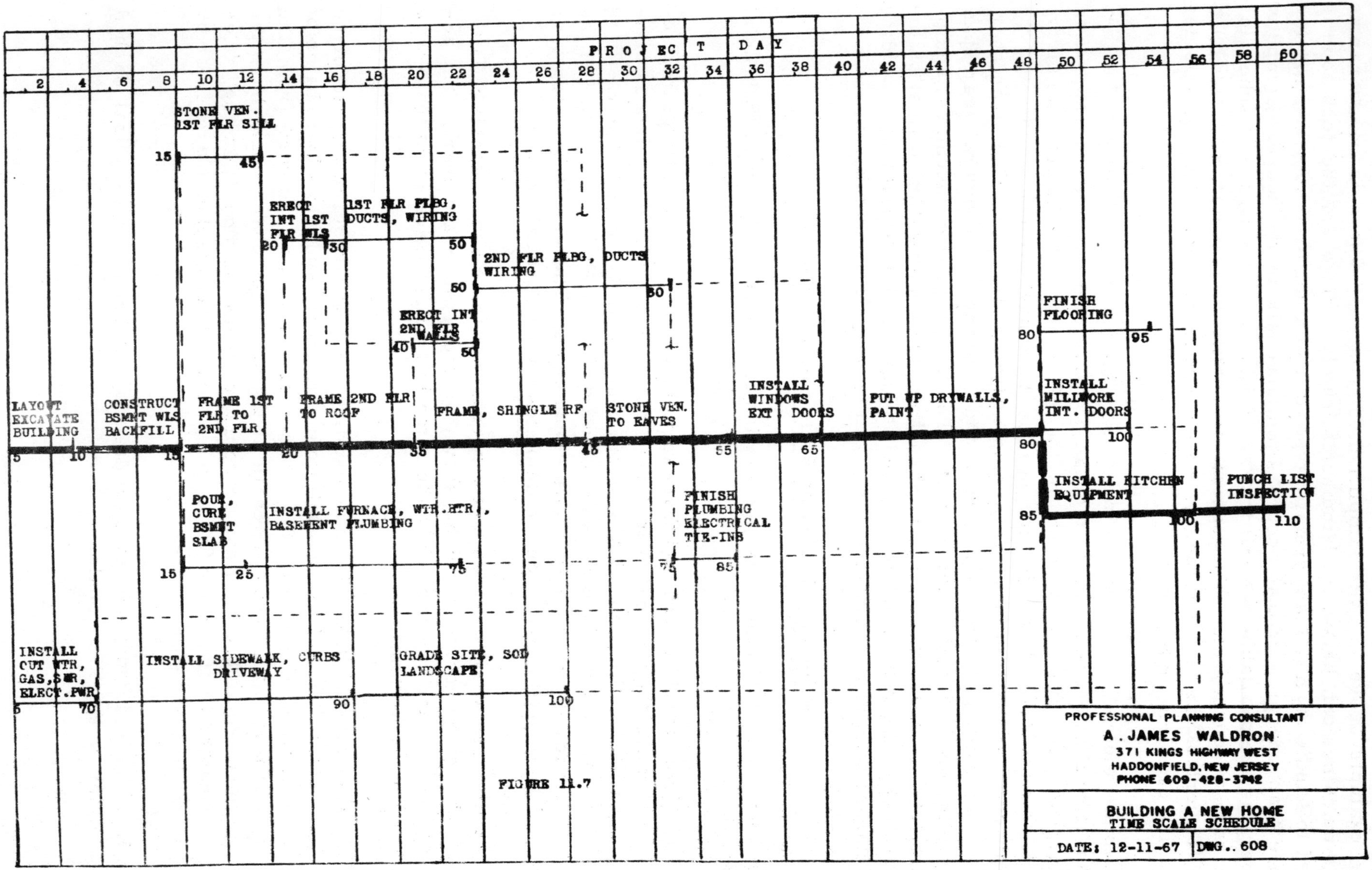

LAYOUT EXCAVATE BUILDING

STONE VEN.- 1ST FLR SILL

CONSTRUCT BSMT WLS BACKFILL

FRAME 1ST FLR TO 2ND FLR

ERECT INT 1ST FLR WLS

1ST FLR PLBG, DUCTS, WIRING

ERECT INT 2ND FLR WLS

2ND FLR PLBG, DUCTS WIRING

FRAME 2ND FLR TO ROOF

FRAME, SHINGLE RF

STONE VEN. TO EAVES

INSTALL WINDOWS EXT DOORS

PUT UP DRYWALLS, PAINT

FINISH FLOORING

INSTALL MILLWORK INT. DOORS

INSTALL KITCHEN EQUIPMENT

PUNCH LIST INSPECTION

INSTALL FURNACE, WTR. HTR. BASEMENT PLUMBING

FINISH PLUMBING ELECTRICAL TIE-INS

POUR CURB BSMT SLAB

INSTALL SIDEWALK, CURBS DRIVEWAY

GRADE SITE, SOD LANDSCAPE

INSTALL CUT WTR, GAS, SWR, ELECT. PWR

PROJECT DAY — 2 4 6 8 10 12 14 16 18 20 22 24 26 28 30 32 34 36 38 40 42 44 46 48 50 52 54 56 58 60

5 10 15 45 80 30 50 40 50 50 35 20 25 55 65 45 60 75 85 75 80 85 95 100 100 110 90 100 70 16

FIGURE 11.7

PROFESSIONAL PLANNING CONSULTANT
A. JAMES WALDRON
371 KINGS HIGHWAY WEST
HADDONFIELD, NEW JERSEY
PHONE 609-428-3742

BUILDING A NEW HOME
TIME SCALE SCHEDULE

DATE: 12-11-67   DWG. 608

Fig. 11.7 is the Time Scale Chart for Building a New Home. Note that this is another format of Fig. 6.1 of Chapter 6--indeed, it is constructed from the logic and Event Times of Fig. 6.1. The arrow diagram is needed for the path sequences and interconnections; the schedules for the individual activity Early Start and Early Finish Times. The student should study this schedule of Fig. 11.7, comparing each activity with its activity arrow on Fig. 6.1, and its surrounding Event Times. Be sure how each Start and Finish Time was ascertained.

The position of any arrow on the Time Scale Chart, in relation to the top or bottom of that chart, is purely arbitrary. In order to emphasize the Critical Path, it has been shown here in the middle of the chart, with the other paths of activities above or below in relative accordance with their vertical position on the original arrow diagram of Fig. 6.1. As was previously noted, this could have been portrayed with the Critical Path at the bottom of the chart, and each path, in increasing order of Total Float ranked above the Critical Path.

Fig. 11.7 is our basic Time Scale Chart. We are going to simulate an actual Project Control Chart development by adding some typical project conditions to the basic project conditions of Problem 1.1 in Chapter 1. From this information we will go through the steps of setting up our Project Control Chart.

PROJECT CONDITIONS FOR BUILDING A NEW HOME

1. Project Calendar

   The Project will work a five day work week, with Saturdays and Sundays non-work days. Two holidays will be observed, if necessary: April 4th and May 30th. The Project will start on a Monday morning, February 3rd. We will use a 1969 calendar.

2. Inclement Weather

   From meteorological records at the site, inclement weather is predicted to have the greatest possibility of occurring during the last two weeks in February (heavy precipitation, low temperatures, deep frost, etc.). Past records and experience indicate that work could proceed in this period, but usually 50% of productive time will be lost on weather affected activities. (Note: on any project, if a Winter shutdown is elected, a hiatus or gap in the calendar dates is shown).

3. Weather Affected Activities

   The following activities are adjudged to be weather affected. They will become 50% efficient if they fall into the limiting Inclement

Weather period. These activities would have been "flagged" or noted on the arrow diagram as being weather affected.

Activity No.	Description
5-70	Install Outside Sewer, Gas, Water Lines
10-15	Construct Basement Walls, Backfill
15-25	Pour, Cure Basement Floor Slab
15-45	Stone Veneer to First Floor Sill
35-45	Frame and Shingle Roof
45-55	Stone Veneer to Eaves
70-90	Install Sidewalk, Curbs, Driveway
90-100	Grade Site, Sod, Landscape

Table 11.0

Construction of the Project Control Chart

Once the arrow diagram for the project is completed (remember, this is the first step, the project logic diagram), the basic Forward and Backward Pass made, and all Event Times established, we are ready to proceed to the construction of the schedule on the Project Control Chart. Since we have the result of the initial Forward and Backward Pass, we know the basic Project Duration. Knowing the Limiting Weather Periods, and any other constraint periods or situations, we can easily estimate a realistic overall Project Duration. We will set up a Time Scale Chart, with vertical divisions representing work time units. The author has found a column representing one work day effective on projects less than six months, and a column representing two work days suitable for any project up to two years in overall duration, and a column representing a work week for projects over two years in total duration.

On this Time Scale Chart, we block out the Limiting Periods. Once this has been done, we then start transposing the activities, chain by chain, by their Early Start and Finish Times, onto the Time Scale Chart. In the initial Forward Pass, we note those "flagged" weather affected activities which fall into any part of the Limiting Weather Period. If they do, that part of their duration in the Limiting Weather Period is extended by the anticipated efficiency; in our case, we will double that duration, since we expect 50% efficiency. IN ESSENCE, WE ARE MAKING A SECOND FORWARD PASS WITH WEATHER ADJUSTED DURATIONS FOR WEATHER AFFECTED ACTIVITIES.

The conversion of the Project Day to the Calendar Date is shown at the top of the chart. Since the adjusted schedule due to weather conditions does not extend beyond the end of April, the May 30th holiday does not pertain on our Project Control Chart of Fig. 11.8.

Note the anticipated weather adjustments: activities #5-70 and 10-15 will

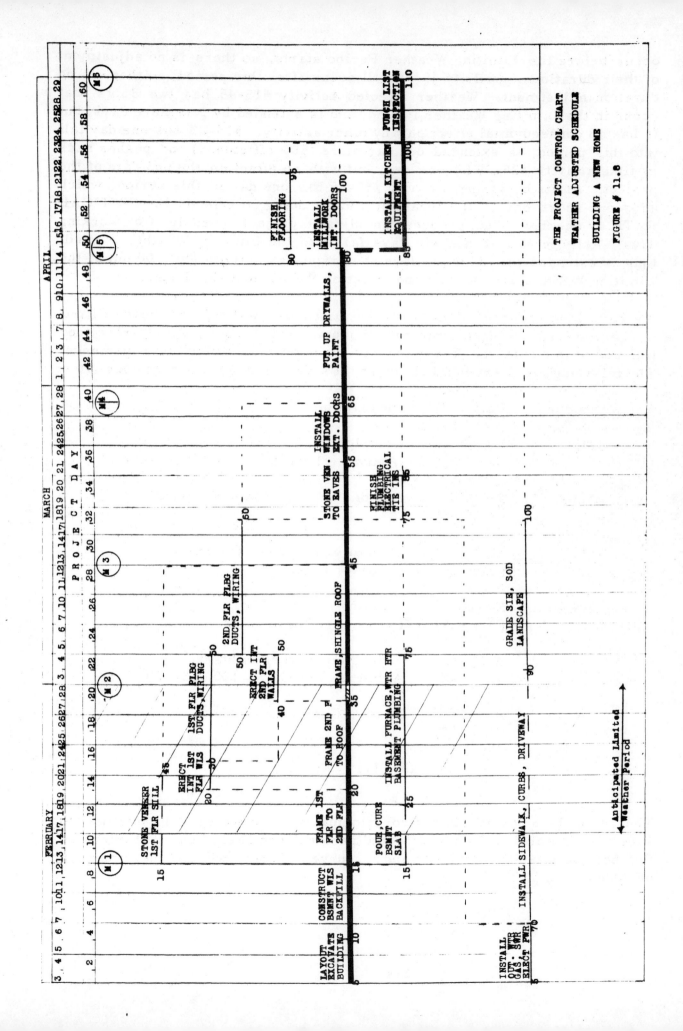

THE PROJECT CONTROL CHART
WEATHER ADJUSTED SCHEDULE

BUILDING A NEW HOME

FIGURE # 11.8

occur before the Limiting Weather Period starts, so there is no adjustment of their durations. Activity 45-55 will occur after this period, so there is no duration adjustment. Weather affected activity #15-45 has two days that occur in the Limiting Weather Period, and is extended by two more days, but it has no consequential effect on any other activity. #15-25 has one day fall into this period, is extended one day more (50% efficiency) and pushes back in time activity #25-75 by one day, but with no effect on the end date of the project. At its beginning, activity #35-45 has one day in this period, which adds one more day to its duration, and by doing so extends the CRITICAL PATH by one day. This produces an adjusted project schedule of 60 working days. Activity #70-90 has six days fall into the Limiting Weather Period, thus is extended for six more days, but it pushes activity #90-100 out of the Limiting Weather Period, with no effect on the Project End Date.

We have also shown Project Milestone Dates on the Project Control Chart. These are dates when meaningful amounts, or phases of work, are either finished or to be started. These are determined by judgement and experience. The following are the rationalization of the selected target, or milestone dates:

Milestone	Date	Description
M1	Feb. 12 (Day #8)	Completion of Foundations (Activity #10-15)
M2	Feb. 28 (Day #20)	End of Limiting Weather Period
M3	Mar. 12 (Day #28)	Structural Frame is Complete (Activity #35-45)
M4	Mar. 27 (Day #39)	Building is Closed In (Activity #55-65)
M5	Apr. 14 (Day #49)	Final Finish Work to Start Activities 80-95, 80-100, 85-100
M6	Apr. 29 (Day #60)	Project Completion

Table 11.1

Thus, Fig. 11.8 has all the control information necessary for the manager to supervise his project from a time and performance basis. Later chapters in this book will explain how costs and resources of manpower and machines can be added to it.

ONCE THE PROJECT CONTROL CHART IS OBTAINED, IT IS RECOM-
MENDED THAT THE ARROW DIAGRAM AND ITS DIGITAL SCHEDULES BE
CAREFULLY FOLDED AND PUT AWAY! IT IS THE AUTHOR'S CONTENTION
THAT THE ARROW DIAGRAM IS A WORK SHEET, AND SHOULD NOT BE
THE DOCUMENT ISSUED ON A PROJECT. THOSE UNFAMILIAR WITH
NETWORK PLANNING WILL HAVE NO PROBLEM READING AND USING THE
PROJECT CONTROL CHART.

Control of a project and updating a project's schedule are not necessarily
related. Control of a project goes on continuously day after day. It is exerted
in the same manner that a thermostat is part of an environmental temperature
control system. In the thermostat there is a reference temperature, the dial
setting. The actual temperature is measured by a bi-metallic element. Any
deviations between reference and actual is detected, and energy is exerted in
some form to correct the environment's temperature. So, too, in project
control. The Project Control Chart is the reference document. Actual performances are observed and compared. Once deviations are detected, energy
in some form is to be exerted to bring the project back on schedule.

Updating a schedule is discussed in a later chapter. It is essentially a re-
estimate of the remaining work on a given project.

Note one advantage of the manually drawn Project Control Chart versus the
computer drawn Time Scale Chart. Manually, the Project Control Chart can
be organized and "banded" (see Chapter 2) to show smooth work flow in
physical areas, or by trades, or by sub-contractors, or in any combination
thereof. The computer drawn Time Scale Chart is only horizontally oriented
against the Time Scale--the vertical orientation is pre-determined by the computer's program seeking available space for printing the description of the
activity in a clear space on the chart. Thus, natural sequential relationships
may not fall on the same horizontal row, but become discontinuous "bars" on
a bar chart type of presentation.

The bar chart, which almost every reader is familiar with, is another form
of schedule presentation. Fig. 11.9 is such an example of our project, with
each activity on a separate row. The shortcoming here is that if for any
reason an activity is adjusted or extended or moved, the consequential relationship cannot be easily seen. For example, if activity #5-70 is delayed,
extended or moved in any manner, it automatically affects the starting time
of activity #70-90, but you cannot readily determine that relationship. If the
reader studies the Project Control Chart of Fig. 11.8, he will readily note
the logic relationship.

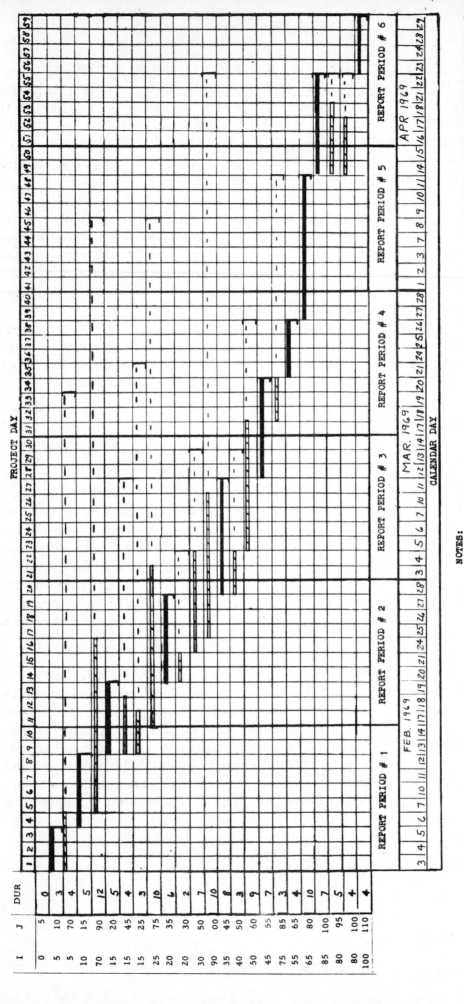

PROFESSIONAL PLANNING CONSULTANT
A. JAMES WALDRON
371 KINGS HIGHWAY WEST
HADDONFIELD, NEW JERSEY
PHONE 609-428-3742

BUILDING A NEW HOME
TIME SCALE CHART
(BAR CHART)

DRAWN: 12/22/68 | FIGURE # 11.9

NOTES:
1. Activities listed in chronological
sequence in order of importance
(Early Start Time - Major)
(Total Float Time - Minor)

2. ▬▬▬▬ Critical Activity

3. ▭▭▭▭ - - - - [Latest
Float/Slack Activity   Finish Time

CALENDAR BASIS:
Projects Start Date: FEBRUARY 3, 1969
Five (5) Day Work Week
Holidays Honored:
Memorial Day (May 30, 1969)

MANPOWER ALLOCATION

We will start assigning resources to a network. On the five arrow diagram, Fig. 11.10 we have assigned various manpower requirements. In order that we crawl before we walk, we shall set up the following restrictions on this project: (1) we cannot start the particular operation until all the men required are available, and (2) once an operation is started, it must be worked upon without interruption until it is completed.

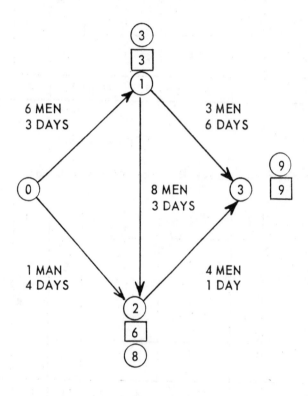

Fig. 11.10

Fig. 11.11 will indicate the time scale representation of Case I, all activities at their Early Activity Start Time. Fig. 11.12 presents the Case II, leveling by using the float activities to minimize the peak manpower. The activities with float are maneuvered to minimize overlap. Note that in all cases, the logical sequence of the activities always pertains; no activity may be scheduled before its logical predecessor is completed.

By assigning the manpower requirements to each activity arrow, we are able to add up the daily force requirements. In Case II, we see that by sliding the

Case I All Early Starts

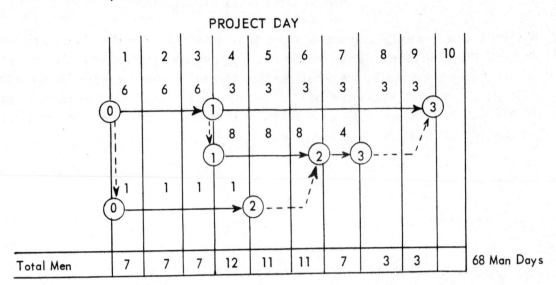

Fig.11.11

Case II Using Float to Level Resource (Manpower) Usage

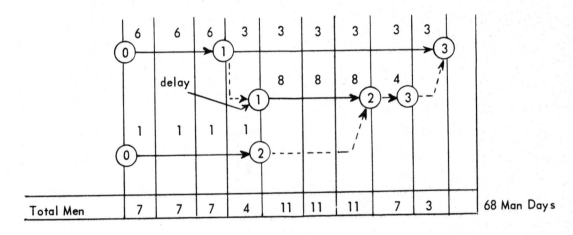

Fig.11.12

floater job arrows back and forth, we are able to take advantage of the float and reduce the peak from 12 to 11, and still maintain the normal project duration of 9 days.

Next, the project shall be "forced", by establishing a limit below that minimum of 11 in Fig. 11.12 No matter how the floater activities are manipulated, the minimum usage in Fig. 11.12 will not go above 11 men on any particular project day. So Case III will be investigated, wherein every other restriction

118

remains (must have specified crew size available before scheduling, and once started, an activity must be scheduled to completion before scheduling the next one), except now we only have a maximum of 10 men available on any one day.

At this point, a change in methodology is in order. A table will be established, listing the activities in a certain sequence as an ordinate, and the project day sequence as an abscissa. The assignment shall be entered each day, and under a slant bar, the remainder from the available force. Thus, by looking up any columns, (project day), the available number of men for the next assignment can be quickly noted. Fig. 11.13 indicate Case III - 10 men daily limit. Again, the logical sequences of the plan must be kept; no activity can be started ahead of the completion of all of its predecessors.

Case III

ACTIVITY	PROJECT DAY	1	2	3	4	5	6	7	8	9	10	11	12	13	14	15
	LIMIT	10	10	10	10	10	10	10	10	10	10	10	10	10	10	10
0 - 1		$6/4$	$6/4$	$6/4$												
0 - 2		$1/3$	$1/3$	$1/3$	$1/9$											
1 - 2					$8/1$	$8/2$	$8/2$									
1 - 3					DELAY			$3/7$	$3/7$	$3/7$	$3/7$	$3/7$	$3/7$			
2 - 3								$4/3$								
AVAIL		3	3	3	1	2	2	3	7	7	7	7	7			
USAGE		7	7	7	9	8	8	7	3	3	3	3	3	68 MAN DAYS		

Fig.11.13

119

Note two interesting aspects of the above schedule. First, the limit of 10 men available each day still extended the duration out from the normal of nine (9) days to a new length of twelve (12) days by delaying the critical activity 1 - 3. Remember the constraint mentioned above; we cannot start an activity until all the men required for that operation are available. Later we shall start activities at a "threshold", when an acceptable minimum crew size is available. Then we shall extend the duration, based on the proportion of a constant man-days figured for that operation.

Note also that although we had ten men available every day, we never used more than nine men on any single project day.

Next, we shall investigate an extension of Case III. The sequence of listing the activities in Case III was arbitrary. Why not see if the order in which we listed the jobs has some bearing on a schedule that is generated by this method? We shall consider Case IV which is the same as Case III except we shall list the jobs in order of criticality, within the logic of the network. Activity 0-1 and 1-3 have zero float, activity 1-2 and 2-3 have 2 days of total float, and activity 0-2 has four days of total float. In figure 11.14, we shall enter the remainder from our limited available resources. In this way we just note the last number in any day's column and immediately learn how many men are available that day for the next assignment. This is a simplified bookkeeping method.

Project Day	1	2	3	4	5	6	7	8	9	10	11	12	13	14	15
Limit	10	10	10	10	10	10	10	10	10	10	10	10	10	10	10
Activity Float															
0-1 0	4	4	4												
1-3 0				7	7	7	7	7	7						
1-2 2						delay				2	2	2			
2-3 2													6		
0-2 4	3	3	3	6											
Avail	3	3	3	6	7	7	7	7	7	2	2	2	6		
Usage	7	7	7	4	3	3	3	3	3	8	8	8	4		

Fig.11.14

Case IV gives interesting results. Although it is the same as Case III, 10 men available, all other criteria the same, just the order in which we listed the jobs caused the project schedule to go out another day to day 13; and while ten men were always available, no more than eight on any one day were assigned to this project.

From experience, there is no consistently best way of listing the sequence of activities. Since the "i - j" numbers were assigned at random, there is no clear cut pattern. For best results, three separate sequence listings should be investigated: "I" Major, "J" Major, and Early Start Time Major with Total Float Minor. Fig. 11.15 indicates the different sequence of each listing. In general, the "Early Start Time" Major, "Total Float Time" Minor sequence should be investigated first. In a good percentage of actual projects, it produces adequate information to allow decision making.

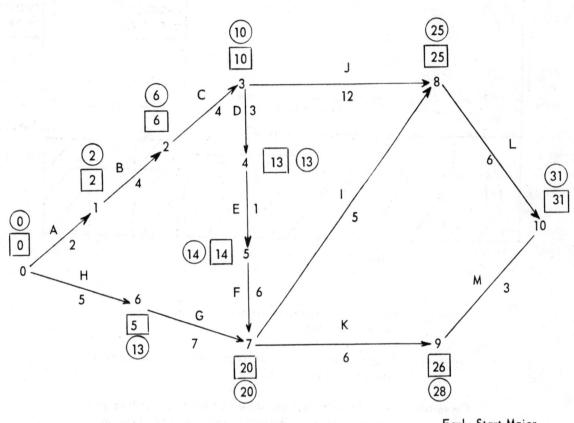

I Major Listing	J Major Listing	Early Start Major Total Float Minor
0-1	0-1	0-1
0-6	1-2	0-6
1-2	2-3	1-2
2-3	3-4	6-7
3-4	4-5	2-3
3-8	0-6	3-4
4-5	5-7	3-8
5-7	6-7	4-5
6-7	3-8	5-7
7-8	7-8	7-8
7-9	7-9	7-9
8-10	8-10	8-10
9-10	9-10	9-10
I in ascending order	J in ascending order	Early Start Time determines ascending order

Fig. 11.15

121

Realize that the man-hours or man-days stay the same, since these were put into the system as estimates necessary to accomplish the work in normal time. The leveling aspect keeps the same area under the man-time curve, but flattens it out. Fig. 11.16 graphically shows the effect of leveling.

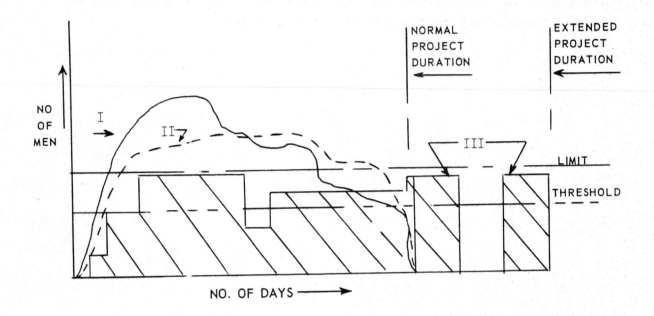

Curve I: Manpower Distribution, all activities starting early, unlimited resources.

Curve II: Manpower Distribution, utilizing project Float to level within normal project duration.

Curve III: Manpower Distribution, showing effects of limiting an available resource. Project Duration extended due to limits. Difference between "Limit" and "Threshold" called Reserve Force, or Elasticity. Threshold can be used to start activities when a certain minimum number of a resource is available, E.G. If an activity calls for six men for two days (12 man-days), and threshold is four, the method will schedule that activity for three days with the four available men.

Area under Curves I, II, and III are the same.

Fig. 11.16

Multicraft Scheduling

The same approach is extended to the situation where more than one resource (craft) is assigned to a particular activity arrow. The limit for each craft is listed under the project day row, and a sequence listing of activities shown in a column. Fig. 11.17 is a project diagram, with crafts A, B, C requirements assigned to each arrow. The daily limits of each craft are set as follows:

Craft A - 5
Craft B - 3
Craft C - 4

We shall now take the next step, and allow "Splitting" of an activity, i.e. each activity is scheduled a day at a time. If the availability of a particular craft is less than a specific day's requirement, then the activity skips that day, and goes on to the next day, seeking adequate availability. Thus in a 5 day duration activity, if after the second day of consecutive scheduling, the availability of a particular craft is less than the requirements, the next day's work is delayed until the availability is attained.

Again, taking the technique in steps, the example shown in Figures 11.17 and the schedule in 11.18 has a constraint - the team assignments of crafts must be maintained. Thus, if activity 4-8 requires two men of craft B, and two men of craft C, for a total of 6 days, both two men of B and two men of C must be found on the same day, before a days work for that activity can be scheduled.

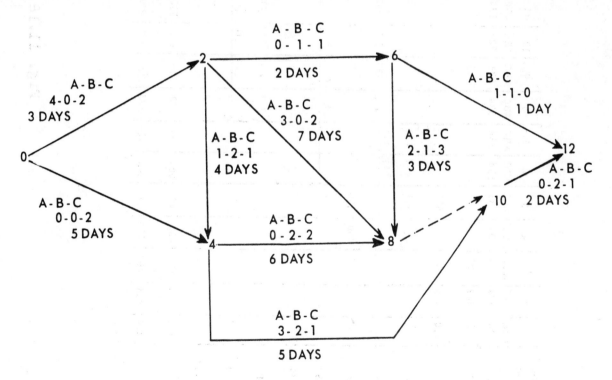

THREE CRAFT PROBLEM

Fig. 11.17

The schedule will be generated by an Early Start Major, Total Float Minor Activity Sequence Listing. Only remainders, after assignment, are shown.

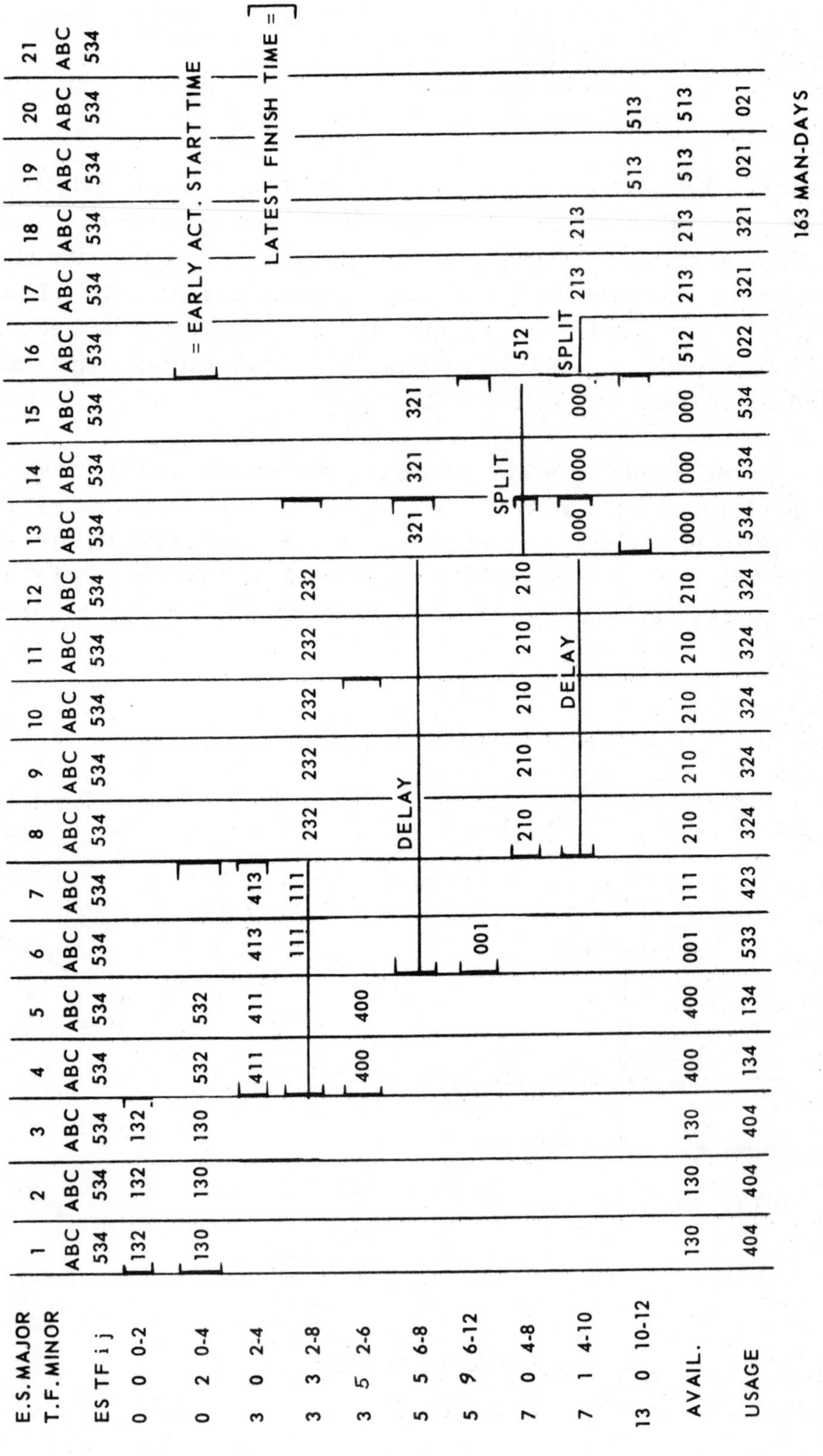

Fig. 11.18 — Early Start Major, Total Float Minor resource leveling schedule.

Activity listing (left columns):

E.S. MAJOR	T.F. MINOR	i j
0	0	0-2
0	2	0-4
3	0	2-4
3	3	2-8
3	5	2-6
5	5	6-8
5	9	6-12
7	0	4-8
7	1	4-10
13	0	10-12

Resource schedule (each period column headed ABC = 534 available; values are remainders after assignment):

Period	1	2	3	4	5	6	7	8	9	10	11	12	13	14	15	16	17	18	19	20	21
ABC	534	534	534	534	534	534	534	534	534	534	534	534	534	534	534	534	534	534	534	534	534
	132	132	132	532	532	413	413	232	232	232	232	232	321	321	321	512	213	213	513	513	
	130	130	130	411	411	111	111	210	210	210	210	210	000	000	000		213	213	021	021	
			130	400	400	001		210				210	000								
AVAIL.	130	130	130	400	400	001	111	210	210	210	210	210	000	000	000	512	213	213	513	513	
USAGE	404	404	404	134	134	533	423	324	324	324	324	324	534	534	534	022	321	321	021	021	

Chart annotations:
- ⎡ = EARLY ACT. START TIME
- ⎤ LATEST FINISH TIME =
- DELAY
- SPLIT
- 163 MAN-DAYS

124

The limits are listed for each day. The remainders are listed in the table so that the next activity for scheduling can be compared to the availability of each craft, in the last number in the project day column immediately above it.

Thus, activity 2-8 could start immediately after the completion of its logical predecessor, 0-2. A search for the craft requirements (3 men of craft A, 2 men of C) can start on day #4, the day after the scheduled completion of 0-2. Unfortunately, there is an inadequate supply of craft C (2 men of craft C having been previously scheduled for 0-4, and 1 man of craft C previously scheduled for 2-4). This activity must be delayed until day #6, when both craft A and C have adequate men available.

Activity #4 - 10 is normally scheduled to start on day #8 (the end of day 7), after its predecessors 2 - 4 and 0 - 4 have been completed. But a search on day 8 for the 3 men of craft A, 2 men of craft B, and the 1 man of craft C reveals a shortage of one man in each craft. These do not become available until day 13. On day 16, the previous assignment to activity #4 - 8 results in a deficiency in craft B. Thus activity 4 - 10 must be split, it cannot be scheduled to work on day 16. The remaining two days of work must be scheduled on days 17 and 18.

Resource Thresholds

In the normal planning of a project, the determination of a particular crew size is sometimes arbitrary. Thus, it may originally have been assumed that a reasonable crew for a particular activity was four carpenters and two laborers. However, if at the time that activity could start, the foreman finds two carpenters and one laborer available, chances are he will start that activity with what he has. As additional men become available, they will be added to that crew. So as a matter of practicality, we should approach scheduling in this manner, rather than artifically delaying an activity when a complete crew is available. The approach is obvious; we will determine a reasonable minimum crew size for starting. We know the original man-day estimate. Thus, we are using X men on an activity that originally called for 3X men, and we have expended Y men days so far. The original estimate called for E man days total. The remaining duration of the activity is $(E-Y)/X$. If X is increased any future day, then the consumption of man day rises, and a new remaining duration calculated by the division of $(E-Y)$ by the new value of X, the number of men working.

Resource Contours

The technique, and some available computer programs, will allow the limit to be contoured. Thus a "step" increase can be shown, if on a particular project day, a resource (men, machines, money) are released from another project, and are indicated available for this particular project. Resource limits can be "contoured" at will, so that advantageous rates of expenditure of resources can be applied or tested, on a particular project. The availability table for the project in Fig. 11.17 might conform to the one shown in Fig. 11.19.

PROJECT DAY	1	2	3	4	5	6	7	8	9	10	11	12	13
CRAFT	ABC	ABC	ABC	ABC	ABC	ABC	ABC	ABC	ABC	ABC	ABC	ABC	ABC
AVAILABILITY	5-3-4	5-3-4	5-3-5	5-6-5	5-6-5	5-6-5	5-6-5	6-5-5	7-4-5	8-3-5	9-2-5	10-2-5	10-2-5

Fig. 11.19

Each craft availability may be contoured separately, as well as its particular "Threshold" value.

In order to develop a most powerful aspect of this technique, the manpower assignments in Fig. 11.20 below will be scheduled by this method, with the following rules applied to teach a methodology.

1. The Plan of Fig. 11.20 shows the "Normal" crew and duration. Each activity may start with a minimum crew of two less (-2) than the Normal shown, and executed with three more (+3) as a maximum crew. The man-days of Normal crew/Normal duration will be the determining factor, with crews ranging between minimum and maximum as specified, and compared to a contoured limit.

2. The only exception allowed to the minimum crew size will be the last day of the particular activity, where the remainder of man-days will dictate a smaller than minimum crew. This will be considered a "clean up" function, with any size crew needed less than minimum allowed.

3. The work will be scheduled on a man-day basis, in increments no smaller than a day.

4. The resources will start with 8 men (necessary for the initial activity, 0-2), and will build up one more man a day until a maximum of 12 men is reached. A limit of 12 will be kept on this project until completion. The build up starts on day #4.

5. Any activity may be "split" (one day at a time, not necessarily consecutive days). Work cannot be scheduled until the minimum crew availability is present, except for the last day "clean up" function.

126

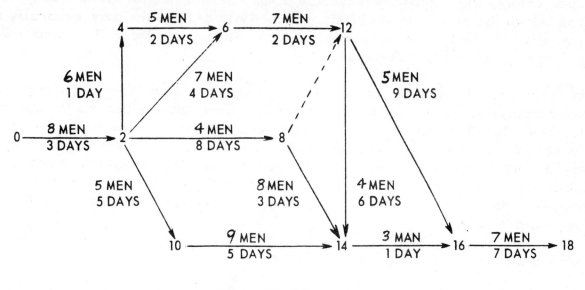

Fig. 11.20

Fig. 11.21 is the Project Schedule for the above limitations. The activities have been ranked by Early Start (ES) Major, Total Float (TF) Minor. Note how the ES-TF Column goes in ascending numerical order from 0-0 to 20-0. The sum of the Man-Days column must be equal to the total of the Usage Table (the sum of the men per day for the 31 days).

The logic of the Plan is always adhered to, Activity 12-16 follows Dummy 8-12 and Activity 6-12. It (12-16) cannot start until the 16th day because its prede- cessor 6-12 is scheduled to finish on the 15th day. Although we start looking for men on the 16th day for Activity 12-16, we find no one available until the 18th day.

Another advantage of the Sequential Event Numbering method is apparent here. When we start scheduling 12-16, we note its "i" number to be 12. Looking above it in the Activity List, in the "j" column, we note that the dummy 8-12 and the Activity 6-12 must precede it (its "i" follows a previous "j" in the list). The dummy tells us we must also examine all "j's" of 8 as predecessors.

The resultant schedule produces valuable information. The usage is fairly constant, except for days #4 and 14. There is a big dip in usage of originally available resources on day #26. There is not much that can be done about the availability of 9 and 10 men on days #26 and 27. This may have to be written off as an inefficiency factor. The table also indicates that our "Resource Contour" could start dropping off to 11 men on days #23, 24, and 25, and down to 9 men on the last day of this schedule.

If the "Normal" Forward and Backward Pass is made on Fig. 11.20, the end date will be day #27. The schedule for the normal durations is shown in Table

8.0, Chapter 8. Since we are reducing some durations by utilizing maximum crews, the significant schedule dates are the Latest Finish (LF) dates. It is left to the reader to ascertain why Activity 8-14 is the most critically affected by the limited resources. 16-18 has an improved duration, and makes up some of the time lost by 8-14.

At least one organization is establishing rates of expenditure of its capital (a basic industrial resource) on its company projects, that represents the optimum return to the organization. It is used in conjunction with the selection of key milestone events on a project arrow diagram. The Early Event Times for the key events are selected, and an optimum rate of capital commitment is established. Once a key milestone is reached, the rate of resource availability (management commitment) may have a different slope or contour to the next key milestone event.

Some features of existing computer programs are:

1. It may handle over 20 crafts. Assign up to any 7 crafts to any one arrow; 15 crafts scheduled in one day.

2. The computer can be instructed to split duration of jobs in unit times.

3. The start day of any arrow may be specified.

4. It may specify a "start delay threshold": If an activity is postponed indefinitely, or to an impractical date due to limits of its resources, a constant may be put in representing the latest acceptable starting date. If the resources limit reaches this constant, the computer will reset this particular activity up to its earliest start date, and specifically "flag" the particular activity as having an unrealistic resource limit.

5. The limit is above the threshold creating an elasticity in the system, or a reserve force.

6. Contour of limits by day or periods of time may be set into the program.

7. Usage threshold may be different for each craft.

8. Space limitations effects can be considered a resource limit.

9. Dollar expenditures can be considered a resource limit.

PROJECT SCHEDULE

Fig. 11.21 PROJECT SCHEDULE WITH CONTOURED RESOURCES AND BASED ON MAN DAYS WITH MIN-MAX CREWS.

Note:
$$\frac{8}{0} = \frac{\text{Assignment}}{\text{Remainder}}$$

* = Minimum Crew Not Available

Activity data

Activity I-J	Rank ES	Rank TF	Crew Size Min	Crew Size Norm	Crew Size Max	Normal Duration	Man Days
0-2	0	0	6	8	11	3	24
2-8	3	0	2	4	7	8	32
2-6	3	2	5	7	10	4	28
2-4	3	3	4	6	9	1	6
2-10	3	6	3	5	8	5	25
4-6	4	3	3	5	8	2	10
6-12	7	2	5	7	10	2	14
10-14	8	6	7	9	12	5	45
8-12	11	0	DUMMY				
12-16	11	0	3	5	8	9	45
12-14	11	2	2	4	7	6	24
8-14	11	5	6	8	11	3	24
14-16	17	2	1	3	6	1	3
16-18	20	0	5	7	10	7	49

Daily schedule (Assignment | Remainder)

Project Day / Resource Limit header:

Day	1	2	3	4	5	6	7	8	9	10	11	12	13	14	15	16
Limit	8	8	8	9	10	11	12	12	12	12	12	12	12	12	12	12
0-2	8|0	8|0	8|0													
2-8				7|2	7|3	7|4	7|5	4|8								
2-6					*	*	5|0	8|0	10|2	5|7						
2-4					*	*			2|0							
2-10					3|0	4|0				7|0	8|4	7|5	1|11			
6-12														8|0	8|0	
10-14													11|0	10|2	4|8	12|0

Day	17	18	19	20	21	22	23	24	25	26	27	28	29	30	31	32
Limit	12	12	12	12	12	12	12	12	12	12	12	12	12	12	12	12
10-14	12|0	2|10														
12-16		8|2	8|4	8|4	8|4	8|4	5|7									
12-14		2|0	4|0	4|0	4|0	4|0	6|1									
8-14								11|1	11|1	2|10						
14-16											3|9					
16-18												10|2	10|2	10|2	10|2	9|3

Availability Table / Usage Table

	1	2	3	4	5	6	7	8	9	10	11	12	13	14	15	16	17	18	19	20	21	22	23	24	25	26	27	28	29	30	31	32
(remainder)	0	0	0	0	2	0	0	0	0	0	0	0	0	2	0	0	0	0	0	0	0	0	1	1	1	10	9					3
AVAILABILITY TABLE	8	8	8	7	10	11	12	12	12	12	12	12	12	10	12	12	12	12	12	12	12	12	11	11	11	2	3	10	10	10	10	9
USAGE TABLE																																

Again, a note of caution is due. This is still just an information system. Judgment must be exerted on the numerical answers received. The schedule may present an answer such as this; 23 pipefitters are needed for a three month period, then there is no requirement for this craft for the next six weeks, then some 14 pipefitters will be required for the next 10 weeks. Is the manager to dismiss this craft for the six week interim, and then attempt to rehire a lower number? The answer to this problem is in the area of labor relations, and beyond the context of network planning. Again, this technique does not solve "people problems".

One very effective usage of the manpower leveling or Contouring aspect, is in the scheduling of a hard to get resource, so that it is scheduled as early as possible, and continuously scheduled until those affected operations are completed.

As a case in point; a contractor building several dormatories for a large university, found himself competing with a large amount of residential development construction in the area. A particular craft, bricklayers, was in short supply. If he received a schedule that indicated a "split" in bricklaying activities, he had no guarantee of getting this craft back on the project when needed. His approach; schedule bricklaying activities as soon as possible, and insure that they were continuously scheduled to completion of that chain of bricklaying activities. This criterion delayed other activities, resulting in a six week extension of the project duration. To the contractor, this was an acceptable compromise, since there was no guarantee that he could regain the dismissed craftsmen within six weeks of dismissal.

A noteworthy point for the reader! When using the above described matrix approach to resource countoured schedules, the reader must exert judgement as to when he should go above the "Normal" crew or resource assignment towards the "Maximum" crew or resource assignment. There are going to be some networks that, by their configuration, it will be wise to keep the early activities at normal crew sizes. This will leave some of the available resource (or resources, if multi-resource matrix is being used) available for further scheduling. Inevitably, if an early activity is brought up to maximum resource utilization, there may be little or no resources of that class available for subsequent activities. Note in Fig. 11.21 that a non-critical activity, #8-14, is extended furtherest past its Latest Finish Date. As a general practical approach, use only up to "Normal" crews or resources for the first run. Observe how much over the desired Project End Date this result gives. Study those activities that went past their Latest Finish Date. Make a second run, going to "Maximums" wherever necessary to improve the end date. There is no simple generalization that will produce the best schedule for any project; it usually takes two or more studies.

11.1 In the diagram below, lay out the plan on a time scale, and (a) start every activity at its early start time, and note the peak (the amount and the working day or days) man-power load. Then (b) manipulate the floater activities, within the normal duration of the project, to minimize the peak man-power requirements. Note the reduction of maximum man-power requirements, and on what day or days the reduction occurs. Hints: Review Figures 11.8 and 11.9. (c) What is the schedule of this project if the maximum availability of men on any one day is 11.

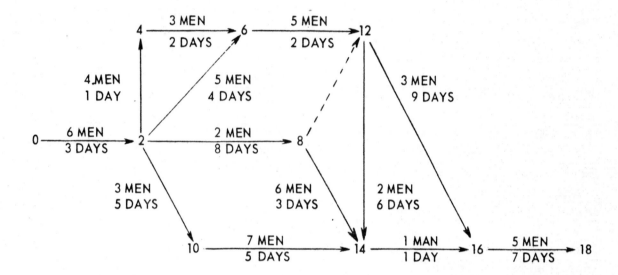

11.2 List the activities of Fig. 11.17 page 124, in a J Major sequence, and schedule with the craft limitations as shown in Fig. 11.18 (b) List the activities of Fig. 11.17 in an I Major sequence, and determine the schedule. Is there any difference in the completion dates? Compare all three schedules, and note different peak usage points.

11.3 In the contractor's planning of the construction of the house, Problem 1.1, he has based his working time estimates on the following craft assignments:

Craft #1	Carpenters
Craft #2	Laborers
Craft #3	Plumbers
Craft #4	Electricians
Craft #5	Masons
Craft #6	Painters

ACTIVITY	Duration (Working Days)	1	2	3	4	5	6
Sign contract	0						
Layout and excavate for building	3		2				
Construct basement walls and back-fill outside	5		2			2	
Install outside water, gas, sewer lines and electrical power	4		3	2			
Frame first floor to second floor	5	4					
Stone veneer to first floor sill	4					2	
Stone veneer to eaves	7					4	
Frame and shingle roof	8	2					
Frame second floor to roof	6	4					
Pour and cure basement floor slab	3		3				
Install furnace, water heater, basement plumbing	10		2				
Erect interior first floor walls, and rough flooring	2	3					
Install first floor plumbing, ducts, wiring	7			2	2		
Install windows, exterior doors	4	2					
Install sidewalk, curbs, driveway	12		2				
Erect interior second floor walls, and rough flooring	3	3					
Install second floor plumbing, ducts, wiring	9			2	2		
Put up inside drywalls and paint	10	2					2
Install kitchen equipment	7		2	1			
Grade site, sod and landscape	10		2				
Finish flooring	5	3					
Install millwork, interior doors	4	2					
Finish plumbing, electrical tie-ins	3			2	1		
Punch list fix up and inspection	4						

(a) What is the peak requirement for carpenters on the normal schedule with the craft #1 assignments above?

(b) If the threshold for carpenters is 1 man, and a limit of 4 carpenters available any one day, can the normal 59 day completion date be met? (When using thresholds, you are working in man-days, and as men become available the remaining duration is reduced by the formula $(E-Y)/X$.)

(c) If the limit on laborers is 2 men available on any one day, and a laborer threshold of 1, can the 59 day completion date be met?

(d) With a maximum of two plumbers used in the original estimates, will the conditions in (c) above affect the plumber schedule, and the completion date? (Handle this as a two craft problem, making sure both laborers and plumbers are available on the same day.)

In all the parts of problem 11.3 above, every activity may be split, i. e., scheduled one day at a time.

11.4 Construct the Time Scale Project Control Chart for the project of Problem 2.6, Chapter 2. "BUILDING A COIN OPERATED LAUNDRY." Correct your arrow diagram to the Network Plan solution of 2.6 before proceeding. The weather affected activities are:

#5	Form Pour Footings
#8	Install Outside Sewer, Water, Gas Lines in Trench
#9	Form, Pour, Cure Floor Slab
#10	Erect Cinder Block Walls
#11	Erect Roof Rafters and Purlins
#12	Waterproof Exterior Walls
#13	Interior Plumbing
#21	Cover Roof
#35	Grade and Pave Outside Parking Area

Inclement weather can be expected between the Start of the 11th Day and the end of the 25th Day. It is expected that two-thirds of the working days will be lost in this period (activity production is 33% efficient). The start date of the project is December 1, 1968, and a six day work week followed. Sundays, and December 25th and January 1st are non-work days.

Show the following Milestones:

1. The Day the Floor Slab has been Poured and Cured
2. The Day the Roof is Covered
3. The Start of Final Plumbing Tests
4. The Start of Final Electrical Hookup and Tests

(SPEND AS MUCH TIME AS YOU NEED ON THIS PROBLEM, IT REPRESENTS THE CRUX OF THE FUNDAMETALS OF PROJECT PLANNING AND CONTROL.)

REVIEW OF FUNDAMENTALS

Pause and Consider

At this point in the material, the reader has been exposed to the thesis of separation of Scheduling from Planning; how this logic system relies on sequences of activities in a plan, their estimates on a disassociated basis, a Forward and Backward Pass to establish time boundaries which are Event Times, the establishment of the schedule, and some comments on how to analyze that schedule.

As noted previously, it is fully expected that the initial schedule will probably be an unhappy one, not meeting the project Time requirements. A second "Pass", made after decisions are reached from the information generated, is to be expected in 80% of the projects so planned by this technique.

Again, it might be wise to return to Chapter 3 and review the material on the Transfer Event. This is the shorthand approach, with very pertinent remarks to the powerful use of the Dummy in adjusting the plan, once the initial schedule proves inadequate.

At this point in the material, the reader is where 95% of the "State of the Art" is--planning and time scheduling. From here on in the book, further sophisticated manipulations of the data bring more information. The next dimensions of the technique now consider time and resources, and time and costs, as simultaneous parameters. But the fundamentals, as hereto developed, must be understood before any benefit can be obtained. Perhaps a review is in order?

One other aspect of the technique is the rapid and explicit evaluation of alternate strategies. It should be obvious that there is usually more than one way of executing a project. You have learned the tool to quickly test more than one way of attacking your project. It is extremely easy to change just a section of a plan, and obtain new information. Here is a case in point:

Alternate-Strategy Techniques

One tremendously powerful aspect of Network Planning is that the technique may be used to quickly evaluate the feasibility of alternate strategies. Any section of a network can be lifted and replaced by another sub-network, representing a different strategy to be evaluated. As an example, let us consider the following case history.

Addition of a New Product Line

The ABC Company, large manufacturer of electronic products, has a small line of nuclear instruments. The ABC management feels it would be advantageous to diversify. The technical capability of design and application engineers, combined with existing production facilities should be a base upon which to build a line of industrial process-control instruments. A market survey is made, establishing current total sales of $2 billion annually, and more than 100 producers currently in this market.

A three-pronged study is made. The first considers a cross-licensing approach. Manufacturers in the industrial instrument market do not have a competitive line of nuclear instruments, and are not strong in the nuclear field. It is believed that a good possibility exists of combining their production techniques with ABC circuit techniques, enabling ABC to extend its operation, first in the nuclear industry with a line of non-nuclear instruments, and eventually into the domestic process-instrument market. A company team of engineering, marketing, legal and production personnel is formed to investigate cross-licensing possibilities, particularly with European manufacturers.

Concurrently, authorization is given to the Engineering and Marketing Departments to prepare a proposal for a company-built facility for design and production of its own line of industrial instruments. Also, a management consultant firm is retained to study direct acquisition of existing manufacturers. The sales and distribution organization, suppliers, and advertising and sales literature, are all investigated simultaneously (See Fig. 10.8).

The three major proposals (cross-licensing, company-built facilities and direct acquisition) are presented for management review and decision. The Network Plan for this review and decision has three possible terminating sub-networks. In this case, terminations have similar structure, but there are differences in timing. When costs and other resources are applied, management receives the necessary timely and accurate information to decide and execute the optimum course of action.

In this case (a true history), the actual costs, and the resultant management decision cannot be revealed. The salient point is that part of a logic network can be replaced by an alternate sub-network, and new information on schedules, key events, and other resources allocation, presented quickly to management for its use in decision making.

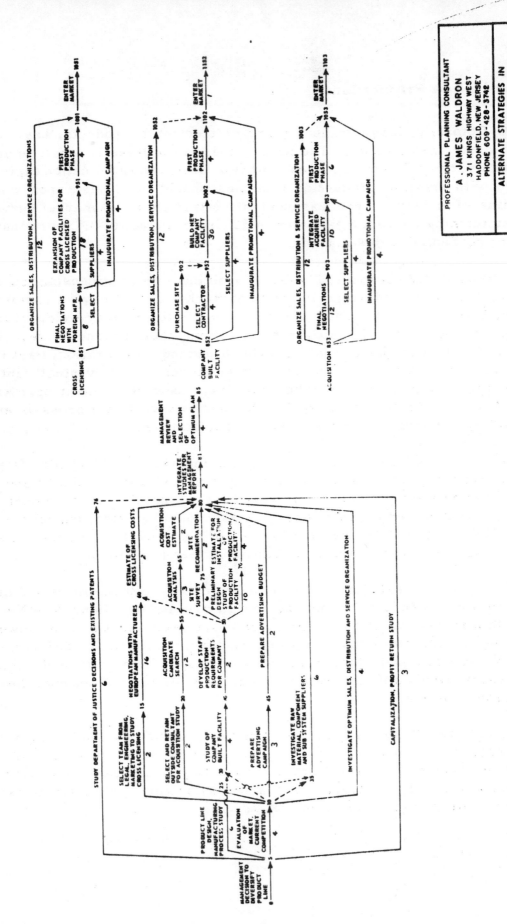

Fig. R1.0

Quite often it is difficult for the Planner to determine the sequences of activities, or to obtain the co-operation of those who are to plan sections of the overall Project Network Plan. A technique from Industrial Engineering is available that allows sequential relationships to be obtained rather rapidly.

A numerical table is set up similar to the one in Fig. R2.0 below. The number of diminishing rows will be determined by the number of activities to be evaluated for sequences. Essentially, the activities are assigned a number, in any random fashion whatever. This will serve to identify the activity. Thus, the activity list that the neophyte planner should compose before starting his first plan, may be numbered. Then each activity is compared with every other activity, and the basic question asked, "Does this activity come before that activity?" Numerically, the technique consists of comparing first the activity with number 1, with all the others; then number 2 with all the others; then number 3, etc. The comparison is done on the Rank Table. The first row, all number 1's, is compared with all the remaining activities. If No. 1 is to be done before No. 2, then No. 1 is circled. If No. 2 is to be done before No. 1, then No. 2 is circled. If it can't be decided which should precede the other, then neither is circled. Then proceed to the third and fourth rows, where No. 2 is compared, one by one, with all the remaining activities (No. 2 has already been compared with No. 1 in the previous exercise). Then No. 3 is compared with the remaining activities, then No. 4, etc.

After all comparisons have been made, then a count is made of all the circled number 1's, the number 2's, etc. The activity with the highest score is the initial activity in the sequence, the next highest score is second, the next third, and so on. Ties, or those activities that have been circled the same number of times, are concurrent activities. Ties at the highest score indicate initial concurrent activities. As a case in point, the Reproduction Machine Project in Chapter 2 will be examined. The Activity List has been random number identified as follows:

8. Authorization
3. Issue P.O.
7. Delivery Time of Equipment
1. Installation of Equipment
5. Hire Operator
2. Train Operator
6. Bring Inspector
4. Inspect the Installation

Fig. R3.0 indicates the comparison and the scoring. In the first row, Activity No. 1 (Installation of Equipment) was compared to No. 2 (Train Operator). Since the problem statement in Chapter 2 specifically notes the training is to take place on the machine, then Activity No. 1 must be done before No. 2,

Activity Ranking

```
1   1   1   1   1   1   1   1   1
2   3   4   5   6   7   8   9   10

    2   2   2   2   2   2   2   2
    3   4   5   6   7   8   9   10

        3   3   3   3   3   3   3
        4   5   6   7   8   9   10

            4   4   4   4   4   4
            5   6   7   8   9   10

                5   5   5   5   5
                6   7   8   9   10

                    6   6   6   6
                    7   8   9   10

                        7   7   7
                        8   9   10

                            8   8
                            9   10

                                9
                                10
```

Rank, Lowest to Highest

No. of Times Circled	0	1	2	3	4	5	6	7	8	9	10
Activity Number											

Fig. R2.0 Activity Rank Table

consequentially, No. 1 is circled in preference to No. 2. Comparing 1 to 3 (Issue P.O.), obviously No. 3 must be done before the Installation represented by No. 1, the planner circles No. 3 in preference to No. 1. There is no logical sequential relationship between No. 1 and No. 5 (Hire Operator), or between No. 1 and No. 6 (Bring the Inspector), so neither of these combinations are circled. No. 7 (Delivery of Equipment) must preceed the Installation of No. 1, so No. 7 is circled. And No. 8, Authorization, must preceed No. 1, so No. 8 is circled.

The planner continues down to the next row of activity combinations, and circles precedence in the same manner. After comparing all the activities, a count is taken of how many times each activity was circled. The count shows activity No. 8 was circled the most times, seven. It is the initial activity of the plan. If the Arrow Diagram Plan of Fig. 2.7 of Chapter 2 is studied, the number of arrows that follow any one particular arrow will be the same as the number of times that particular arrow has been circled. For example, Activity No. 7 has been circled 3 times. This Activity (Delivery Time of Equipment) is followed by three activity arrows in Fig. 2.7; "Installation", "Train Operator", "Inspect Installation." Note the zero circled activities have no successors.

This technique does not replace the logic of the diagram; it helps the planner obtain logical sequences, but not all of the logical interrelations. It is particularly limited in ascertaining subsequent concurrent activities. The Arrow Diagram is still the basic mechanism of planning.

The Ranking technique is fairly good for small sections of a plan. It is recommended that the Rank Table be kept to 50 or less activities for maximum effectiveness in Network Planning assistance.

One application is the obtaining of sequences from other persons responsible for executing some of the work represented on the Arrow Diagram Plan, and who either do not know how to construct the network for their portion of the work, or who are "too busy" to sit down with the planner. A few quick questions of "Which activity would come first," notations on the Rank Table, and the planner will be able to retire and construct the arrow sequences.

The Decision Tree

Another use of the arrow diagramming technique is the construction of a "Decision Tree", which is nothing more than a graphical portrayal by arrows, of all the options available in any particular circumstance. By including probabilities and estimated costs, the choice of activities to the most desirable event can be evaluated for time, cost and probability, and that chain can then be developed into a closed arrow diagram strategy for that option.

The technique is widely used in Marketing Strategies, Research and Develop-

Activity Ranking

(1)/2 1/(3) (1)/4 1/5 1/6 1/(7) 1/(8) 1/9 1/10

2/(3) 2/4 2/(5) 2/6 2/(7) 2/(8) 2/9 2/10

(3)/4 3/5 3/6 (3)/7 3/(8) 3/9 3/10

4/5 4/(6) 4/(7) 4/(8) 4/9 4/10

5/6 5/7 5/(8) 5/9 5/10

6/7 6/(8) 6/9 6/10

7/(8) 7/9 7/10

8/9 8/10

9/10

Rank, Lowest to Highest

No. of Times Circled	0	1	2	3	4	5	6	7	8	9	10
Activity Number	2 4	5 6	1	7	3			8			

Fig. R3.0 Ranking of Activities, Reproduction Machine Project

ment, and Political Analysis. It produces an open network (a "tree"), with the individual branches terminating in a desirable or undesirable event. The probabilities along any branch are multiplied, while the costs and time durations are summed, generating the numerical information of probability, length of time, and costs of attaining that event.

To illustrate the use of the arrow diagram in the "Decision Tree", the following hypothetical case will be explored: You are considering your forthcoming summer vacation. There is a campaign on within your family, particularly with the children, for a long cross-country trip. You have an eight year old car which is beginning to make strange noises after your struggle to start it. The original tires are now completely bald. You realize that the probability of their rupturing on the trip is rather high, with the consequence of an expensive purchase of new tires while out of town. On top of this there is the probability of an accident if the tires blow out while traveling high speed on the highway. There is no guarantee that the engine will last for the trip, and major motor overhauls while away from home are exorbitant.

The first development of the Decision Tree for the Vacation project looks like that in Fig. R.4.0. below.

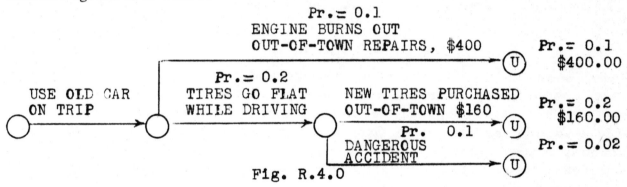

Fig. R.4.0

The more you consider the ends of the branches, the more undesirable the trip becomes to you. The next obvious option is added to the Decision Tree in Fig. R.5.0.

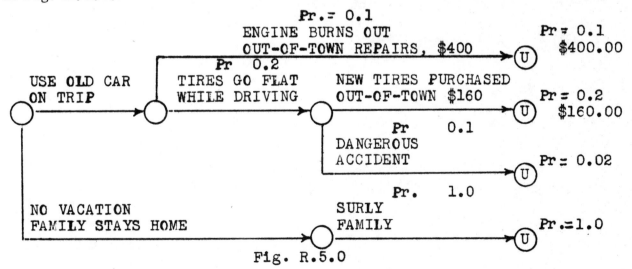

Fig. R.5.0

Since that last branch looks most undesirable, you consider further options:
(1) buy a new car, which is out of your budgetary plans, or (2) rent a car,
which proves an acceptable alternate. These are added to the Decision Tree
in Fig. R.6.0.

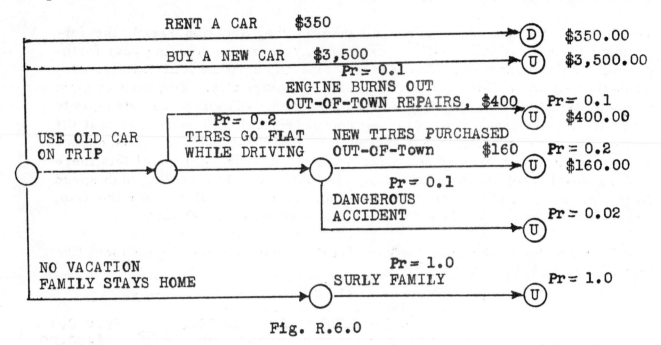

Fig. R.6.0

Further options, as you think of them, may be added to the Tree; (1) go by
public transportation (bus, train, or airplane). Each would be its own branch
in the Decision Tree, or (2) send the children to camp, and you stay home, or
visit your wife's relatives, etc. Once a chain leading to a desirable event
within your parameters of acceptable cost, probability and duration is found,
that branch is then developed into the complete closed arrow diagram for that
strategy, and the schedule obtained.

The nodes in the Decision Tree may be considered decision points.

Note: Ⓤ = Undesirable Event

Ⓓ = Desirable Event

CHAPTER 12

AN "IDEAL" PROJECT SCHEDULE

The general schedule generated from the arrow diagram shows the limits of time for each activity. The Early Activity Start Time (ES) points out that if all goes well and according to both the logic of the plan, and the validity of the duration estimates, this is the earliest possible time that the activity could start. The Latest Activity Finish Time (LF) points out that if all goes wrong, this is the latest permissible finish time for that activity if the resultant completion date is to be maintained. In essence, the schedule for a network plan shows the range of starting and finishing times for each activity, set by the Event Time boundaries. In addition, the nature of "Total Float" is that of lengths of non-critical paths, parallel to but shorter than the network Critical Path. Although the network technique assigns the Total Float or slack to each activity by formula, no one arrow enjoys that latitude in time scheduling by itself. If any arrow on a path loses Total Float or Slack, then every arrow on that path, or an interconnected path, loses the same amount of Float or Slack. Even if an activity is a future one, in terms of reference as to the date of analysis, and loses float, it affects its predecessors as well as its successors. Thus, an operation "Deliver MG Set #103", that is not scheduled to occur for another eight months, receives a new extended duration because of new information from the manufacturer, it will affect every one of its predecessors and its successors schedules by removing from its interrelated paths that amount of float represented by the incremental increase of its new duration.

Distributed Float

To increase the value and utility of the network technique of planning and scheduling to the manager, a method will be developed herein that allows the line manager to evaluate the importance or controllability of each activity in his plan. By assigning an Activity Weight, or Activity Value Number, representing the managers' estimate or value judgement of an activity's importance or controllability, a weighed or "ideal" schedule will be generated. This will (a) still find the original Critical Path, and (b) produce a specific starting and finish time for each activity, in accordance with that activities Weight or Value Number. In essence, all of the float in a network will be distributed to each activity, so that each activity obtains a specific block of time, which will be independent of its predecessor or successors block of time. These specific milestone dates will give the manager another set of information with which to monitor the progress of his project, will serve as the basis of a simplified but powerful cost control system, and become the basis of a simplified simultaneous project scheduling system.

The value judgement number may be on any arbitrary scale; 1 to 9, 1 to 999, etc., with the more indefinite, or least controllable activity getting the higher activity weight factor, in order to receive more of the path Float/slack that is distributed.

For the CPM user, and the single time estimate PERT user, the activity weight factor is added in the planning stage, reflecting the judgement of the planner who estimates the particular activity. Since this method divides up the path Float/slack and distributes it percentagewise to all the activities on that path, no matter how complicated that path, in accordance with the assigned weight factors, any range of weights may be used. Therefore, for the three time estimate PERT user (Optimistic, realistic, pessimistic) the range of these three elapsed time estimates (range equals pessimistic time minus optimistic time) will serve as the activity weight factor, since it reflects the uncertainty involved, proportionately, for that particular activity.

The weights used, in whatever range desired, may represent a judgement, or history of past poor performance or uncertainty. Such factors as an uncertain new state of the art operation, past poor performances of a group or craft, past failures of a supplier or sub-contractor to live up to his dates, etc. are reflected in the weights assigned. The more of the Distributed Float it is desired to obtain for an uncertain activity, the higher the weight or Value Number assigned. Fig. 12.1 below, using a 1 to 9 range, indicates the approach.

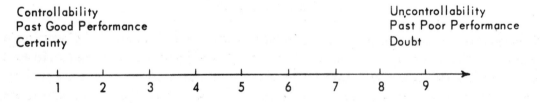

Fig. 12.1
Range of Activity Value Numbers

In the arrow diagram below in Fig. 12.2, the manager or the planner has assigned weight factors next to the single time estimate of elapsed time for each activity. These are shown as encircled numbers. The network will be recognized as that of Fig. 7.2, page 41 of Chapter #7.

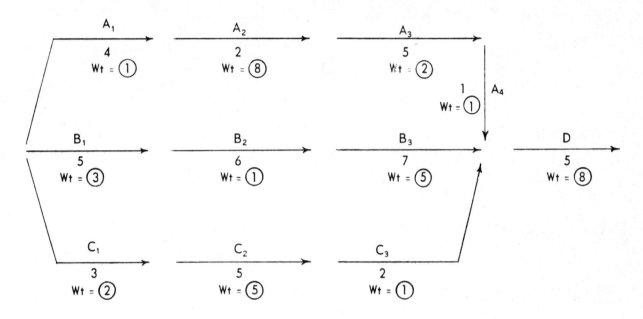

Fig. 12.2
Weighed Network Plan

The next step is the execution of the Forward and Backward Passes, the establishment of the Early and Late Event Times, and the ascertainment of the Critical Path. Inspection will reveal that the chain of A arrows has a Float/slack of 6 time units, the C chain Float/slack of 8 time units, and the B chain and operation D are critical. From the time bounded plan indicated in Fig. 12.3, a simple distribution of Float/slack can be made by allocating that proportionate path Float/slack in ratio to the activity weight/path sum of weights. For example, the weights along path A add up to 12 (1+8+2+1). Thus A_1 would receive 1/12th of the path Float/slack of 6, or 0.5 time units. A_2 = 8/12ths of 6, or 4 time units; A_3 obtains 2/12ths of 6, or 1 time unit, and A_4 is 1/12th of 6 or 0.5. The "Distributed" Float/slack then adds up to the path Float/slack, 0.5+4+1+0.5 = 6. If the "Distributed" Float/slack for a particular activity is added to that activity's Early Event Time, $\boxed{T_E}_j$, we will then obtain a "Distributed" Event Time for each Event. The "Distributed" Event Time at a Critical Event is the same as the other Event Times. Symbolically, we have shown the Distributed Events in Fig. 12.3 in triangles. It shall be identified as T_D.

145

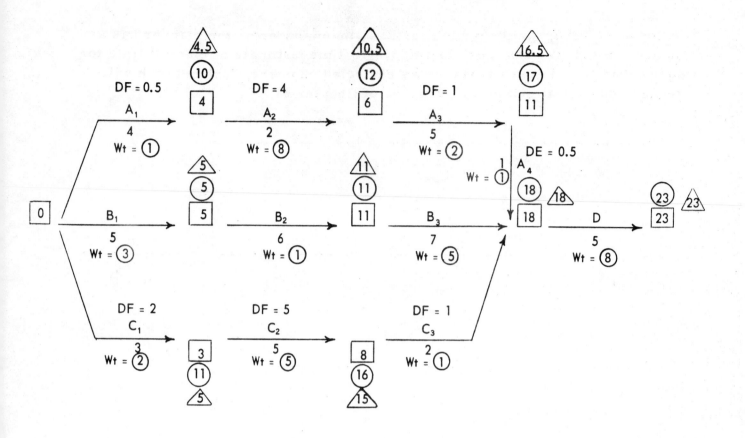

Fig. 12.3

Network With Early, Late and Distributed Event Times

The above is for purposes of example. It is not a formal method of establishing Distributed Event Times, and cannot be feasible on a complicated, interconnected real project diagram.

Event Weight Factor, "X"

After the establishment of Early and Late Event Times, and the Critical Path, the following methodology is employed to establish an Event Weight Factor, "X", which will be assigned to each event.

1. Dummies are considered to have an activity weight of 0.

2. Critical Events $(T_E = T_L)$ are assigned an Event Weight Factor of 1.

3. At each non-critical event, an Event Weight Factor, "X" will be established. This is simply the highest activity weight of all the arrows that enter that event. The Event Weight Factor is not an accumulated number, just that highest activity weight, of all the arrows entering that event.

4. A dummy transfers the Event Weight Factor, "X" of the event at its tail, to its head. This transferred weight is considered with all the other activity weights entering that event at the head of the dummy.

Fig. 12.4 is an illustrative example of how the Event Weight Factor, "X" is obtained for each event.

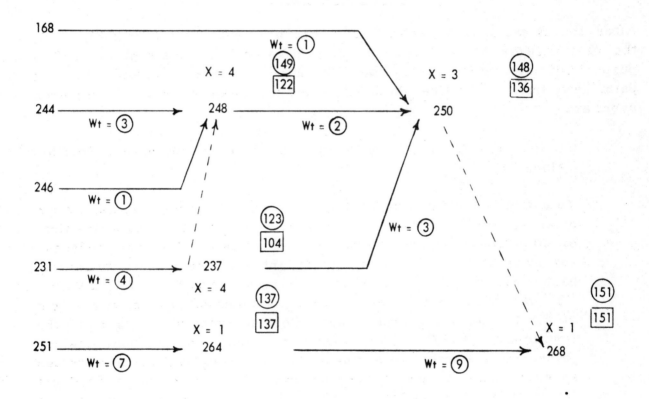

Fig. 12.4
A Section of a Network With Event Weight Factors Established

In Fig. 12.4 above, Event #250 has an Event Weight Factor of 3, since the highest weight factor of the three arrows entering Event #250 is ③ on activity 237-250. Note that Event #237 has an Event Weight Factor, "X" of 4, but that this has no bearing on the "X" or Event Weight Factor of Event #250. The Event Weight Factor "X" is not an accumulated number, just the value of the highest activity weight factor or Value Number of all the particular arrows entering that specific event, each event is evaluated separately for only those

147

arrows that enter it (those arrows that have that Event Number as a "j" number). Note that Event #248 has an "X" of 4, set by the dummy 237-248. The dummy is a transfer agent, and it transfers the "X" or Event Weight Factor at its tail to its head, for evaluation along with the weight numbers of those activity arrows that also enter the event at the dummy's head. Finally, note that Event #268 is a Critical Event, thereby receiving an Event Weight Factor of "X" = 1. This is regardless of the dummy 250-268 that also enters Event #268 with an Event Weight Factor of "X" = 3. Thus each event receives an X or weight value. Since the origin and final event of a closed network are on the Critical Path, they automatically receive an "X" = 1 from Rule #2 above.

Position Value, "Y"

After the Event Weight Factor, "X", has been established for each event, the next step is the setting of a Position Value, "Y", for each event. The purpose of this number is to count the number of events on a non-critical path, away from the Critical Path. The rules for establishing a "Y" for each event are:

1. Critical Events, including origin and final event, have a Position Value, "Y" of 0.

2. To establish the Position Value, "Y", for non-critical events, a form of an accumulative backward pass will be made. Starting at the final or objective event, and proceeding backwards into the next predecessor event that is not critical, the weight factor of the activity leading backwards into the predecessor will be added to the Position Value, "Y", at that activity's head. The highest sum of Y_j + activity weight of all the arrows coming back into that event will be taken as the Position Value "Y" of the event under consideration. Dummies are paths, and are activities having a 0 activity weight. They are treated as every other arrow on the diagram. Fig. 12.5 indicates the Event Weight Factor, X, and the Position Value, Y, for each arrow of the diagram of Fig 12.4.

148

Since this part of the methodology is a form of a Backward Pass, note the effects. The Position Value at Event #250 is the "Y" Value at the head of the dummy, 250-268, plus the weight value assigned to that dummy. In this case, it is a sum of 0+0=0. The "Y" value at Event #237 is a comparison of the two paths back into Event #237, or from 248 back to 237 and from 250 back to 237. Thus, before the "Y" of Event #237 can be found, the "Y" of both Event #248 and 250 must be established. The Y_{237} will be the higher of the two sums, Y_{248} plus $wt._{237-248}$, compared to Y_{250} plus $wt._{237-250}$, or 2+0 compared to 0+3. The "Y" of Event #237 is set by the Y_j plus wt_{i-j} from Event #250. The Position Values, or "Y" is thus accumulated all the way back to the origin Event.

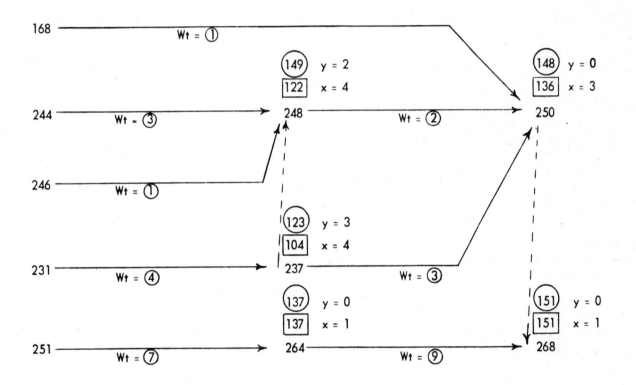

Fig. 12.5

The Event Weight Factor, "X", and the Event Position Value, "Y", for each event in Fig. 12.2 is shown below in Fig. 12.6.

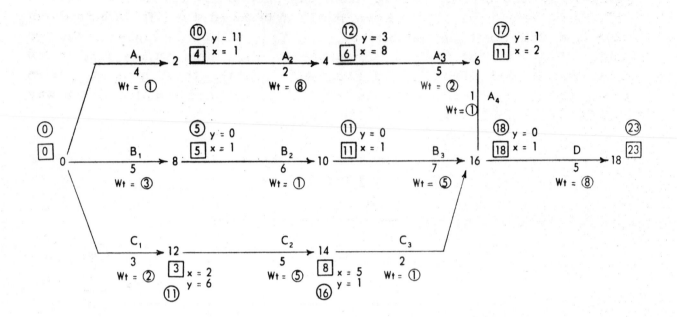

Fig. 12.6
X and Y Event Values Assigned to Network Plan

Distributed Event Times

The Distributed Event Time, designated T_D, and shown symbolically in triangles at each event, is found from the formula:

$$\triangle T_{Dj} = \frac{\left[\bigcirc T_{Lj} \times X_j \right] + Y_j \left[\triangle T_{Di} + D \right]}{X_j + Y_j} \text{ (max at j.)}$$

(12.1)

where D is the duration of the activity

The Distributed Event Time, T_D, at Event j is found by the addition of two quantities. The sum of these two quantities is then divided by the sum of the Event Weight Factor, X_j and the Position Value Y_j, The first quantity is the

150

Latest Event Time, T_L, of that event multiplied by the Event Weight Factor X of that event. The second quantity is the evaluation of all the activities coming into that event. Each arrow entering the event under consideration has its duration, D, added to the prior established Distributed Event Time at its tail, or $\triangle T_D$ i + $D_{(i-j)}$. The combination of $\triangle T_D$ + $D_{(i-j)}$ of all arrows entering this event that gives the highest sum is selected for this value. This quantity is then multiplied by the Position Value Y of that event. The sum of these two quantities is then divided by the sum of X_j and Y_j. Note since the formula requires a prior established $\triangle T_D$ i, Distributed Event Times are calculated in a forward pass. Table 12.1 tabulates the quantities, and the Distributed Event Times for Fig. 12.6.

Event No.	X	Y	T_L x X	$Y_j \left[T_{D\,i} + D \right]$ max	T_D
0	1	0	0	0	0
2	1	11	10x11	11(0+4)	4.5
4	8	3	12x8	3 (4.5+2)	10.5
6	2	1	17x2	1 (10.5+5)	16.5
8	1	0	5x1	0	5
10	1	0	11x1	0	11
12	2	6	11x2	6 (0+3)	5
14	5	1	16x5	1 (5+5)	15
16	1	0	18x1	0	18
18	1	0	23x1	0	23

Table 12.1

Note that the formula works for Critical Events as well as for Float/slack events. Distributed Event Times may be rounded off. A computer program will truncate, or drop the decimal part of the number.

An "Ideal" Schedule

The individual activities shall be given just one starting and finish time, based on the Distributed Event Times of the network.

1. "Ideal" Activity Start Time = $\triangle T_D$ j - $D_{(i-j)}$

2. "Ideal" Activity Finish Time = $\triangle T_D$ j.

151

3. The activity that is last on a Float/slack path, that is, the activity that enters a Critical Event, shall have its ideal starting and finish times set by the following formula.

   a. "Ideal" Activity Start Time = $\underset{D}{\triangle T}_i$

   b. "Ideal" Activity Finish Time = $\underset{D}{\triangle T}_i + D_{(i-j)}$

Table 12.2 indicates the Distributed Event Times for the network of Fig. 12.7. Fig. 12.7 is the statistically averaged network of the PERT network of Fig. 10.4, page 68. The activity weight factors are the ranges of the three time estimates of the arrow. Table 12.3 shows the "ideal" schedule, in digital form, for the network plan of Fig. 12.7. Fig. 12.8 shows this same information in bar chart form.

Distributed Float/slack

The formula for finding the Distributed Float/slack for an activity is:

$$\text{D.F.} = \underset{D}{\triangle T}_j - \underset{D}{\triangle T}_i - D_{(i-j)} \qquad (12.2)$$

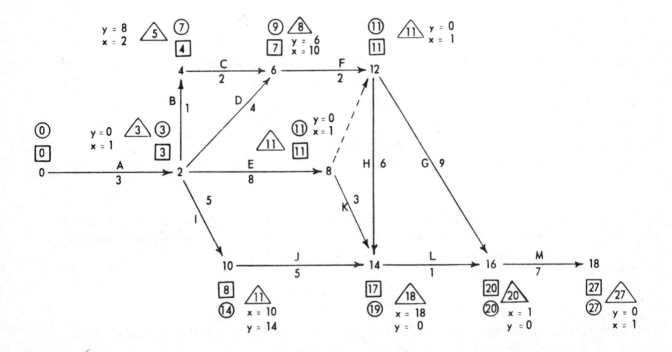

Fig. 12.7
Distributed Event Times for Non-Critical Events

Event No.	X	Y	$T_L \times X$	$Y_j \left[T_{D_i} + D \right]_{max}$	T_D
4	2	8	7x2=14	8(3+1)	4.6=5
6	10	6	9x10=90	6(3+4)	8.3=8
10	10	14	14x10=140	14(3+5)	10.5=11
14	18	0	19x18=342	0	18.0

Table 12.2

Ideal Schedule for Fig. 12.7

ACTIVITY	START	FINISH
A	0	3
B	4	5
C	6	8
D	4	8
E	3	11
F	8	10
G	11	20
H	12	18
I	6	11
J	13	18
K	15	18
L	18	19
M	20	27

Table 12.3

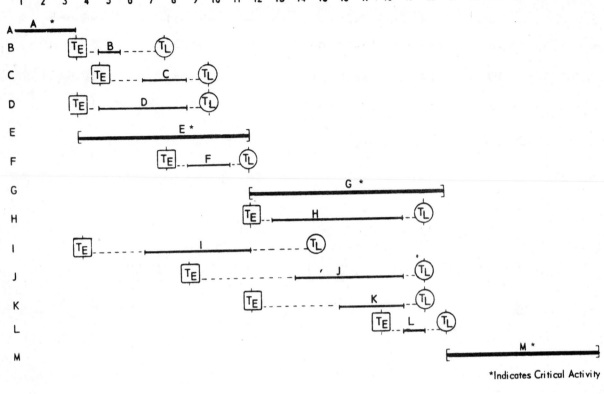

Ideal Schedule

Fig. 12.8

Distributed Float in Cost Control

The success of any control system is the timely, explicit and accurate collection and transmission of input information. With the network logic plan as the basis, the assignment of direct costs and labor to each activity is the next step. This is done either by a specific cost code that is compatible with the network plan, and definition of the activity, or by the evolving "work package" concept that is correlated to the network plan.

There have been several cost reporting and control systems relating to PERT and CPM described in contemporary literature, that appear to have a common shortcoming. The network plan, and the definition and fineness of detail ("level of indenture") of the activity arrows have been based on an extant cost code. This is unfortunate, because it is a case of the cart pulling the horse. When the plan accommodates an established cost code system, the plan becomes extremely limited. Consequently, the information generated will be artificially and impractically limited. Generally speaking, an existing cost code and reporting system will have to be modified, through an evolutionary period of time, to suit the installation of network planning techniques in the company. It is a pretty certain bet that the work and cost categories in a prior established code system will not fit the categories shown on the activity arrows of the network plan.

A project cost domain may be constructed for the basic network plan, from the schedule. The boundaries of the cost and man-hour domain take the general form of a hysteresis curve. The first time-dollar, or time-man-hour boundary is the curve based on the accumulation of direct costs or man hours if all the activities started as early as possible. The second expenditure boundary curve is based on all activities starting at their latest date (everything critical). Within this domain, the "ideal" single curve will be based on the "Distributed Schedule". Note in all cases there is a single total Project Cost Figure. This is a more meaningful guideline than the standard S curve (the Gompertz Curve) of the form, $y = (1 - e^{-at^2})$. Against the "ideal" "Distributed Schedule" cost or man-hour curve, the actual accumulation of project costs and man-hours can be shown. Trends towards over or under runs can be rapidly detected, and predictions made quickly as to when the cost domain of that project will be penetrated - a milestone to be avoided if reasonably possible. Cash Flow and Fiscal control techniques may obviously be structured on the "ideal" schedule. Fig. 12.9 indicates the presentation of the cost domain and the accumulated costs against the "ideal" guide line.

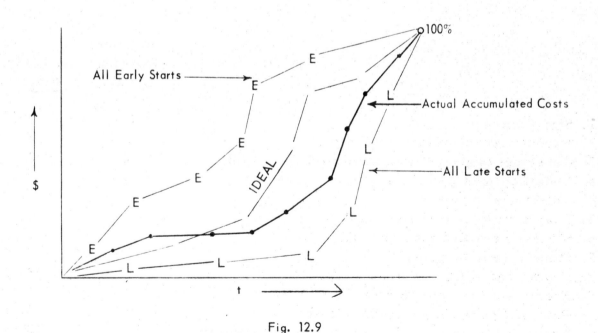

Fig. 12.9
Project Cost Domain and Guide Lines

155

Other Applications

Many uses of the Weighed schedule become noteworthy. The outside activities on a construction project would receive higher weight numbers, in order to take advantage of any float and set up a protected schedule for bad weather possibilities. The case mentioned previously in the Calendar Dating section of Chapter 8 might be handled in this manner: The bad weather date is selected as a Calendar date at that site, for example, November 1st. Outside activities would become more inefficient at that date. Activity weights could be assigned to any outside activity whose Early Start Time occurs after the equivalent of November 1st, thus distributing any float to operations that could use it. Since the method described works with any range of weights or values, it might be feasible to use the duration estimate itself as a weight factor. Thus, on a float path, those activities with the longest durations would receive the most Distributed Float. Remember, an actual project network plan is complicated, wit h many interrelated paths, and it is very difficult to predict the allocation of path float, and then the ultimate distribution of that float.

Again, it must be pointed out that this is only an information system, manipulating estimated data and logic to produce data upon which an experienced and forthright manager is willing to act.

PROBLEMS

12.1 Activity weights for the activities involved in the construction of the house, Problem 1.1, are listed below. Calculate the ideal schedule for the house.

Activity	Duration (Working Days)	Activity Weight
1. Sign Contract	0	0
2. Layout and excavate for building	3	4
3. Construct basement walls and backfill outside	5	2
4. Install outside water, gas, sewer lines and electrical power	4	1
5. Frame first floor to second floor	5	4
6. Stone veneer to first floor sill	4	3
7. Stone veneer to eaves	7	3
8. Frame and shingle roof	8	2
9. Frame second floor to roof	6	1
10. Pour and cure basement floor slab	3	1
11. Install furnace, water heater, basement plumbing	10	2
12. Erect interior first floor walls, and rough flooring	2	5

Activity	Duration (Working Days)	Activity Weight
13. Install first floor plumbing, ducts, wiring	7	3
14. Install windows, exterior doors	4	2
15. Install sidewalk, curbs, driveway	12	1
16. Erect interior second floor walls, and rough flooring	3	5
17. Install second floor plumbing, ducts, wiring	9	3
18. Put up inside drywalls and paint	10	5
19. Install kitchen equipment	7	6
20. Grade site, sod and landscape	10	2
21. Finish flooring	5	2
22. Install millwork, interior doors	4	4
23. Finish plumbing, electrical tie-ins	3	5
24. Punch list fix up and inspection	4	1

12.2 Calculate the ideal Schedule for the Inertial Guidance System project, Fig. 10.3 of Chapter 10. Use the range [optimistic minus pessimistic] of the three time estimates as the activity weight.

12.3 The same contractor who is building the house of Problem 12.1 also receives the contract to build the Car Wash Station of Problem 5.3. The activity weights are listed below for the Car Wash Station project. He starts work 12 working days after the start of the house project, calculate the ideal schedule for the Car Wash, and place it on the same time scale as the house, with the 12 days delay shown (This will be a multi-project schedule, shown on the same time scale sheet).

Activity	Duration (Working Days)	Activity Weight
1. Obtain building permit	5	4
2. Layout and excavate building	3	2
3. Award Mechanical and Plumbing subcontract	10	3
4. Award Structural Steel, Roofing subcontract	5	5
5. Award Electrical subcontract	10	2
6. Order, deliver Wash Pit railing, steel grating	10	2
7. Order, deliver hot water tank and spray headers	10	5
8. Order, deliver Motor Operated Overhead Doors	15	3
9. Order, deliver Air Heater and Blower	20	4
10. Install outside underground water, gas, sewer lines	5	2

Activity	Duration (Working Days)	Activity Weight
11. Form and rebar Wash Pit Walls, East column footings	2	1
12 Form office-locker room section slab, West column footings	1	2
13. Pour and cure wash pit walls, East column footings	4	1
14. Pour and cure office slab, West column footings	2	2
15. Mechanical rough in (underground)	3	3
16. Pour and cure Wash Pit Slab	2	1
17. Deliver steel columns, trusses, galvanized roofing	15	6
18. Construct light standards, sign foundations	2	1
19. Install light standards, sign	4	4
20. Erect Steel columns	4	3
21. Install chain drive in pit	3	1
22. Erect roof trusses, ridge pole, and plumb steel	2	3
23. Cover roof	6	4
24. Erect cinder block walls	4	1
25. Install windows, doors	2	2
26. Install plumbing in Wash Pit	3	3
27. Install Hot H-2⁰ tank, spray headers, plumbing fixtures	6	5
28. Install railing, steel grating over Wash Pit	2	2
29. Install motor operated overhead doors	2	4
30. Install air heater-blower	2	3
31. Hang acoustical ceiling, office and locker room	4	2
32. Hang interior ceiling light fixtures	2	3
33. Install millwork, office and locker room	2	3
34. Final plumbing connections, and test	3	4
35. Grade, compact and pave outside area	6	2
36. Final electrical connections, and test	3	4
37. Paint interior	3	3
38. Install floor tile	2	2
39. Final Inspection	1	5
40. Install interior partitions	5	3

CHAPTER 13

OPTIMUM SCHEDULES BASED ON TIME AND COST

In the previous sections of this text, the criteria for obtaining duration estimates for each activity was discussed. The estimate was based on the use of the practical resources available; this represented our best intuitive estimate of man hours divided by the available men, in consideration of practical limitations, such as: the physical space for the accomplishment of the activity; safety requirements; labor regulations, etc. This original estimate, the "usual" or best estimate will be considered the "Normal Time", and the direct costs so associated (direct labor, material costs) will be considered "Normal Costs". Normal Time and normal cost combined make the starting point on an Activity Cost Curve, Fig. 13.0.

This point is defined as the "Normal Case" and corresponds to the lowest cost, least time situation. To define it as the lowest cost point alone may not be sufficient to pinpoint the required duration. Consider the case of the painting of a room which will require 80 man-hours. It is possible to have two painters working simultaneously with no decrease in efficiency or increase in cost. Therefore, the duration may be 80 hours or 40 hours (but it is still an 80 man-hour job at the same cost). The Normal Duration would be then 40 hours, which is the shortest time at the lowest cost, if the "Normal" resource of two painters is available. As more men, more equipment, or more expensive equipment are used to do the job, the duration time is reduced but the cost increases. Similarly, if inadequate manpower and/or equipment is used inefficiently, the duration is increased and the cost also goes up.

As a case in point, consider the example of a draftsman working on a "B" size drawing (30" x 40"). Only one draftsman can work on the drawing at a time. Let us start by assuming that it takes 40 man-hours to complete this drawing. Since only one man can work on it at a time, we have a 40 hour duration. Based on an eight (8) hour day, we have 5 working days as the "Normal Time". If the draftsman's salary is $2.00 per hour, the direct normal labor costs are 80 dollars ($80) for five days. If management decides to expedite this activity they may authorize a second shift.

For the purpose of exaggerated example we shall assume the overtime shift is reimbursed on a time and a half basis. This operation would then be completed in 3 calendar days at a total cost of $96 direct labor. If this activity becomes critical, it could be "crashed" by the third shift. In this example, we shall assume a time and a half premium for third shift. We would then have our drawing after 2 calendar days at a total crash cost of $104. Study the development of Fig. 13.0.

The "Crash" position on the curve represents the absolute minimum time in which the activity may take place. Any additional manpower and/or equipment would result in an increase in cost without a corresponding decrease in time. Thus the "crash" case is defined as the minimum <u>additional</u> cost over normal cost, at the irreducible minimum time.

The cost curve for this activity is shown in Figure #13.0.

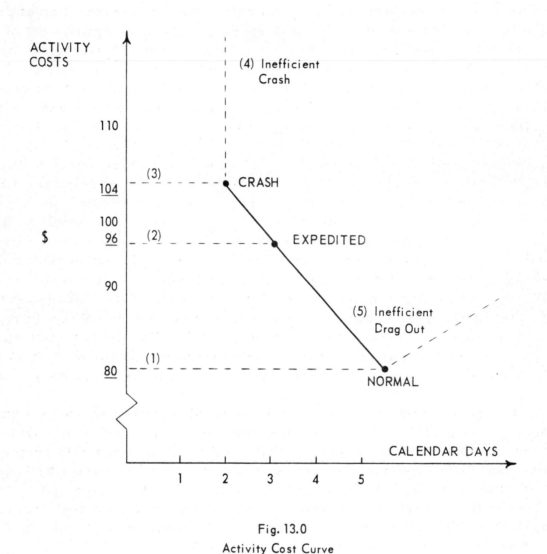

Fig. 13.0
Activity Cost Curve

Point (1)   One draftsman works for five 8-hour day shifts over a period of five days. (Normal Case)

Point (2)   Two draftsmen work on two shifts and complete the operation in three calendar days distributing the work over three day shifts and two second shifts. Because of shift premiums, the cost is higher. (Expedited Case)

160

Point (3)   Three draftsmen work on three shifts and complete the activity in two calendar days distributing the work over two day shifts, two second shifts, and one third shift. The cost incurred is higher than before.(Crash Case)

Point (4)   More than three draftsmen are assigned. The cost burgeons, but the calendar time can not be less than two calendar days. The additional people can only look on and not contribute to the reduction of time.

Point (5)   One man drags out the job for more than five calendar days due to poor supervision, etc. Cost rises, time increases, but only 40 manhours of work are effectively completed, in a much longer period of time. A graphic example of "Parkinson's Law" - "The work will fill up the time allowed available".

The network planning method deals with the following:

1. The normal duration and normal cost
2. The crash duration and crash cost
3. The rate of increase in cost for a unit decrease in time: that is, the slope of the straight line joining normal and crash points.

This last value is equal to the slope of the activity cost curve and it is defined as:

$$\frac{\text{Crash Cost} \quad \text{Minus} \quad \text{Normal Cost}}{\text{Normal Time} \quad \text{Minus} \quad \text{Crash Time}} = \text{equals slope}$$

$$S = \frac{\$_C - \$_N}{D_N - D_C} = \$/\text{Day} \qquad (13.1)$$

Thus, the slope of the activity cost curve becomes a premium rate, in terms of dollars per time unit. This is an additional information "number", to help a manager make efficient decisions.

Depending on the number of duration, and associated direct cost, points for an activity, we can then construct an activity cost curve. An obvious need is the availability of information to set the Normal Duration, Normal Cost Point; and the Crash Duration and the Crash Cost. The cost curve can be drawn as a straight line between these two points.

Note that although the slope of this curve is mathematically negative, in this context we shall consider it positive. The Normal Case is the starting point.

The slope represents the premium rate that must be paid, above the Normal Cost to reduce the duration of the activity.

Sometimes more information is known about the cost curve for an operation concerning the upward cost trend as the duration decreases. As indicated thus far, we may reasonably take the cost curve for an activity to be the straight line connecting its normal and Crash Costs. When intermediate costs are known it is sometimes desirable to make use of this information and obtain a more precise cost curve.

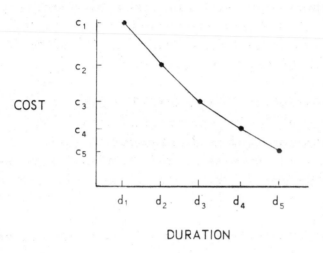

Fig. 13.1

Curvelinear Activity Cost Curve

Where formerly we represented the cost curve by one straight line, we now represent it by several smaller straight line segments. Here, d_1 corresponds to Crash Duration and d_5 corresponds to Normal Duration.

The cost curve can be generated from several cost-time points, if they are known, by successive linear approximations. However, the more segments the curve is broken into, the longer the calculation effort. Another requirement of this technique is the activity cost curve must be either linear or concave upwards. Following the curve from Normal Time to Crash Time, the mathematical value of the slope must increase. The curve shown in Fig. 13.2 below cannot be used. The mathematic support of this technique breaks down for such a curve, and is not valid. A simple intuitive analysis of this curve will also be demonstrated later to show why it is not logically feasible.

In this technique, the slope of the curve is considered positive; it is regarded as the premium that must be paid to reduce time.

An appropriate analogy is this: Just as energy must be expended to compress a gas, so must capital (money) be expended to compress the time duration of an activity.

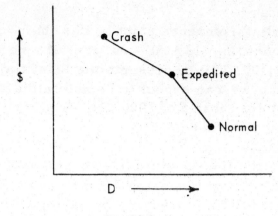

Fig. 13.2

An Invalid Activity Cost Curve

Some activities are of such a nature that a normal time-dollar point, and a crash time-dollar point can be ascertained, but there can be no interpolation in between. Examples typical of this situation would be:

(1) A supplier, who quotes a "Normal Cost" of X dollars, for a Normal Delivery Time of Y days, will "Crash" his delivery to 1/2 Y days, for a premium cost of 2X dollars. However, he will not allow an interpolated delivery time, pro-rating reduced time by a proportionate increase in cost. His is an "Either or" proposition.

(2) In certain applications of load and environment, a large concrete pour takes 28 days to cure, at a certain cost. This is the "NORMAL TIME-NORMAL COST" point. In this case, it may be possible to add an accelerator, which at a premium cost, will reduce the curing time to 7 days. This last would be the "CRASH TIME-CRASH COST" point. In this activity, it is physically impossible to add just half the admixture, and obtain a 17 day cure.

This technique will handle the both cases, that of an Activity Cost Curve running continuously upward from Normal to Crash, or to that of an either Normal or Crash Case, with no interpolation between.

The Use of Activity Cost Curve Slopes

Fig. 13.3 indicates a path of activities from a project plan

$100/D : SLOPE
$10,000 : NORMAL
            COST

i_1 ————————→ i_1

6 Days Normal
3 Days Crash

$1,000/D : SLOPE
$1,000   : NORMAL
               COST

i_2 ————————→ j_2

12 Days Normal
7 Days Crash

$10,000/D : SLOPE
$100   : NORMAL
                COST

i_3 ————————→ j_3

8 Days Normal
4 Days Crash

Fig. 13.3

163

The "datum" or starting reference point of this chain of activities is the sum of the normal durations $(6 + 12 + 8) = 26$ days at a normal cost of ($10,000 + $1,000 + $100) of $11,100. This represents the least time least cost situation. Using these estimates we can expect to complete the series of operations in 26 days, and for not less than $11,100. This is our Normal Time -- Normal Cost point -- we start from here.

If we wish to reduce the time to complete the chain of activities i_1-j_1, i_2-j_2, i_3-j_3, we will use the activity slopes ($/time unit) of those activities as the criterion. We shall apply time pressure by paying the least premiums. If we wish to reduce the overall duration of this chain of activities, we find on investigation it costs the least premium to reduce i_1-j_1. There is no sense, from a minimum additional expenditure of money criterion, in reducing any of the others, since their slope (Premium rate) is higher. Once we reduce i_1-j_1 to its irreducible limit of Crash Time (3 days, at a total Crash cost of $10,000 plus $100/D times 3, or $10,300), further reduction in this chain of activites will be at the activity with the next highest slope; i.e., activity i_2-j_2. Thus, if we want an overall duration of this chain to be 20 days, we would reduce i_1-j_1 to its crash limit first, reducing 3 days from the normal duration of 26 days, then we would reduce activity i_2-j_2 to 9 days, at an additional premium $1000/D x 3 day reduction; or $3000 additional. The total premium for a 6 day reduction from normal is 3 x $100/D($i_1$-$j_1$ reduced to its crash limit), plus 3 x $1000/D for i_2-j_2 (expedited from 12 to 9 days but not to its crash limit) or $3300 extra to gain 6 days.

Consider now a combination of paths. Fig. 13.4 below has added a path parallel to the first two arrows, i_1-j_1, and i_2-j_2.

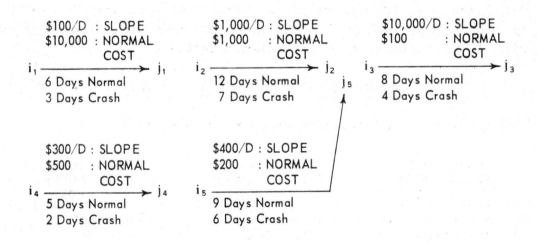

Fig. 13.4

It can be seen from the configuration and durations, that the first reduction in the parallel path plan of Fig. 13.4 will be up to 3 days at the $100/Day premium rate activity, i_1-j_1. Then the next highest slope, on the longest path of the plan, is that of i_2-j_2. The fact that the parallel activities have lower premium rates doesn't help; they cannot reduce the project until the longer path they are in parallel with has been reduced to their sum. Thus, a 4th day may be taken from the completion date at j_3 by reducing the next highest slope activity on the longest path; that at i_2-j_2. Once this has been done, any further reduction must now be taken on the parallel path along with the originally longest path, because at the end of the 4th day's reduction of the completion date (from Day 26 to Day 22 at j_3) the parallel paths are now equal in overall duration, 3+11 and 5+9. The next day in reduction will now add the least slopes of the parallel paths, in this case the $1,000/Day of i_2-j_2 plus the $300/Day slope of i_4-j_4. It now costs $1300/Day to go from completion date Day 22 to Day 21 at j3.

Making Up Lost Time

The same approach, based on compressing the least cost slopes of paths between given or fixed time boundaries, is used in making up slippages in a schedule. Fig. 13.5 indicates the last section of a project plan for the construction of a school. The plan is being reviewed as of project day #167. There have been slippages on various activities in the prior part of the project. The plan as shown has the remaining "Normal" durations shown and the original Event Times T_E and T_L. Those activities that have been completed by project day #167 are so noted. This particular project contract has a penalty clause: the Contractor will be assessed $500 a day for every day after project day #178 that he has to remain on the site performing work.

The objective of the Contractor is to determine (1) how can he attain the key event #640 by day #176. This is his real goal, since the last activity, #640-645, is under the control of the owner and cannot be expedited; and (2) determine whether the premium he must pay is worth it. In other words, does it pay the Contractor to expedite certain activities or does it pay him to continue on the original normal schedule and absorb the $500 a day penalty.

Thus, there are 9 working days left to the Contractor (Project day #176, goal, minus project day #167, time now) to make up the lost time. A study of the network plan, Fig. 13.5, reveals the following:

a. The chain of activities, 590-595, 595-635, 635-638, and 638-640 (ceiling tile, electrical work) will take 11 days normally. Two (2) days must be eliminated in this path.

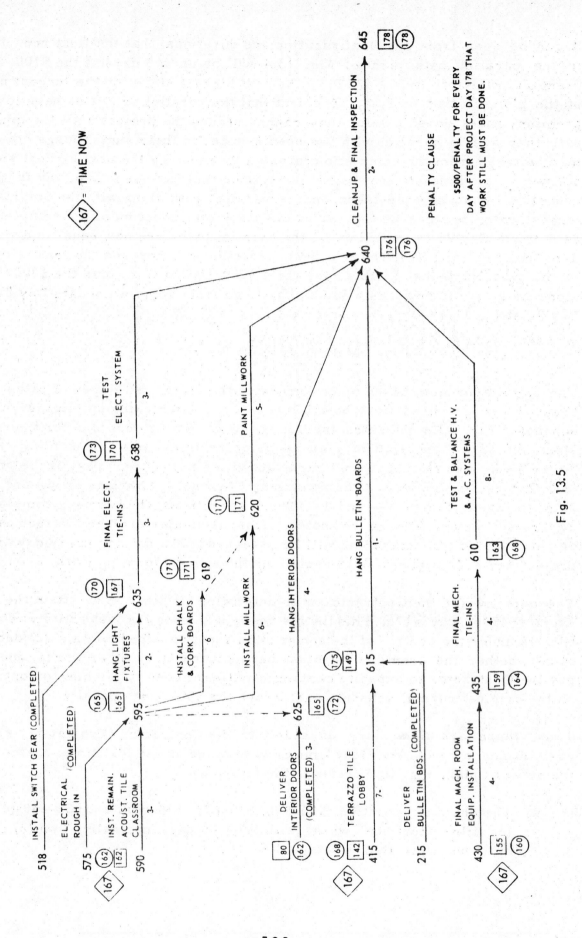

Fig. 13.5

166

b. The chain of activities (millwork) 590-595, 595-619, and 595-620 in parallel, 620-640, will take 14 days normally. This path was the original Critical Path. Five (5) days must be removed from this path.

c. The chain of activities (mechanical) 430-435, 435-610, and 610-640 will take a normal time of 16 days; Seven (7) days must be dropped from this path.

d. Hanging the interior doors (625-640) and the lobby work (415-615, 615-640) can be normally accomplished in the 9 remaining working days and presents no problems.

After a review, the Contractor estimates the premium rates (slopes) for each activity. These are shown on Fig. 13.6 as Normal Time-Crash Time under each arrow, and Normal Cost-Crash Cost above each arrow.

From the analysis above, the slopes of the activities to be expedited are:

590-595	$140/Day	620-640	$135/Day
595-635	$110/Day	415-615	$115/Day
595-619	$ 70/Day	430-435	$125/Day
595-620	$ 90/Day	435-610	$110/Day
635-638	$120/Day	610-640	$145/Day
638-640	$120/Day	625-640	$105/Day

The additional cost of expediting will be:

Along the 590-595, 595-635, 635-638, and 639-640 path the least slope activity is reduced first to "Crash", then the next least slope to obtain the desired 2 day reduction in duration.

595-635	1 day	$110 additional
635-638 or		
638-640	1 day	$120

(either one, since both have the same slope. Other considerations, such as crew size, or type, will then influence the selection)

Along the path 590-595, 595-619, and 595-620 in parallel, 620-640, the analysis indicates two important features: 595-619, and 595-620 are in parallel. If one is reduced without the other, then no reduction is obtained along that path. Remember, the analysis is always made on the basis of the logical sequence of the arrows. The cost analysis may reveal that a different logical sequence should be considered. Here, after installation of some of the millwork, some painting may start. This might allow some reduction in the normal duration of the activity 620-640 "Paint Millwork" without any premium payments. To accomplish this, the logic of the network plan must be reviewed, and a new sequence of re-defined activities drawn.

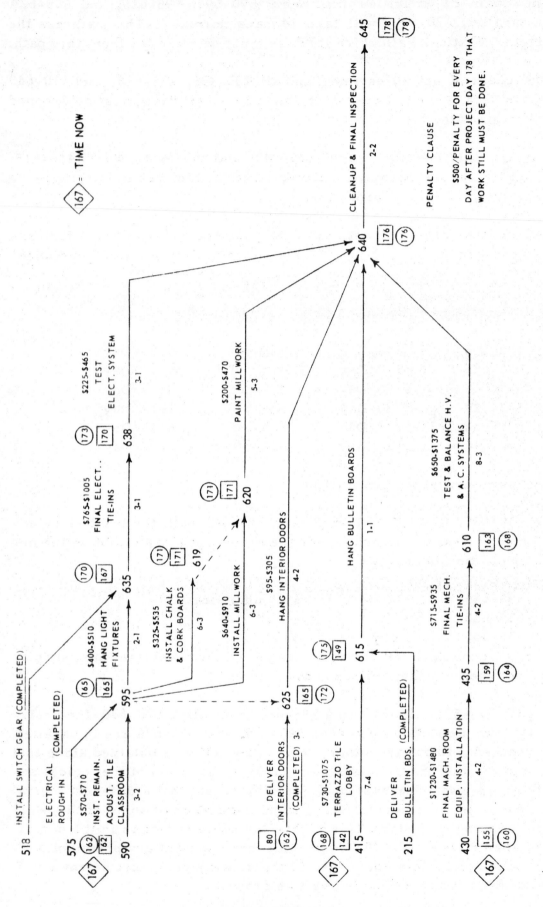

Fig. 13.6

168

Considering the logic of Fig. 13.5 as acceptable, the parallelism of 595-619 and 595-620 creates an "equivalent" slope, the sum of the slopes of the parallel arrows. To accomplish any reduction between events #595 and 620 both activities must be reduced concurrently for a total slope of $70 plus $90 a day, or $160/day equivalent.

The first reduction will be at 620-640, to its crash limit

620-640	2 days @ $135/Day	$ 270

The next highest slope on this path is 590-595, at $140/day.

590-595	1 day @ $140/Day	$ 140

We have made up 3 of the 5 days sought. The remaining 2 days will be taken from the parallel 595-619 and 595-620:

595-619) 595-620)	2 days @ $160/Day	$ 320

The reduction of 590-595, mandatory for the millwork activity, now affects the previous path analyzed, the electrical activities. Since 590-595 is common to two concurrent paths of the millwork and electrical work, a reduction in its time forced by one path, changes the distribution of the premiums on the other path. It is no longer necessary to reduce either 635-638 or 638-640 (the higher premium rate activity on that path), since a 1-day reduction in 590-595 benefits that path by 1 day.

The remaining path, 430-435, 435-610, and 610-640 will be reduced as follows:

435-610	2 days @ $110/Day	$ 220
430-435	2 days @ $125/Day	250
610-640	3 days @ $145/Day	435

The total minimum premium to be paid, to reach event 640 by day #176 is:

595-635	1 day	$ 110
620-640	2 days	270
590-595	1 day	140
595-619) 595-620)	2 days	320
435-610	2 days	220
430-435	2 days	250
610-640	3 days	435
		$1745

The second part of the analysis is the comparison of the premium rates with the penalty rates. Using the target event date as a goal, the sum of the premium rates per day along the parallel paths into the objective event must be compared with the penalty rate per day. The premium rates will vary from day to day as the project is extended past the target event time, due to various premium rates and crash durations on the parallel paths. The penalty rate usually remains the same. For example, if it is decided to reach event #640 by day 177, which activities would be relaxed back towards normal? It would be the highest cost activity on the path for the millwork, the electrical, and the mechanical paths of activities. Thus, if 595-619 and 595-620 are allowed to go to a 5 day expedited duration, the event #640 would be extended a day and the Contractor would not pay $160 premium for these activities. If a 1 day extension is going to occur, then the electrical path could be relaxed a day also and the premium for the highest slope on that path could be saved, or $110 on 595-635. The highest premium on the mechanical path 610-640 or $145 day could be saved from a day. Thus, a 1 day extension on event #640 to day 177, the Contractor would not pay:

$160   for 595-619 and 595-620 in parallel
$110   for 595-635
$145   for 610-640
$415   he would not pay $415 in premiums, but he would pay
$500   penalty.

A path is extended by allowing the highest cost slope activities first to relax back to normal. It is the SUM of the premiums of parallel paths, per day, that is compared to the penalty rate. Once a high premium activity is relaxed to normal, then the next highest slope is considered. Table 13.0 indicates the order of diminishing highest premium activities.

Table 13.0

Project Day at Event #640	Activity Not Expedited	New Duration	Premium Savings	Penalty
177	595-619) 595-620)	5 (E)	$ 160	
	595-635	2 (N)	110	
	610-640	6 (E)	145	
		total	415	$ 500
	(All other activities previously crashed are still crashed)			
178	595-619) 595-620)	6 (N)	$ 160	
	610-640	7 (E)	145	
		total	$ 305	$ 500

Table 13.0
(continued)

179	590-595	3 (N)	140	
	610-640	8 (N)	145	
		total	$ 285	$ 500
180	620-640	4 (E)	$ 135	
	430-435	3 (E)	125	
		total	$ 260	$ 500

This method applies to the evaluation of
bonus rates, by comparing the premium
rates to be paid, versus the bonus to be
gained.

Note from the table that if the penalty rate had been $300 a day, the analysis
would reveal that it would be feasible to let the project extend to project day
#178, since the premium rate is higher than the penalty. After project day #178
the premijm rate is less than the penalty rate of $300 a day and is to the Con-
tractor's advantage to expedite those indicated activities in Table 13.0.

The above approach may be more readily seen in a graphic Cost Curve. The
Cost Curve for the Network Plan of Fig. 13.5, shown in Fig. 13.7 below,
starts at day #183. This is the new completion date of the plan in Fig. 13.5
as of the new start base line day #167. It represents the remaining 16 days of
the Mechanical Sub-contractor's chain of operations, activities 430-435, 435-
610 and 610-640. This is the new Critical Path as of day #167.

To reduce from day #183 to day #182, a one day reduction, the least slope activity on the Critical Path is activity #435-610. This can be reduced one day
at the least premium rate of $100. This can be reduced another day at the same
rate, then it reaches its crash limit. Now we have come back to day #181, and
at this point the Millwork chain of operations (590-595, 620-640) becomes
parallel Critical. To reduce the end date of this project another day, back to
day #180, we must reduce the remaining least slope activity on the Mechanical
operations path and the least slope activity (620-640) on the Millwork path.
Thus, we can pick up another two days at the sum of their slopes per day. At
the end of these two days, coincidentally, these activities have reached their
Crash Time Limits, and the next highest premium rate activities on the
parallel Critical Paths must now be selected. Thus it goes, at an ever increasing premium rate until the target date is reached. The variable rate
curve is compared to the constant penalty rate curve (or liquidated damages
curve), and the point of intersection noted for the optimum schedule; the least
time at the least total cost. The sum of the daily premiums in Fig. 13.7 totals
to $1745.00, the least cost to bring this project home on time.

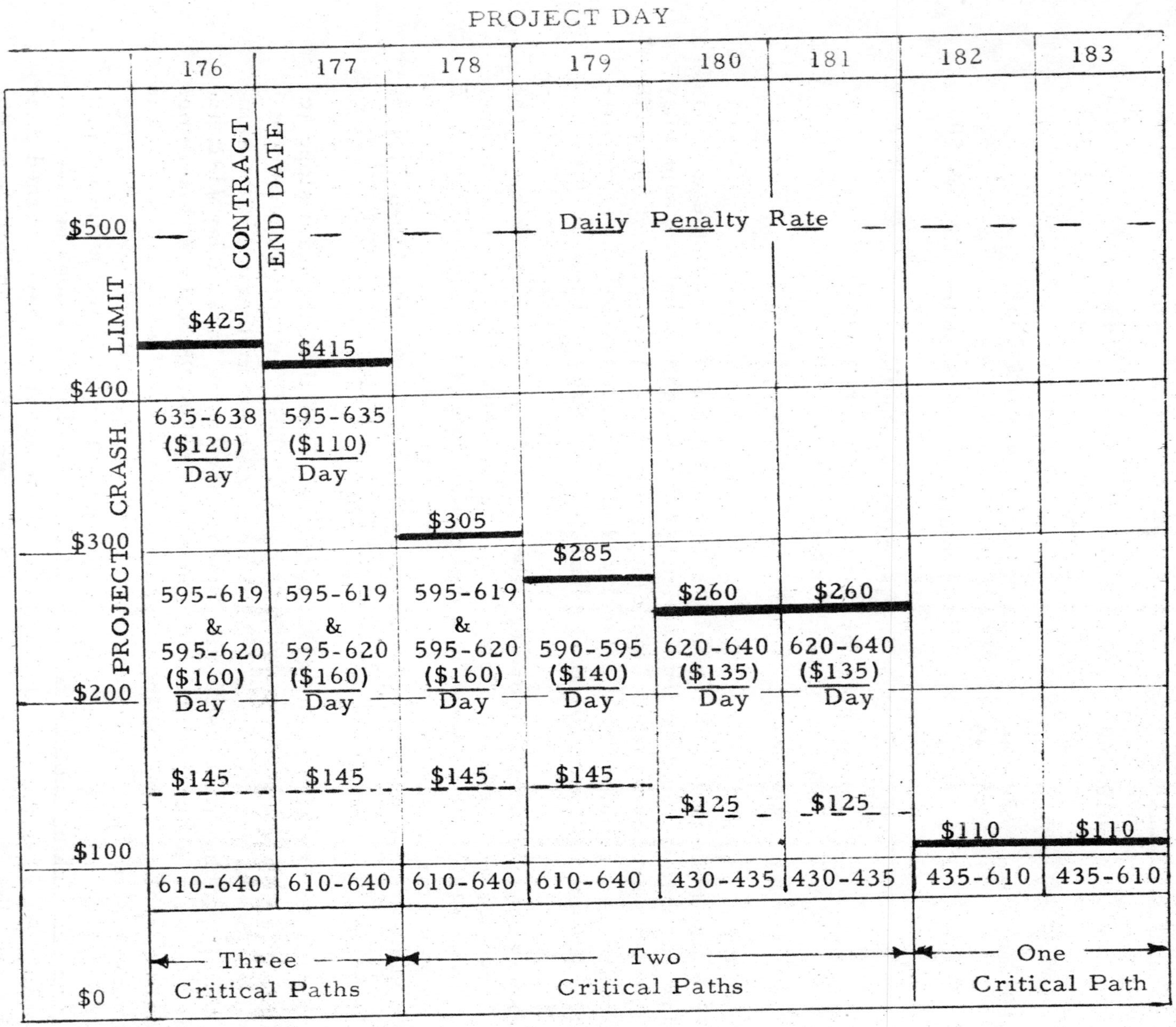

	176	177	178	179	180	181	182	183
$500 LIMIT	$425	$415						
$400 PROJECT CRASH	635-638 ($120)/Day	595-635 ($110)/Day						
$300			$305	$285	$260	$260		
$200	595-619 & 595-620 ($160)/Day	595-619 & 595-620 ($160)/Day	595-619 & 595-620 ($160)/Day	590-595 ($140)/Day	620-640 ($135)/Day	620-640 ($135)/Day		
$100	$145	$145	$145	$145	$125	$125	$110	$110
$0	610-640	610-640	610-640	610-640	430-435	430-435	435-610	435-610

CONTRACT END DATE

Daily Penalty Rate

Three Critical Paths — Two Critical Paths — One Critical Path

Fig. 13.7 Making Up Lost Time--Cost Curve

If a Bonus Clause is also available to justify the Penalty Clause, this approach is continued until the particular premium rate exceeds or is greater than the bonus rate.

Note that according to the sequence of the plan of Fig. 13.6, and the Crash limits assigned to the activities, this project could be reduced to day #175. At this point, the Millwork chain is at its crash limit for each activity, and cannot be reduced further. This becomes the limiting chain of operations, or the Critical Path, in this plan. If further reduction is required, re-evaluation of the plan and its logic and durations would be required.

An analogy to the cost optimization approach above is to consider the plan similar to two parallel plates connected by springs of various spring constants, in a series-parallel connection, as shown on Fig. 13.8 below.

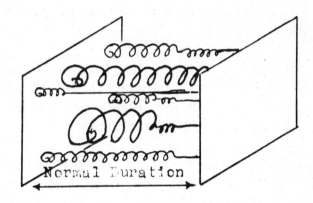

Fig. 13.8 Cost Optimization Analogy

The "at rest" distance between the two plates corresponds to the normal Critical Path project duration. The distance between the plates may be reduced by applying a force on the outside of the plates. The spring constants are analogous to the premium rates, or cost slopes of the activities. The force applied to the plates to reduce the distance between them is slight at first, representing the force to compress those springs on the longest chains which have the smallest resistance to compression. The force is analogous to the incremental sum of the cost rates per day times days reduced; that is, the force is the integral of the rates summed up per day of reduction.

The approach of re-evaluating the logic, and perhaps changing it to suit a target event time, is called a "Time-Risk" option in the PERT literature. Unfortunately, this literature does not define the word "risk". The "risk" that the manager takes in a fixed price contract is very much different from the "risk" taken on a cost-plus-fixed-fee, or a cost-plus-incentive fee contract. In essence, it is a suggestion of evaluating alternate strategies, and comparing their direct costs only.

The Optimum Crashing of a Project

When it is desired to reduce a project to its minimum duration, or make up time due to slippages, there will be an optimum combination of activities that represents the minimum additional expenditure of money to reach the desired crash schedule.

A. "Crashing" A Project

   1. Find the CRASH Critical Path, by the usual forward and backward pass using the Crash durations of all activities, to establish the Crash Early Event Times and the Crash Latest Event Times. Those activities on the Crash Critical Path determine the minimum crash time duration of the project. You must pay the premium crash dollars for these activities.

   2. Compare the difference between the Normal Duration of the project and the Crash Duration of the project. Those activities whose Normal Total Float is greater than this difference will remain normal even in the Crash case. Enter their normal costs.

   3. Next, those activities which are not on the Crash Critical Path, but whose Normal Total Float is less than the numerical difference between the Normal Project Duration and the Crash Project Duration, must be examined. The activities, or paths of activities, must be fitted into the Crash Event Time Boundaries. Where there is such a path, the "fitting" of the crash non-critical paths is done by first setting all crash non-critical paths at their Normal Duration, then testing to see if that Normal Path will fit between the Crash Event Time Boundaries. If so, the path stays normal. If not, start reducing in duration those activities with the least slope. Reduce them first to their crash duration. If more reduction is required, then select the next highest slope activity on that path until all crash non-critical paths are exactly fitted between Crash Event Times. On non-critical Crash paths, the Crash Event Time becomes the time boundary into which the non-critical Crash paths must be fitted by reduction.

B. Example of Optimum Crashing of a Project

   1. You are to crash the project as planned below, and to reach the shortest project duration possible at the least additional investment. What is the minimum crash cost schedule?

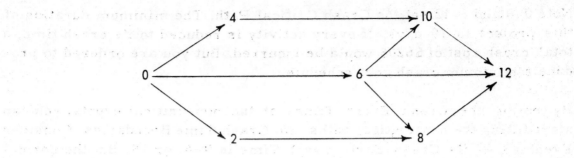

i-j	Normal Time	Normal Cost	Crash Time	Crash Cost	Slope
0-2	3	$ 200	1	$ 320	$ 60/D
0-4	8	100	5	400	100/D
0-6	12	400	9	580	60/D
2-8	8	400	4	1200	200/D
4-10	6	120	3	240	40/D
6-8	14	500	6	900	50/D
6-10	6	540	4	580	20/D
6-12	12	300	10	450	75/D
8-12	8	750	3	900	30/D
10-12	5	475	3	635	80/D
		$ 3785		$ 6205	

SOLUTION

i-j	Selected Time	Cost $
0-2	3-(N)	200
0-4	8-(N)	100
0-6	9-(C)	580
2-8	8-(N)	400
4-10	6-(N)	120
6-8	7-(E)	850
6-10	5-(E)	560
6-12	10-(C)	450
8-12	3-(C)	900
10-12	5-(N)	475
		$ 4635 (Δ = $850)

Fig. 13.7

175

C. Analysis of Solution

Note 0-6 and 6-12 set the Crash Critical Path. The minimum duration of this project is 19 days. If every activity is reduced to its crash time, a total crash cost of $6205 would be incurred. But you are ordered to produce a minimum crash cost schedule.

By noting the Crash Event Times at the non-critical events, you can start fitting the non-critical paths into Crash Time Boundaries. Consider Event #8 -- its Crash Early Event Time is 9+6, or 15. But the normal durations of 0-2 and 2-8 equal 3+8, or 11 so that chain of activities will stay normal: there is no sense in spending a premium that will not affect a reduction in project time.

Note 0-4 and 4-10 have a Normal Total Float of 15 days, which is the same as the difference between the Normal Project Duration (34 days) and the Crash Project Duration of 19 days. Thus, they will stay at normal time -- normal cost.

The only remaining activities are 6-10, 10-12, and 6-8 and 8-12. Consider 6-10 and 10-12. They must be fitted between the Crash Critical Event Times of day 9 at event #6 and day 19 at event #12, a 10 day crash time boundary. The sum of their normal times is 6+5, or 11. One day must be eliminated from this path to make it fit into the 10 day Crash Time boundary between events 6 and 12. The slope of 6-10 is $20/D; the slope of 10-12 is $80/D. Thus, we will reduce 6-10 one day at a $20 premium and leave 10-12 normal.

The same approach is used on path 6-8-12. The total normal duration is 14 (6-8) + 8 (8-12) or 22 days. This is to be compressed into the same 10 day CRASH Critical Event Time Boundary, day 9 at event #6 and day 19 at event #12.

6-8 has a slope of $50/D and 8-12 has a slope of $30/D. Thus, we will reduce 8-12 first to its crash duration of 3 days. We have to find an additional 7 days (trying to fit a 22 day normal path into a crash boundary of 10 days). We have picked up 5 of the 12 by reducing activity 8-12 from 8 to 3, and we find those remaining 7 days in the activity with the next highest slope - activity 6-8.

PROBLEM

13.1   The original plan for the construction of the house, problem 1.1, produced a 59 day schedule. The plan is shown in Fig. 13.8. (For a discussion of the logic, refer to the answer book.) There is increasing pressure from the wife of the owner to move into the new

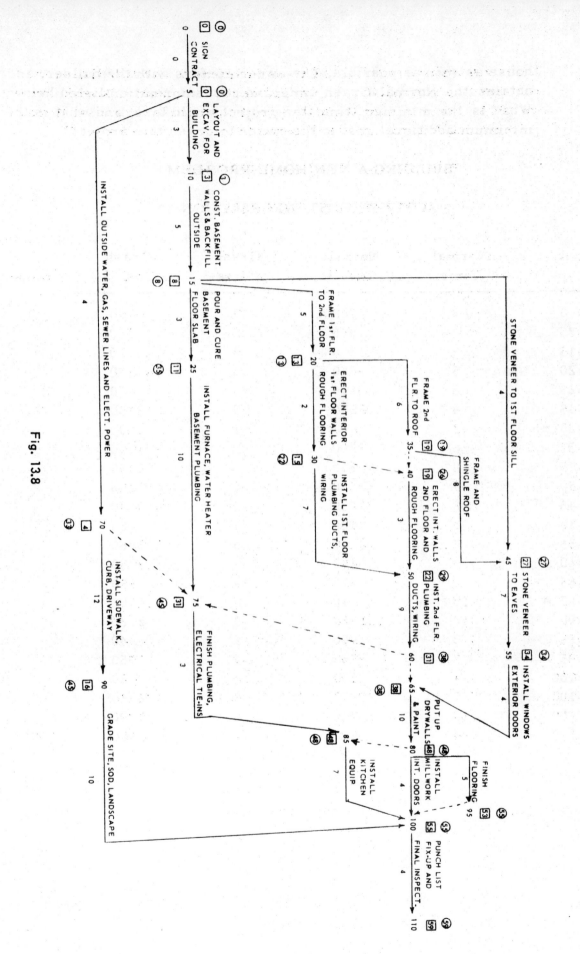

Fig. 13.8

house as soon as possible. The owner confers with the builder, and obtains the Normal-Crash time and cost information listed below. What is the minimum time the project would take, and what is the minimum additional cost to the owner to crash this project?

BUILDING A NEW HOME PROBLEM

ACTIVITY COST INFORMATION

Activity i j	Normal Time	Normal Cost $	Crash Time	Crash Cost $	Slope
5-10	3	700	3	700	0
5-70	4	950	2	1350	
10-15	5	1320	2	1770	
15-20	5	850	3	1580	
15-25	3	470	3	470	0
15-45	4	2200	3	2450	
20-30	2	430	1	605	
20-35	6	1200	2	4000	
25-75	10	350	5	1350	
30-50	7	650	4	1250	
35-45	8	1130	3	3630	
40-50	3	490	2	665	
45-55	7	2060	3	3060	
50-60	9	480	5	1280	
55-65	4	370	1	850	
65-80	10	840	6	1560	
70-90	12	1320	3	2220	
75-85	3	490	2	840	
80-95	5	730	4	950	
80-100	4	490	2	770	
85-100	7	1870	3	2830	
90-100	10	750	6	1390	
100-110	4	500	4	500	0

CHAPTER 14

THE LEAST TOTAL COST SCHEDULE FOR A PROJECT

The Heuristic Approach to a Project Direct Cost Curve

In the previous section we noted the value of the cost slope of the activity in ascertaining a minimum cost project "crash" schedule, or making up lost time between rigidly fixed time boundaries. We will now extend the concept of the activity cost curve and investigate a series of optimum schedules; that is, for each incremental reduction in time of the project, we will find the best combination of activity times and costs. This will be based on the criterion that we are only interested in the least additional amount of cost for each incremental reduction of project duration from the normal duration.

Within the limits of the crash and normal durations of each activity, management has considerable choice for selecting how long each operation in a project should take. Each set of selections will lead to a different schedule and, in consequence, a different project duration. Conversely, there are generally many ways to select activity durations so that the resulting schedules have the same duration. First, in making a choice, management must have some way of evaluating the merits of each possibility. One criterion of choice, and the one we will focus upon here, is the basis of cost.

Consider the following network:

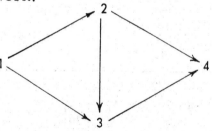

Activity	Normal Time	Normal Cost	Crash Time	Crash Cost	Slope
1-2	5	$180	4	$230	$50/day
1-3	11	75	8	255	$60/day
2-3	7	165	6	205	$40/day
2-4	12	48	10	108	$30/day
3-4	6	312	4	452	$70/day
Total	18	$780			

Table 14.1

Case #1

You are ordered to shrink this project one day, to a 17 day duration. What is the optimum schedule?

Since the critical jobs here form a single chain, we are in a situation on analyzing just those activities on that critical path. Reducing any other activity avails nothing in reduction of project duration. Applying time pressure to this chain, we see that only activity 2-3 should decrease in duration because it has the smallest resistance to the pressure. That is, Slope (2-3) is less than Slope (1-2) and (3-4). We may thus reduce the project duration by at least one time unit and obtain a 17 day schedule on a minimum additional investment of $40 for a total cost of $820.

Case #2

You are now ordered to reduce this project to a 16 day duration. What is the optimum schedule?

Some possibilities are:

Activity	Cost
(1-2)	$180
(1-3)	75
(2-3)	165
(2-4)	78
(3-4)	452
Total	$950

Schedule 1

Activity	Cost
(1-2)	$230
(1-3)	135
(2-3)	205
(2-4)	48
(3-4)	312
Total	$930

Schedule 2

Activity	Cost
(1-2)	$230
(1-3)	75
(2-3)	165
(2-4)	48
(3-4)	382
Total	$900

Schedule 3

Table 14.2

180

Each of these schedules satisfies the normal and crash limits of activities and takes 16 days to execute. However, each schedule involves a different cost. This situation is typical. In general, management would select the least costly alternative. In the present case this is Schedule 3.

Note that job (2-3) has been "relaxed" back to its normal duration, in Schedule 3, even though for the 17 day project duration it had been expedited to its crash point.

The problem of computing the project cost curve in a valid way involves mathematical considerations beyond the scope of this text. However, a simplified valid method will be develcped in a following section.

This is a combinatorial problem; a "heuristic" solution is necessary. Heuristic means "seek and discover," and applies to a system that looks at all possible combinations, including the relaxing of certain operations and selects that schedule which represents the least additional investment for the incremental reduction in the project schedule. In essence, the method will first examine the critical path and reduce those jobs on the critical path with the most attractive slope (the least slope) until we use up the slack or float in other paths. We will stop after we have created a new parallel critical path and select the attendant schedule. Then the system heuristically starts to shrink that new set of critical paths until it again causes more zero float or slack paths, stops, selects that optimum schedule, and proceeds to the crash schedule.

Note that there is even an optimum crash schedule. The all "Crash" condition is shown below:

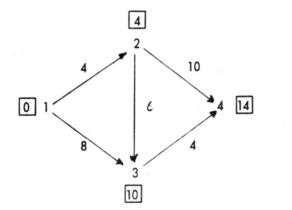

Activity	Cost
(1-2)	$ 230
(1-3)	255
(2-3)	205
(2-4)	108
(3-4)	452
Total	$1250

Table 14.3

From the foregoing it is apparent that the project cannot be completed in less than 14 days. No activity can be reduced any further. Activity 1-3, however, has two days Float in it under this scheme. Extra investment has been made to crash this activity without affecting in any way the completion time of the project. Obviously, to achieve the minimum Project Duration it is not necessary nor economical to expedite all the activities to their crash position. In this case, the extra investment with no return amounts to $120, or approximately 10% of what should be the project cost. Hence, the minimum Project Duration may be attained with a variety of investments, but again only the minimum investments should be considered. Note that for a project duration reduction, more activities become "critical" by the definitions in Chapter 6. This is now an academic classification, the activities to be reduced in time of execution become the new goal of management.

For the present example, we obtain the following spectrum of schedules:

Schedule	1	2	3	4	5
(1-2)	5	5	4	4	4
(1-3)	11	11	11	11	10
(2-3)	7	6	7	7	6
(2-4)	12	12	12	11	10
(3-4)	6	6	5	4	4
Duration	18	17	16	15	14
Total Cost $	780	820	900	1000	1130

The total cost curve for this project is as follows:

Project Cost

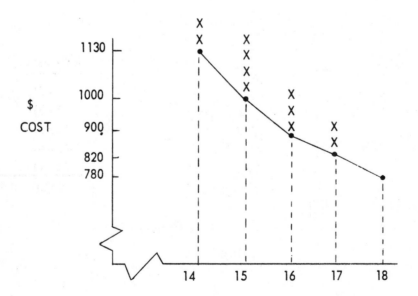

Table 14.4

In the graph the X's represent typical schedules of given duration plotted versus their costs. The vertically aligned x's are cost schedules of the same duration. The lowest one in each column is the least costly for its duration. If a curve is drawn to connect adjacent minimum cost schedules, we obtain a function of project cost versus project duration. Such a function will be called the Project Direct Cost Curve.

Note that on the Project Cost Curve only one "schedule" is depicted for the longest project duration. This represents what may be considered as the "all-normal" case. For this project duration all activities are at their normal condition. From the "activity cost concept" the normal case represents the lowest direct investment for that activity, so that the sum of all normal costs represents the lowest direct investment for the project. Now consider what may be defined as the "all-crash" schedule; i.e., all activities are at their crash duration. This represents the highest direct project cost, but is a minimum of all possible cost schedules at this point.

The Project Direct Cost Curve, by itself, is a limited set of management information. To continue the criterion of the best schedule at the least cost, we need additional information. To collect this information we shall investigate the Indirect Cost Curve, and the Utility Curve.

The Project Indirect Cost Curve

From the regular accounting procedures of the organization, we shall collect information as to the indirect costs assigned to the particular project.

An Indirect Cost will be defined as those costs that are definitely attributable to the project, but cannot be specifically assigned to a particular activity, or group of activities, on that project. Such items as rental of temporary equipment (welding machines, telephone, communications, office equipment); temporary services for the project (clerical, secretarial, drafting, etc.); temporary heating in order to maintain inside construction; utilities used on the project; gas and oil for the pickup trucks, the company cars; trailer for the construction office, etc. Quite often the cost allocation is an accountant's dream, but an owners nightmare. The accounting procedures used by the company must be evaluated for a realistic assignment of Indirect Costs. The salary of the General Superintendent may be considered an Indirect Cost for that project - only if it stops at the completion of the project. Living expenses certainly will come under the Indirect Cost category. The guide rule for differentiating between Indirect Costs, and Overhead and Burden is this: Indirect Costs end with the completion of the project. The Overhead, Burden, G and A (General and Administrative) costs that go on whether the company is working on a project or not belong in a separate accounting category. If the Overhead and Burden Costs are proportioned to a project, it may well give a misleading picture. Generally, depreciation, by tax law, will go on over a calendar period, and is independent of work on a project. However, if accelerated depreciation, due to use on a project, of equipment, can be legitimately and legally shown, then the amount of accelerated depreciation, above the normal, can be considered an Indirect Cost.

In addition to the Indirect Costs listed above, another category should be included in the Indirect Cost curve. These are Liquidated Damages, that start to accrue after a specific Target Date (T_S). If there is a Penalty-Bonus contractual clause, the Penalty of X dollars per day is added to the Indirect Cost curve, accumulating from the Target Date (T_S) to the Normal completion Date. The Bonus is considered a negative cost (profit), and is shown below the abscissa, starting at T_S, and accumulating negatively to the Crash completion Date.

The Indirect Cost curve starts at O at time O, and accumulates to the Normal Duration of the project. Fig. 14.1 displays the Indirect Cost Curve.

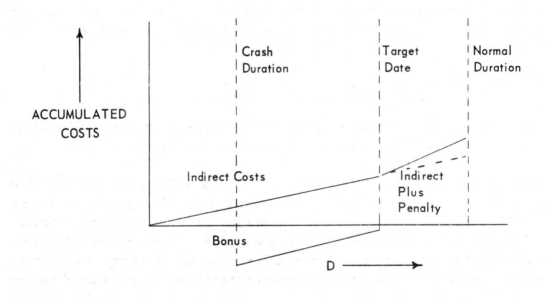

Fig. 14.1

The Project Utility Cost Curve

For the owner, or the person or organization that enjoys the profit or income generated by the purpose of the project, an account should be made on the loss or deprivation of that income that could accrue between the Crash Project Duration and the Normal Project Duration. In other words, a method can be established to indicate whether the income that could be enjoyed in the period between the Crash Duration and the Normal Duration of the project is worth a premium rate effort. As a case in point, consider the erection of a motel. The network plan is drawn, duration times, costs, and premium rates are collected. A Project Direct Cost Curve is established. The Utility Curve would be that income that would accrue to the owner if the motel enjoyed a certain occupancy from the period from Crash to Normal Project durations.

A Contractor may use the Utility Curve to indicate the interest earned on his final payments, if he crashed the project, that would accumulate in that period between Crash and Normal.

The contractor may be able to place a value on his bonding capacity that would be tied up on the project, while he is working in that period between the possible Crash Duration of the project to the Normal Duration of that project. In general, contractor Utility Cost curves are rather flat, and do not influence much the Total Cost curve. The Utility Cost Curve is more applicable to the owner of the project under execution.

The Utility Cost Curve starts at Crash Duration, and accumulates up to the Normal Project Duration.

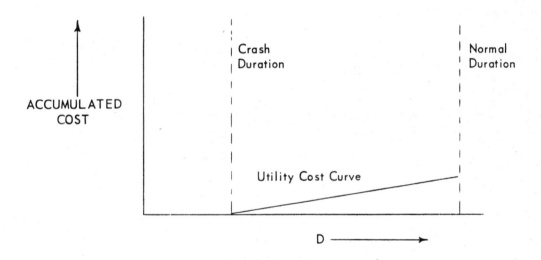

Fig. 14.2

A Total Project Cost Curve can then be generated by the summing of the applicable Indirect Project Cost Curve, and/or the Project Utility Cost Curve, with the Project Direct Cost Curve, as shown in Fig. 14.3 below. Quite often this total Cost Curve will indicate an optimum schedule to be followed that represents the least Total Cost for the specific project; an expedited Project Duration, somewhere between the Normal and Crash Durations of the project. Management decisions and directed schedules can be established based on this information.

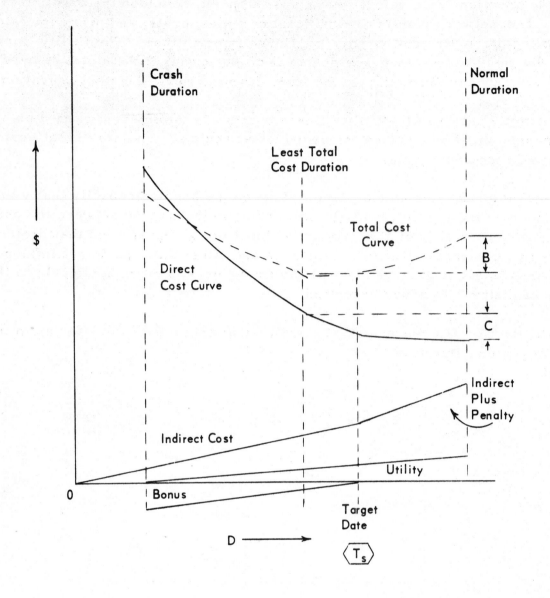

Fig. 14.3
Total Cost Curve
Benefit-Cost Ratio = B/C

It should also be obvious that this technique will allow management to test or investigate various business conditions (acceptable ranges of premium rates, or slopes, for particular activities; various escalation situations; the several accounting methods, both tax and internal, that would create various G and A and OH Indirect Cost assignments, etc.) to help formulate a specific plan of execution for a particular project.

One approach to the use of the Project Indirect and Project Utility Curves, is the extension of Cash Flow accounting techniques to this area. Thus, a needed influx of capital from the project may have a utility value to the organization executing the project, and can be shown as such.

Another use of the Total Cost Curve was in the demonstration that the Overhead Burden carried by a military electronics company was excessive. It was insisted upon that a 300% ratio of Direct Labor costs be carried as an accumulated Indirect Cost, representing that portion of the Overhead and G and A of the company that the particular project was to carry. In every case, the minimum Total Cost Curve point was at Crash - stating in effect that every project should be crashed, in order to cut off the assigned burden costs. What was forgotten was that this very high burden did not stop at the completion of the project, but continued. This was a case of abnormally high staffs, and there was a proof in the Total Cost curves that always started at minimum at Crash - this information should have indicated to the management that there was one area for investigated cost reduction. That area was the large number of staff experts, who did not produce on projects, but were carried as technical advisors.

Benefit-Cost Ratio

More and more fiscal planning studies are producing a number, called the benefit-cost ratio, to substantiate the capital investment, or additional expenditures, on a project to produce an attractive return to the investor. The concept is basic, how many more dollars will be returned by the expenditure of so many dollars. Government agencies are requiring these in studies for authorization of large public works projects.

Fig. 14.3 specifically shows the Benefit-Cost ratio from the network generated Project Direct Cost curve, and the Project Total Cost curve. The reduction in the Total Cost Curve from the higher cost at the Normal Duration point to the lower cost at the minimum of the Total Cost curve, shown as the increment B, is divided by increase in the Direct Cost Curve from Normal to the same point on the Total Cost curve, shown as increment C.

In a study for a particular river basin by the US Army Corps of Engineers, the Utility Cost curve represents the prevention of losses due to floods or droughts in that area. In addition, an estimate of income to the government from the proposed recreational facilities is also factored into the Utility Cost Curve. The Total Cost Curve minimum point is the basis for the recommended schedule to the Congress.

Again, the same reminder! This is just another aspect of an information system. The results produced depend entirely on the validity and practicality of the input information. Establishment of the activity cost premiums, the

Indirect Cost curve, and the Utility Cost Curve will not be easily done. Gathering that information, requires astute, experienced judgement.

A Simplified Method for Mannually Calculating the Project Direct Cost Curve

The "Flow Analogy" Method

In the basic network technology, the time boundaries of Event Times (T_E and T_L) are established by the forward pass (the summation of the longest path forward to each event from the origin), and the backward pass (the negative summation of the longest path back to each event from the terminal event). This concept will be continued, but expanded to include the activity cost slopes. In addition to the standard forward pass, we shall carry along slope (in dollars per time unit) information.

Somewhat of an analogy for this methodology will be that of an hydraulic circuit, consisting of interconnected sections of pipe, each with its own length, and an orifice that limits the fluid flow in that section. This will be the equivalent of a planning network, with each arrow having its durations (pipe length equivalent) and its cost slope (orifice equivalent). The technique will consist of finding the longest length of continuous pipe (the forward pass), and the smallest orifice or flow capacity (slope) in that length. Once that is found, the end point of the Project Direct Cost Curve has been found. The normal project cost is the sum of all of the network's activities normal costs. We also have the slope of the Project Direct Cost Curve from that point. How far that slope extends towards the crash will be determined by the next forward pass which will establish the next point on the Project Direct Cost Curve. After the first forward pass is finished, and the longest time (duration) with the least slope is found, we make a modified backward pass. Along that original longest path only, we remove that least flow capacity from each element. The activity or activities that had that smallest slope now have their slope removed, and their normal durations replaced by either their "crash" duration, if the slope was a straight line from "normal" to "crash," or to the next "expedited" time of the activity has a concave upward curvelinear cost curve. This will be seen in the example below. Once an activity reaches its "crash" time limit, it no longer affects the capacity of the network, since it is at an infinite capacity or slope.

Each forward and backward pass will be called a "phase."

The basic rules for flow capacity (the dollar cost capacity, or the premium rate of the activity cost slope) are:

(1) The flow capacity into a node (event) equals the flow capacity out. Thus parallel, equal length (sum of durations) paths into an event will have their capacities added in the forward pass out of that event. The equalization of flow capacity may not be established until the backward pass. "Flow" in and out of a node applies in both forward and backward passes of a phase.

(2) The flow through any element (arrow) is less than or equal to the capacity of that element (arrow). This applies to both forward and backward passes of a phase.

(3) The backward pass will remove that least capacity along the longest path only (or pro-rated over paths if in parallel), and establish the new activity durations, crash or expedited, resulting from the complete removal of that arrow's flow capacity (slope). At least one activity must have a new duration ("crash" or "expedited") established by the backward pass.

These rules are still subordinate to the basic Critical Path rule that the longest path of durations through a network sets the duration of the project.

Consider the following chain of activities, and their individual cost slopes:

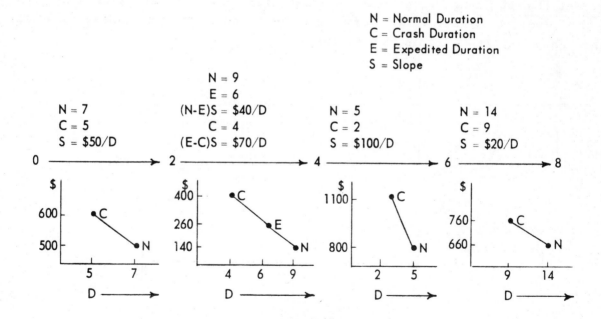

Fig. 14.4

This method is as follows: the first forward pass is made, and the first slope is carried along. As soon as a subsequent arrow is encountered whose slope is less than the forward carried slope, the new lower slope is then carried forward on the standard forward pass. The example is developed to explain the technique; "bookkeeping" methods of entries on drawings, or on forms for a specific project plan, will be described later.

1st. Forward Pass

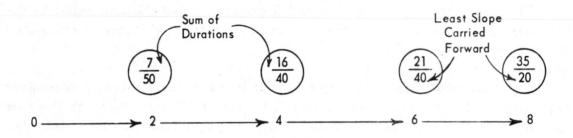

Fig. 14.5

The final duration, and the least slope carried forward to the end event represents the normal duration of the project and the slope of that segment of that Project Direct Cost Curve from "normal" towards "crash." The next point of the Project Direct Cost Curve will be found on the next phase forward pass. Thus far we have this information:

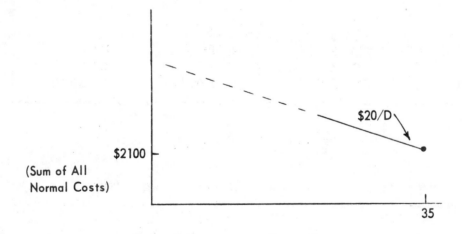

Fig. 14.6

190

The backward pass will consist of (1) removing that capacity, or slope, from each arrow on that longest path. The arrow, or arrows that had that slope have all of their capacity removed, and the new duration (the time or the end of that slope on its activity cost curve) is entered against that arrow. The remaining slope is infinity (∞), indicating that this arrow cannot be reduced any further in time duration. It's slope is no longer considered.

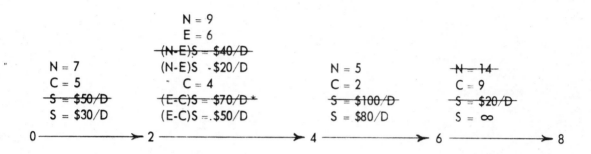

Fig. 14.7

*Note that capacity is removed also from the higher slope section of the curvelinear activity cost curve of activity 2-4. This technique considers a curvelinear cost curve as a series of arrows, with different slopes.

With the corrections of remaining capacity, and new durations made by the first backward pass, the next phase forward pass can be made, and will be:

2nd Forward Pass

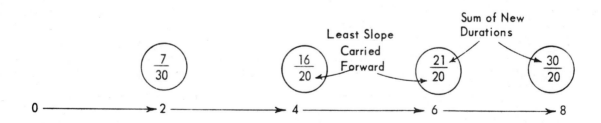

Fig. 14.8

191

The second forward pass establishes the next time point on our Project Direct Cost Curve, that of 30 days. The new Project Direct Cost slope of $20/D is added to the previously established slope, for a total slope of $20/D + $20/D or $40/D. The slopes carried to the final event are added to the previous phase slope of the Project Direct Cost Curve.

The Project Direct Cost Curve now looks like this:

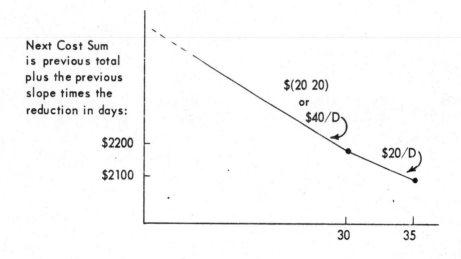

Next Cost Sum
is previous total
plus the previous
slope times the
reduction in days:

$(20 20)
or
$40/D

$20/D

$2200

$2100

30   35

Fig. 14.9

The second backward pass will remove that next additional least capacity that was carried forward in the second forward pass. The arrow diagram now has this remaining information:

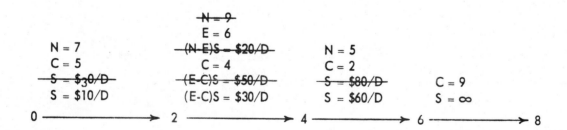

N = 9
E = 6
(N-E)S = $20/D

N = 7   C = 4   N = 5
C = 5   (E-C)S = $50/D C = 2   C = 9
S = $30/D   (E-C)S = $30/D S = $80/D   S = ∞
S = $10/D

0 ————→ 2 ————→ 4 ————→ 6 ————→ 8

Fig. 14.10

192

The third forward pass generates this:

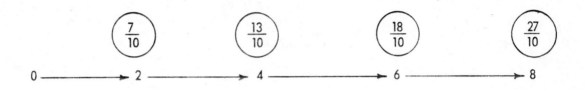

Fig. 14.11

The next point on our Project Direct Cost Curve is:

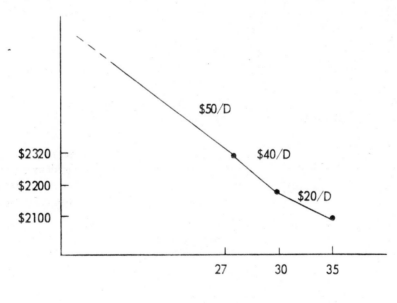

Fig. 14.12

The third backward pass removes the additional slope of $10 from all those activities that have not yet reached their crash limit.

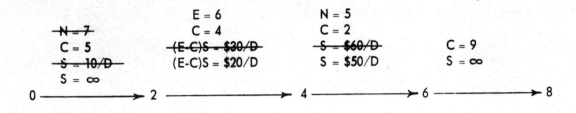

Fig. 14.13

An activity at crash in the beginning of a path is considered to have an infinite slope. This is carried forward until a smaller slope is encountered. The project reaches its crash limit when it can be reduced no further, and is identified when an infinity slope symbol (∞) is carried to the final event.

The remaining forward passes generate the rest of project durations and accumulated slopes, so that the complete Project Direct Cost Curve for the network in Fig. 14.14, once the "crash" duration of the project is found, is:

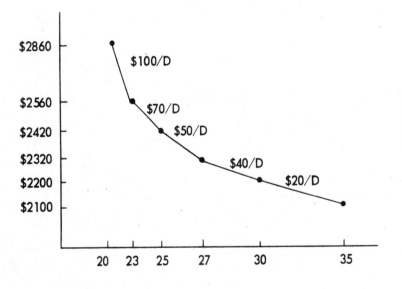

Note that the crash total is equal, in this case, to the sum of the crash costs.

Fig. 14.14

The slopes of the Project Direct Cost Curve are accumulative. The particular Project Direct Cost for each schedule alone has been based on multiplying the differential reduction in project duration by the accumulated slope between the increments of reduction. The particular costs were obtained thusly:

Project Duration	Project Direct Cost
35	$2100 (sum of all normal costs)
30	$2100 + 5 × 20 = 2200
27	$2200 + 3 (20 + 20) = 2320
25	$2320 + 2 (20 + 20 + 10) = 2420
23	$2420 + 2 (20 + 20 + 10 + 20) = 2560
20	$2560 + 3 (20 + 20 + 10 + 20 + 30) = 2860

Note now why an activity cost curve that is concave downward will give invalid answers. Consider the activity curve below:

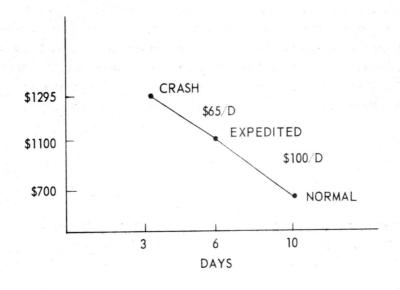

Fig. 14.15

The backward pass, in removing the "flow capacity" in preparation for the next phase forward pass, would start removing it here from the middle of the activity (it would start at the expedited day 6 and reduce it towards crash day 3, and would ignore the normal day 10). This is illogical in that the curve states that we start at a normal time of 10 days (least or "usual" time at least cost), pay a premium rate of $100/D to reduce it to an expedited duration of 6 days. Only after we reach an expedited duration of 6 days can we start crashing at a lower premium rate.

195

In general, this type of activity curve is generated when an activity has been defined too grossly. For example, an arrow is defined:

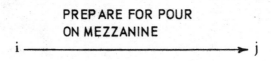

PREPARE FOR POUR
ON MEZZANINE

i ————————————————➤ j

Fig. 14.16

This is to entail the erecting of a plank ramp road with limited capacity of getting buggies to the mezzanine. The maximum effort (highest slope) is in erecting a plank access ramp road. Once the road is finished, the remaining preparatory work can be expedited at a cheaper premium rate. This gross activity actually consists of two activities in series (one cannot start until the other is finished). The obvious solution is to break down the gross activity with the concave downward cost slope into the two properly identified operations.

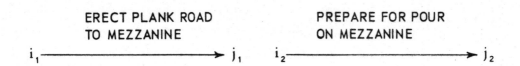

ERECT PLANK ROAD
TO MEZZANINE

i_1 ————————————————➤ j_1

PREPARE FOR POUR
ON MEZZANINE

i_2 ————————————————➤ j_2

Fig. 14.17

Multiple Paths

The plan considered above in Fig. 14.4 is now expanded, and a parallel path added so that rules 1 and 3 may be amplified in the handling of parallel paths. Fig. 14.18 shows the additional parallel path.

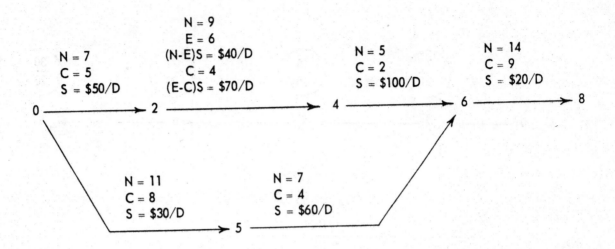

Fig. 14.18

The first forward pass give us this:

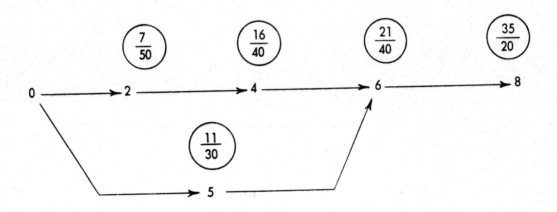

Fig. 14.19

The first backward pass removes $20/D from 0-2, 2-4, 4-6, and 6-8 only, since it was the longest path through the network. Activity 6-8 has gone to "Crash". The second phase forward pass results in:

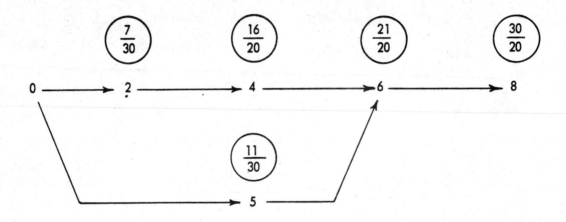

Fig. 14.20

The second backward pass again removes an additional capacity of $20/D from the longest path of the second forward pass; that is, 0-2, 2-4, and 4-6. Since 6-8 was reduced to its crash limits in the previous backward pass, and is now at infinite slope (capacity), it remains unaffected.

The configuration of the network for the third pass is as follows:

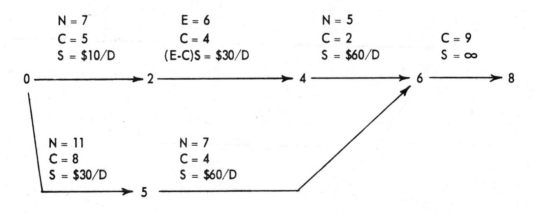

Fig. 14.21

The third forward pass results in a parallel path:

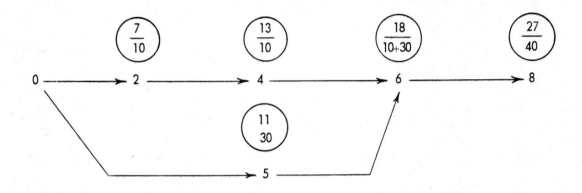

Fig. 14.22

The flow capacities are added into event number 6 and carried to the end of the project. The third backward pass will split the capacities in accordance with the split that came into that event in the forward pass. The path technique is the same, that path flow capacity is taken from each arrow on that path. The network configuration now looks like this, prior to the fourth forward and backward passes:

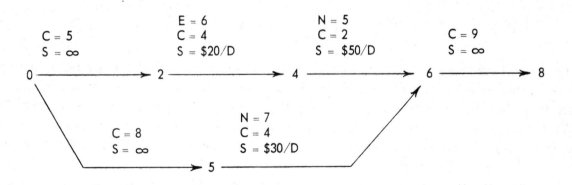

Fig. 14.23

The composite entries on the diagram for the 4th, 5th, 6th, and 7th phases will be as shown below:

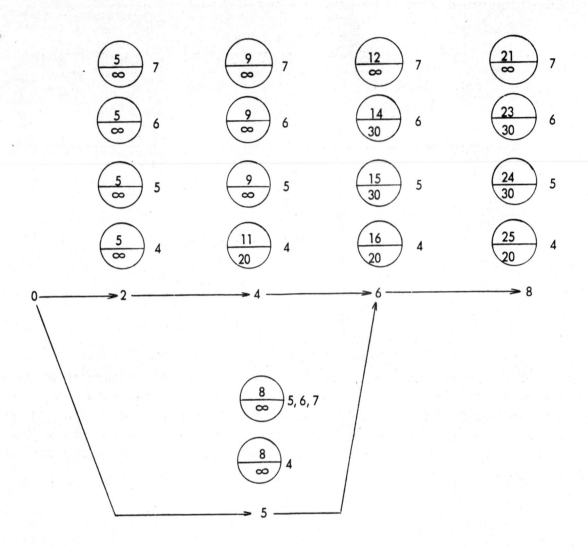

Fig. 14.24

The Project Direct Cost Curve has its crash point established when an infinite slope has been carried to the final event. It may arrive at the final event in this form: 61/90+ ∞, indicating that there is a path parallel to the Crash Critical Path, which path could be reduced further but to no avail in Project Crash Duration, since there is a parallel irreducible Crash Critical Path.

Thus far we have been only dealing with the method of generating the Project Direct Cost Curve. We have not discussed that schedule (individual durations of activities) at any point on the Project Direct Cost Curve. Note that on the third backward pass, on Fig. 14.23, we removed the slope of $30/D from activity 0-5, reducing it to its crash limit of 8 days. On the 4th pass, Fig. 14.24, we found that the longest path of 16 days into event 6 was set by 0-2, 2-4 4-6. Consequently, if we move to accept the 4th project duration of 25 days, the duration of 0-5 would be relaxed from 8 to 9 days, so that the cost would be minimized; that is, we would let path 0-5, 5-6 relax to the time boundary of day 16, which is set at event 6 by another path. We shall establish the rules for the project schedule (activity times) later. It is superfluous to establish the schedule for each point on the Project Direct Cost Curve as we go along. We shall decide by additional cost criteria that one schedule of all activities in which we are most interested.

In general, the technique of generation of the Project Direct Cost Curve does this: it first starts reducing least slope (least premium rate) activities along the critical path, until it creates a parallel critical path. Then, combinatorially, it reduces that parallel configuration until other critical paths are created in parallel and so on, until the project is reduced to its crash critical path case.

It may be difficult to decide how to handle the first rule of "flow capacity" - the flow into a node will equal the flow out. Consider the network below:

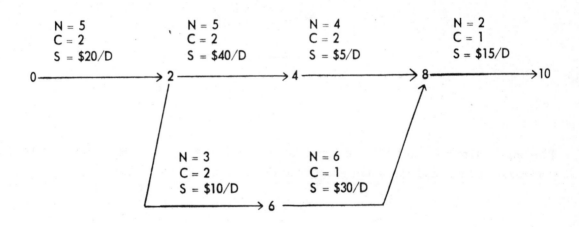

Fig. 14.25

In the first forward pass into event 2, should the carried forward least slope be split evenly into parts equal to the number of paths emanating from that event? Or should that least slope arriving at a "burst" event (that event where several parallel paths start) be carried along each parallel path. On a simple example above quick inspection reveals parallelism, and time concurrency. But a real project diagram may have many parallel and long paths, and it may be difficult to tell if some paths are equal, others shorter in time? If the incoming capacity is split pro-rata, part of it may be lost when one parallel path is longer than another. On the other hand, if that slope arriving at the "burst" event from which several parallel paths start, is carried in toto along each path, then at a "merge" event, (that event where several parallel paths come together), there would be a ficticious summing of that slope, if the parallel paths were equal in duration. The basic rule is for any particular phase, there is only one flow capacity going through the entire network. This is established after the backward pass of that phase has been made. A cut across any section of the network will reveal only one flow capacity—and this cannot be established until after the backward pass of that phase.

Thus in the above network, whether we split the slope coming into event 2 into two equal parts, or use the incoming slope equally, we get this on the first forward pass:

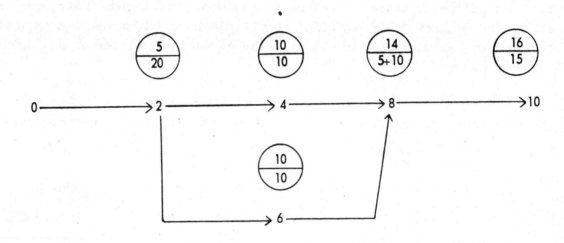

Fig. 14.26

The new slopes, set by the first backward pass, allow us to make the second forward pass. Splitting the incoming slope gives us this:

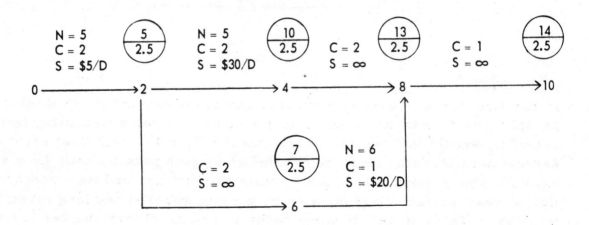

Fig. 14.27

202

This appears erroneous, in that we did not use up the full capacity of one element, activity 0-2. The backward pass, removing the 2-1/2 dollars per day, reduces network capacity, but the duration of the individual activities will not be reduced. The next forward pass will give us the same duration, but a split in the capacity will now give us a duration of 14 days, with a slope of 1-1/4. We find ourselves making several forward and backward passes, coming up with the same project duration, but with ever diminishing slopes.

That the first phase forward pass in this simple network could be made with either the splitting of the incoming slope, pro-rated over the number of paths, or that arriving slope at event #2 carried along each parallel path, and getting the right answer, occurred only because there were slopes on the parallel paths that were less or equal to the split incoming slope. The second phase indicates what can happen if there is only one long path of several parallel paths. Suppose those slopes were higher than the incoming slope, as shown below:

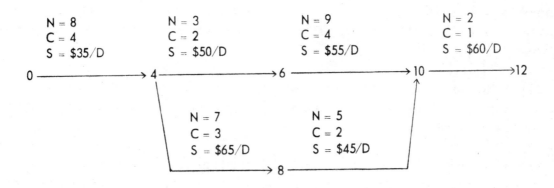

Fig. 14.28

The forward pass, carrying the slope arriving at event #2, into each path without splitting, would give the following results:

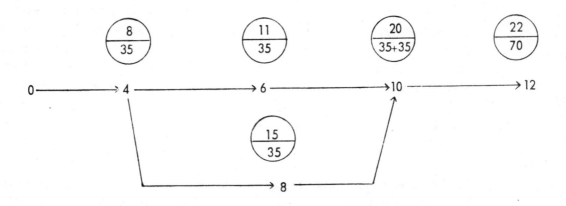

Fig. 14.29

203

Which apparently contradicts the rule that the flow capacity in any phase must be the same anywhere in the network. The backward pass, reducing the capacities of each arrow, would arrive at event 2, and attempt to remove $70/D from 0-2, which only has a capacity of $35/D.

Thus we rely on the backward pass of any phase to:
(1) Check the reliability of the slope arrived at the final event.
(2) Establish the network flow capacity (slope) for that phase.
(3) Establish the remaining durations of the activities for the next phase. It is mandatory in any backward pass to reduce at least one activity to the next point on its activity cost curve, either an "expedited" point on a multiple slope activity, or a "crash" point on a single slope activity.

Practical Rules of Use

Since, in a real, large project, there is no relatively easy way of determining whether parallel paths are equal in time duration; and whether these parallel paths have a least slope equal to or less than a split flow capacity (slope) coming into a "burst" event; it will be necessary to make several generalizations.

(1) In practice, it recommended that the network be reduced, by combining slopes, and eliminating large total float paths. This technique is covered in a subsequent section.

(2) If, after reducing a network, it is still difficult to establish time concurrent paths by inspection, the following steps should be taken:

(A) On the first two phases (normal and first reduction of project) carry the slope arriving at the "burst" event (parallel paths emerge from a burst event) along each path; do not split the capacity. Rely on the backward pass to remove capacities of activities. The "flow in equals flow out" rule applies also to the backward pass. If we find an activity whose slope is lower than that of the backward carried slope, we have encountered equal parallel paths whose individual slopes are higher than that of the slope arriving at a "burst" event. (This situation, of parallel equal paths with higher slopes, may be observed before starting the forward pass, and the decision made then to split the incoming slope).

(B) Since the technique reduces the network by creating parallel critical paths, the third pass should be made with splitting slopes from a "burst" event into pro-rated values along parallel paths.

In all cases, the backward pass must reduce at least one arrow to an end point on a curve slope, either to the crash point, or to an expedited point on a curvelinear activity cost curve, where the cost slope starts to increase over the previous slope.

(C) Make just a forward pass of durations, establish new event times, find parallel paths and mark them. Then split incoming cost slopes.

The Either/Or Case

One final condition should be described. Quite often an activity can be given two points of cost; its normal duration and normal cost and its crash time and crash cost; but there is no way of interpolating a continuous curve between these two points. Such cases were described previously. For this type of activity, we shall give it an "equivalent slope", by the usual formula of

$$\frac{\text{"Crash Cost - Normal Cost"}}{\text{Normal Time - Crash Time}} \left[\frac{\$_C - \$_N}{D_N - D_C} \right]$$ and use it in the generation of

A Project Direct Cost Curve. It is only when we get to the desired schedule of a particular project duration that we must remember it is an either-or case. Then we shall make adjustments in the actual duration.

Establishing the Schedule

Once the technique of generating the direct cost curve of the project has been mastered the next step is the establishing of the schedule (the activity times) for the desired project duration. As mentioned before, it is generally superfluous to calculate the schedule for each point along the Project Direct Cost Curve. There is only one schedule that represents the least Total Cost to the organization executing the project.

As was noted previously, it is the sum of the Indirect Cost Curve and Utility Curve with the Project Direct Cost Curve, that gives the project Total Cost Curve. The Total Cost Curve will have one minimum point. The scheduling method will consist of testing the established points on the Total Cost Curve after each phase (forward and backward pass) of the Project Direct Cost Curve, by adding the newly calculated Project Direct Cost at the phase point on Project Direct Cost Curve, to the sum of the Indirect Cost Curve and Utility Curve (if applicable) at that newly established reduced Project Duration. Once a point on the Total Cost Curve rises in total amount over the amount of the previous point, we have found the minimum total cost Project Schedule. This is indicated below in Fig. 14.30.

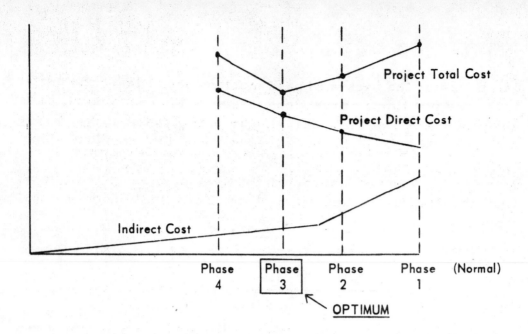

Fig. 14.30

By starting at the Normal Point on the Project Direct Cost Curve, adding to it the value of the indirect Cost Curve and the Utility Cost Curve at that point, we successively generate the Project Total Cost Curve. Once we note that new PHASE Total Cost is higher than the previous PHASE, the previous phase becomes our optimum schedule from the cost criteria.

There can be many shapes of the Total Cost Curve, even though it has one minimum point. It is possible, particularly if the Utility Curve has a high enough slope, that the Total Cost Curve may have its minimum point at the Crash Point. Even in this case, there will be an optimum crash schedule that results in the minimum Crash Cost.

It is also apparent that the Total Cost Curve may have its minimum at the Normal Point of the curve, where the Indirect and/or the Utility Cost Curves are relatively flat. If the 2nd Phase shows an appreciable increase in Total Cost over that of Phase 1, it is a good indication that minimum has already been found. Rember that the Indirect and Utility Cost curves are accumulative, thus their minimum is always at their starting point.

The steps in arriving at the particular optimum schedule for the project are:

1. Establish the Indirect Cost Curve.
   It may be necessary to prior establish both the Normal Duration and Normal Cost of the Project, to obtain complete Indirect Cost Information. This is particularly so if a percentage of the Normal Direct Cost is a factor in the accumulating Indirect Cost curve.

2. Establish the Utility Curve, if applicable. It will be necessary to prior establish the Crash duration of the Project for the starting point of this curve.

206

3. Those activities whose Normal Total Float is equal to or greater than the difference between the Normal Project Duration and the Crash Duration should then be noted. These will never be compressed less than their normal duration. Their schedule times will remain normal, thus their costs. They can be eliminated from consideration in the phases of the Project Direct Cost Curve generation.

4. Once the particular Phase has been determined, examine the event times of that phase. It may be necessary to make the regular Event Time Backward Pass (establishing (T_L) Chapter 5) to set all event times for the optimum phase. Those activities that are in paths shorter than the path that set the event time, can be relaxed back towards normal, to "fill up" the time boundary established by the longest path. Those activities along the shorter path are relaxed by allowing first the activity with the highest slope (highest premium rate) go back towards normal duration. If there is still more time available, then relax the activity with the next highest slope, and so on until the path length coincides with the longest path that set the merge event time.

This is the same technique as described in "Optimum Crashing a Project." page 125. The critical event times of a particular phase become boundaries, and all non critical paths of that phase are relaxed back towards normal, to fill those time boundaries, and minimize the direct costs.

Least Cost Schedule, Phase II, Fig. 14.25

If we analyze the results of the actual Phase II operation for the network of Fig. 14.25, page 152 we would find these resultant durations:

0-2	5	4-8	2
2-4	5	6-8	6
2-6	2	8-10	1

Using these durations, we would find the path into Event #8 that sets its time boundaries. That path is 0-2, 2-6, 6-8, and from the above durations, we note the Earliest Event Time would be time unit 13 for Phase II. Now consider the path 0-2, 2-4, and 4-8. Its time length is 12 time units. Further analysis, via the filling in of established time boundaries, is required. Is there any reason to crash activity 4-8 to 2 time units - it doesn't control the Event Time at Event #8? The answer is no. 4-8 should be relaxed back to 3 time units to fill up that time boundary. Thus our schedule duration for activity 4-8 will be 3 days (expedited), to minimize our direct costs. The sum of the costs of the activities in a particular Phase Schedule must be the same as the cost point at the Phase's location on the Project Direct Cost Curve. The Phase II schedule is shown below in Table 12.5.

i	j	Duration	ES	EF	LS	FL	Float
0 - 2		5 (N)	0	5	0	5	0
2 - 4		5 (N)	5	10	5	10	0
2 - 6		2 (C)	5	7	5	7	0
4 - 8		3 (E)	10	13	10	13	0
6 - 8		6 (N)	7	13	7	13	0
8 - 10		1 (C)	13	14	13	14	0

Table 14.5

5. Those activities that have an "either-or" characteristic (crash or normal points, but nothing in between) and were given an equivalent slope, must now be separately analyzed. If its equivalent slope is the highest in the path, then it must be analyzed to see if relaxation from crash to normal will fit within the time boundaries established by the longer path. If the relaxation to normal adds more time to the path than the event time at the end of the path permits, then it must be kept at the crash time, and the activity with the next highest slope that can be interpolated is relaxed toward normal. Consider the following diagram with the results from its fourth phase shown with their remaining slopes:

PHASE 4 CONFIGURATION

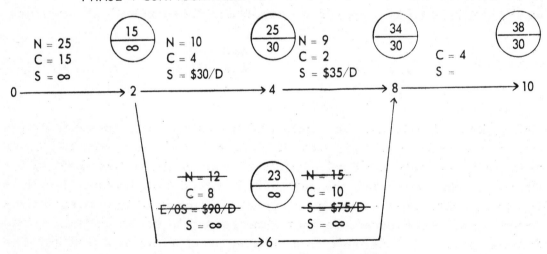

ACTIVITY (2-6) HAS AN EQUIVALENT SLOPE, SINCE IT IS AN EITHER - OR COST CHARACTERISTIC

Fig. 14.31

In this, we see that 0-2, 2-6, 6-8, and 8-10 have been reduced to their crash limits in previous phases. In this example phase IV is assumed to be one that produces the optimum Total Cost schedule.

It is apparent that the path 2-6 and 6-8 can be relaxed, since it has 19 days to go between events 2 and 8. We would first relax the highest original slope activity on this path, which belongs to 2-6. However, it is noted that 2-6 is an E/O (either-or) activity. Relaxation here would mean it must go to a normal time of 12 days, since there is no allowable time in between. If we relax 2-6 to 12 days we are exceeding the phase established time boundary: 2-6 = 12 plus 6-8 at crash of 10 = 22, larger than the optimum 19 days. Hence, we must here leave 2-6 at a crash of 8 days, and relax the next highest slope activity, which is 6-8, to the eleven days available. The schedule durations in this phase are

0-2 = 15 days	4-8 = 9 days
2-4 = 10 days	6-8 = 11 days
2-6 = 8 days	8-10 = 4 days

A Book Keeping System

Naturally, it is imperative that a record of each phase be kept, so that the proper information on the event times and activity durations be available for the desired optimum schedule calculations.

The method shown heretofore, in the development of the technique via simple examples, may be used as a case in point. The block of the "j" event may be used to indicate the longest path, and the lease slope carried forward. The backward pass of that phase will show the removal of slope capacity from that phase's longest path.

The example in Fig. 14.4 on page 140 would look like this, for all its phases.

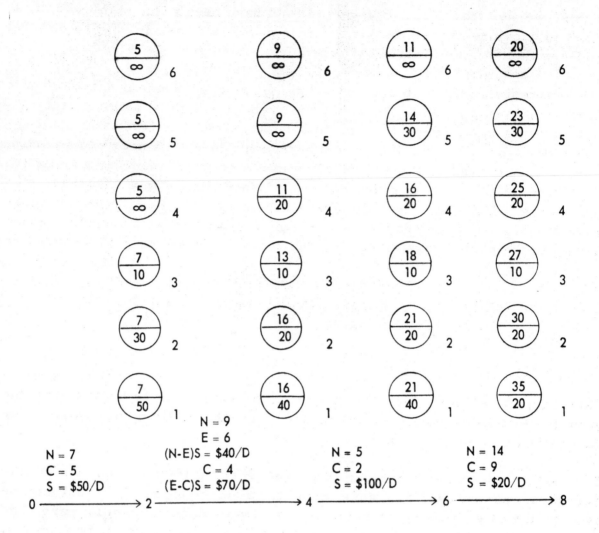

Fig. 14.32

An obvious disadvantage is that the drawing of the network soon becomes cluttered with information. This tends to become confusing, with the attendant possibility of erroneous selection of information. A simple form, shown in Fig 14.33 will help the "Bookkeeping" in a simplified, neat way of keeping the information intact. As the forward and backward pass of any phase is made, the entries are made on an easily read form. When the optimum phase is discovered, the information is rapidly taken off, and the schedule generated.

A listing in a "j" major sequence is recommended. Fig. 14.33 has the phases for the network in Fig. 14.25 indicated. The last two columns in each phase are filled in during the modified backward pass.

ACTIVITY COST SLOPE CALCULATION FORM

I	J	DUR	SLOPE TYPE		PHASE I T_E J	PHASE I LSCF	PHASE I NEW SLOPE	PHASE I NEW DUR	PHASE II T_E J	PHASE II LSCF	PHASE II NEW SLOPE	PHASE II NEW DUR	PHASE III T_E J	PHASE III LSCF	PHASE III NEW SLOPE	PHASE III NEW DUR	PHASE IV T_E J	PHASE IV LSCF	PHASE IV NEW SLOPE	PHASE IV NEW DUR	PHASE V T_E J	PHASE V LSCF	PHASE V NEW SLOPE	PHASE V NEW DUR
1	2	N-5 C-2	20/0 C		5		5/0	5	5	5/0	8	2	2	8	8	2	2	8	8	2	2	8		
0	2	N-5 C-2	20/0 C		5	20/0	5/0	5	5	5/0	8	2	2	8	8	2	2	8	8	2	4	8		
2	4	N-5 C-2	40/0 C		10	20/0	35/0	2	10	5/0	35/0	5	7	35/0	35/0	5	7	35/0	8	2	4	8		
2	6	N-3 C-2	10/0 C		8	10/0	8	2	7	5/0	8	2	4	8	8	2	4	8	8	2	4	8		
4	8	N-4 C-2	5/0 C *		14	5/0	8	2	12	5/0	8	2	9	35/0	8	2	9	35/0	8	2	6	8		
6	8	N-6 C-1	30/0 C *		14	10/0	20/0	6	13	5/0	15/0	6	10	15/0	8	1	5	8	8	1	5	8		
8	10	N-2 C-1	15/0 C		16	15/0	8	1	14	5/0	8	1	11	15/0	8	1	10	35/0	8	1	7	8		

NOTES:
(1) Types of Slope
  C = Continuous
  M = Multiple
  E/O = Either/or
(2) LSCF = Least
  Slope Carried Fwd.
(3) * Slopes sum up into Event #8

Fig. 14.33

When the technique is to be applied to an actual project plan, consisting of several hundred arrows, or more, there is generally a pause for evaluation of the magnitude of the effort to generate the Project Direct Cost Curve.

We have already noted the first step in reduction; the elimination of those activities whose total float in the normal phase is greater than the difference between the Normal Project Duration and the Crash Project Duration. These activities will stay at their normal time and costs.

The next step is to take sections, or groups of activities, out of the overall network, treat them separately as a sub-project, and reduce them down to one equivalent activity of multiple increasing slopes. Consider the network below:

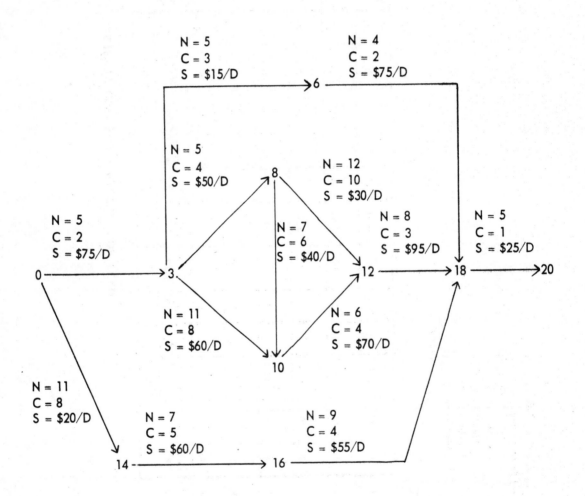

Fig. 14.34

The usual forward pass for the normal durations will give a Normal Project Duration of 36 days. The Crash forward pass for crash durations reveals a Project Duration of 20 days. The difference is 16 days. Activities 3-6 and 6-18 have a Normal Total Float of 17 days, indicating they will never be reduced. They can be eliminated from the Cost Curve phases, and their normal costs entered. Next, the section of the diagram between event numbers 3 and 12 can be handled separately. It will give the following equivalent slope.

DURATION	SLOPE
N = 18	$40/D
E_1 = 17	$80/D
E_2 = 16	$100/D
E_3 = 15	$130/D
C = 14	∞

3 ————————————————→ 12

Fig. 14.35

Thirdly, evaluate the path 0-14, 14-16, and 16-18. The "Normal" duration of this path is 27 days (11+ 7 +9). The first reduction of the chain of activities will take place at the activity with the least slope, 0-14 with 20/D. The next reduction, after reducing 0-14 to 8 days crash, will be at 16-18, along the $55/D slope. Its equivalent activity curve is:

DURATION	SLOPE
27	$20/D
24	$55/D
19	$60/D
17	∞

0 ————————————————→ 18

Fig. 14.36

Thus, a series of arrows is made into one with a multi-slope, curvelinear activity cost curve.

The reduced diagram now looks like this:

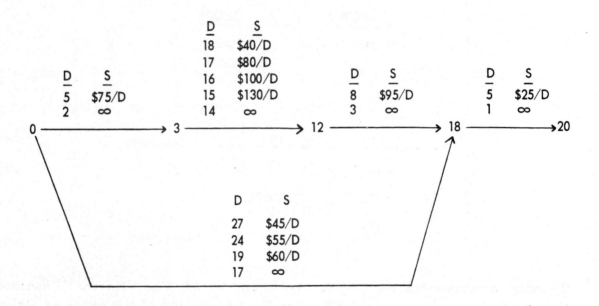

Fig. 14.37

Activities that have an Either/or cost condition (E/O), must be kept as separate arrows in a diagram. They cannot be lumped into a reduced network.

SUMMARY

Rules for Generating a Project Direct Cost Curve

A. The cost slope of an activity, from the formula

$$\frac{\text{``Crash Cost - Normal Cost''}}{\text{Normal Time - Crash Time}} \quad \text{or} \quad \left[\frac{\$_C - \$_N}{D_N - D_C} \right]$$

acts as a "flow capacity" or premium rate, for that activity.

B. The slope of an activity must be linear or concave upward, the slope must increase from normal to crash.

C. An activity that has only the normal time, normal cost and crash time and crash time and crash points, but cannot practically be interpolated anywhere in between, will be given an "equivalent" slope from the formula in (A) above.

D. The basic rule of Critical Path takes precedence, the longest time path through a network, in any phase from normal to crash, determines the duration of the project. The backward pass only removes capacity along that phase's longest path.

E. In a network, the flow capacity (slope) into a node is equal to the flow out of a node. If flow occurs into a node, from which parallel paths emanate, the incoming flow capacity (slope) should be carried forward on each path for the first two phases (forward and backward pass). Thereafter, that flow will be split, pro-rata into the number of parallel paths. This applies to both forward and backward passes.

F. The flow capacity through any element (arrow) is equal to, or less than the flow capacity (slope) of that element (arrow). This applies to both forward and backward passes.

G. The backward pass removes the flow capacity (slope) that arrived at the final event of the project in the previous forward pass. The backward pass follows rule "E". Flow into a node (event) equals flow out. The backward pass must reduce at least one element (arrow) in the path to the shorter duration point on the end of its slope, either the crash point, or the next Expedited Duration, where that activity has a multiple Slope Cost Curve.

H. The backward pass will reveal if a previous forward pass along parallel paths gave an erroneous sum of slopes at a merge event. The slope being carried back must observe rule "F", it cannot be higher than that of any particular activity.

I. The backward pass, along that phase's longest forward pass path, removes that capacity from all slopes of all arrows cost slope activity, and from each separate slope of a multiple slope activity.

J. The slope of the Project Direct Cost Curve is accumulative from the Normal Point up to the Crash Point. Each phase slope at that phases final event is added to the previous accumulated slope.

K. A dummy is a path in the logic diagram, and it becomes an element in our hydraulic analogy. It is considered to have an infinite slope, and passes on the slope entering its start event node.

L. A schedule is derived from the construction of a Total Cost Curve, which is the sum of the Project Direct Cost Curve, the Project Indirect Cost Curve, and the Project Utility Curve. The optimum schedule occurs at the saddle point, or lowest point on the Project Total Cost Curve.

PROBLEMS

14.1 Construct the Project Direct Cost Curve for the network shown in Fig. 13.7.

14.2 Calculate the schedule for the following phases of the Project Direct Cost Curve in problem 14.1 above (assume all continuous slopes)

Phase II	29 day Project Duration
Phase IV	21 day Project Duration
Phase V	20 day Project Duration

14.3 Draw the Project Direct Cost Curve for the following network.

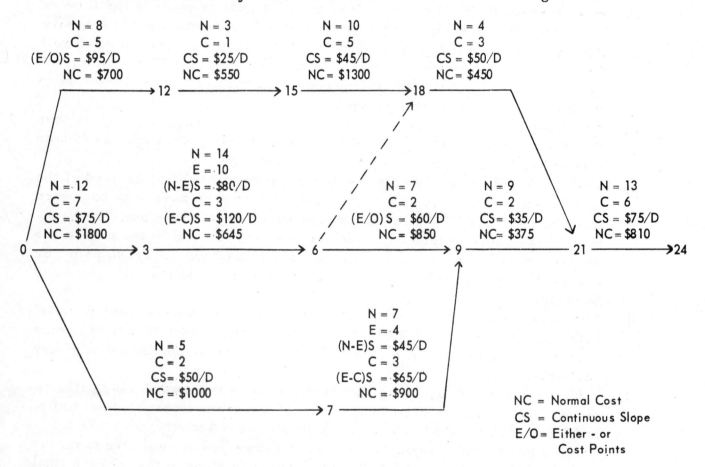

NC = Normal Cost
CS = Continuous Slope
E/O = Either - or
        Cost Points

216

14.4 Calculate the Phase III, Phase VII, and the Crash schedule for the project in problem 14.3.

14.5   OPTIMUM HOUSE SCHEDULE

The Browns, a family of two adults and four children, hold a comprehensive Fire, Theft and Damage policy with the XYZ Insurance Co. The policy offers complete protection and restitution for damages and losses due to fire, storm, civil strife, etc. It also provides for living expenses of $15 per day for each adult, and $10 per day for each child while the damage house is restored.

While the Browns are on vacation, their two story house (Cape Cod) burns to the ground; total destruction. XYZ Insurance Co. obtains bids for rebuilding, and selects the PDQ Construction Co. PDQ furnishes the Critical Path Plan, and the Normal Time, Normal Costs, the Crash Time and the Crash Costs for each activity, as shown in problem 13.1, page 176 . In order to expedite construction, and insure conformance to architectural specifications, XYZ Insurance Co. hires a reputable Engineer as a Construction Supervisor. His fee is $100 per diem, and his travel and living expenses are allowed at an additional $30 per day.

What is the best schedule for XYZ Insurance Co. to rebuild the house?

CHAPTER 15

INSTALLATION OF NETWORK SYSTEMS

The establishment of successful network planning and scheduling systems required careful planning and execution. The main areas of consideration are:

1. Indoctrination of top management.

2. Training of middle-management personnel including such men as project managers, planners and schedulers, estimators and other people who will have direct responsibility for the actual operation of the system.

3. Orientation of those people who must work with the system and report into it, such as engineers, designers, shop supervisors, procurement personnel, and so forth.

4. Analysis and establishment of available data processing and computer equipment that may be required.

5. Establishing the proper organization to conduct network planning and scheduling.

From experience, the middle management people who will use the technique, implement it, and collect the data, and establish the particular systems and procedures, can be given a working skill in five to ten days of intensive training. Such training programs are available from several organizations, either as a general public course, or as a specialized in-plant training program. Today, the majority of companies are taking advantage of the specialized training courses in-plant to build a skilled staff. After working on two or more projects, the planning specialist has sufficient skill accumulated to start developing his own ramifications of the technique.

Another method of implementing network techniques into an organization, is by the retaining of a professional planning consultant to plan a specific project. It is perfectly legitimate and ethical to assign up to three key personnel to stay with the consultant, observe his methods of planning and generating a schedule, and "pick his brain" in the development. This will accomplish two results—the completion of the initial project plan, and the indoctrination of key company personnel in the technique, for further internal planning and up dating of the project. A good consultant should help arrange a computer rental source, if it is required; demonstrate the operation of the computer to the personnel, and leave with the organization a copy of the particular computer program used, and instructions on how to operate the computer.

The use of a professional planning consultant may be mandatory for a firm that finds itself obligated to use CPM or PERT on a large contract, and has no trained personnel.

A word of advice on the selection of consultants. The professional planner selected should be knowledgeable in the network technique, be familiar with the operation of the data processing equipment and computer used, and be experienced in the industrial field of his client. This is a line management science, thus the professional planner should be experienced in the particular industry being served. The field of professional planning consultants abounds with highly competent men who are mathematicians and computer experts, but have little project management experience. A theoretical approach to a real project problem may result in mathematically valid solutions - that may prove impractical. If in doubt, interview the particular planner who will be retained, or assigned to your organization.

Initial Reactions

In general, two attitudes will be encountered in installing Network Planning —that of indifference, and that of hostility. Indifference will generally be encountered in top management. They are involved with the macroscopic problems of management, and may tend to regard this technique as another management fad. In dealing with government agencies, they find they are required to install and use such a system. A one day seminar presentation, replete with case histories, generally is sufficient to demonstrate the potential of the technique and permits top management to read the output data.

Hostility will come from the lower level personnel, usually generated by ignorance of the technique, or fear of a loss of prestige or status by having someone else do his planning for him. Whatever the motivation of the individual, the best way to obtain cooperation is by a short orientation session, describing the technique, the advantages of the information generated, and a general allaying of qualms or doubts. It is mandatory to point out that the plan is always the supervisor's; even though it may originally come from a higher level manager, it is always given to the particular supervisor for his review and concurrence.

This technique is being used down to the foremen level by several firms. It takes a minimum amount of training to have the foremen capable of taking a section of the overall project plan, and from it determine manually their week to week or month to month schedule. This training can be accomplished in two days or less.

Naturally, it is wiser to crawl before you walk. Therefore, the first project to which to apply this technique is one small enough to allow manual calculations. It may be feasible to use the company's traditional methods of planning and scheduling, and to set up network planning and control concurrently, to observe the efficacy of the technique. A good rule of thumb is to limit the number of activities on a manual approach to no more than two hundred. Beyond that the manual system becomes cumbersome and errors by humans are commonplace, and would be carried through the schedule calculations. Manual control on projects of fifty activities, and capital costs of $10,000 has been successfully executed. This appears to be a lower limit use of Network Technique.

With the accumulation of skill, larger projects, and the use of computers and Electronic Data Processing equipment comprise the next step. It is not necessary to procure and install such equipment. Use of such equipment, plus competent advisory help, may be rented at a computer manufacturer's service bureau or from a firm that possesses. This can be done at minimum cost to the company. The actual cost of computer rental is quite low.

A more practical approach is to rent time on a computer from a private company. Many companies have over-estimated their computer needs, or have been oversold on their installation. These companies become good sources of low cost rental, usually, at low demand periods of computer time (after 5 PM, and on weekends). Engineering Companies, banks, and insurance companies have proven a good source of low cost computer rental time.

In the next chapter, #16 "Computer Orientation" the method of instructing the computer to perform the schedule calculations will be described. This is called "programming" the computer. A program must be available for the particular computer selected for use. Most computer manufacturers have available ready made network calculation programs (either PERT, CRITICAL PATH, or some variation), and these may be obtained at no cost from the computer company, by a simple request. However, a cautionary note is very pertinent here. Quite a few of the "canned" programs, freely available, have been written by excellent mathematician and computer experts - but not by project managers. There is a common pitfall to be avoided. It is extremely easy for the line manager to find himself working for the system, rather than the system working for the manager. Limits on the number of arrows, event number digits, lack of a descriptive input for each arrow, can force a large amount of redundant clerical work. A most common trap is the computer program that requires sequential, consecutive "i - j" numbers on an arrow diagram. This means that if any changes occur, or if a chain of activities has to be added to the plan, or if there is deletion, or additions to the size of work of the contract (and these situations are the norm, not the exception in business life today), then the diagram will have to be renumbered to suit the machine.

The manager may have no choice but to live with this situation in the beginning. As the use of the Critical Path Technique increases, there will come a point where it will be feasible to purchase programming service, in order to rewrite a "canned program" to the line manager's specifications.

The Cost of Using Network Techniques

In practice, after Network Planning has been introduced into a company, one specialist and a clerk can handle up to ten networks containing five hundred activities each. After the establishment of network, data handling is primarily a clerical chore.

The initial, non recurring expense, is in the training of personnel (about $350 per man is a good budget estimate). With use, there should be no increase in the number of personnel (train existing people), no increase in time (with experience, the time in planning should take no longer than present planning methods). It is assumed that some formal method of planning is presently employed.

Computer rental time is negligible - a 600 arrow diagram should have its schedule generated for $50 of computer time. The key punching of cards is an initial, rapidly diminishing expense. Key punching, verifying and sorting generally costs 5¢ a card. After the initial pack of input cards is key punched, there is a very small percentage of new cards to be punched for each subsequent up-dating run.

A rough guide of computer time costs, based on normal rental rates, would be as follows:

200 -600	arrows	$50
600 -1200	arrows	$100
1200-2000	arrows	$150
2000-3000	arrows	$200
3000-5000	arrows	$300

This is for computer time only, exclusive of any consultants' time. It should be pointed out that the initial run will 80 times out of a hundred produce errors —that of omissions or incorrect key punching. The above costs do not reflect that situation. You are billed just for the actual time on the machine, so if you are searching for errors, arranging additional key punching, make sure that you are off the computer. The computer usage schedule may be such that you will find yourself waiting until the next day to get back on for a new run. The next chapter will point out some practical recommendations of computer orientation. The rates listed above are from the author's experience in renting time from user's. Generally, the Service Bureaus operated by the computer manufacturers are much higher cost sources. As an example, one computer manufacturer charges 50¢ an arrow for the initial run (including key punching), and 25¢ an arrow for a subsequent updating run.

When computer time is rented, the organization possessing the computer will also furnish an operator, who knows the equipment, and will be able to set up the console, clear the memory of previous programs and problems, and generally assist in the running of the computer. His time may be billed at $10 per hour in addition to the rates above, also pro-rated to actual use time.

Setting Up the First Network

There are several ways of approaching the establishment of the first network. There is one school of thought that advocates obtaining a large sheet of paper, and then starting to draw. The network can be drawn from the final event backwards to the origin event, or vice versa. In practice, the following sequence of planning activities has proven the most effective:

1. Compose a Gross Activity List

    Think of the project, and those aspects of the project that are to be controlled. After listing some thirty to fifty activities (and it will be harder to stop listing activities once you start writing), not necessarily in sequence, prepare to start the network.

    For a contractor, a good source of an activity list is the index of the architect's specifications, and equipment plot plans, which indicate major pieces of equipment and work.

2. Draw the Gross Network

    This is a general picture, made from the gross activity list. It can be made by the Project manager, and/or his planning assistant. One technique that is finding increasing favor, is the development of the first gross plan on a black board in a conference room. At least 12 feet of 3 foot high blackboard is recommended. A group of key people can participate in the logic development. When the diagram reaches the end of the black board, those last events are given alphabetical designators (A, B, C, etc.), to establish a match line.

    Then a picture is taken by a Land Polaroid camera. Once that developed section of the diagram is faithfully recorded, the board is erased, and the logic continued. The photographs can be given to a draftsman, for the initial tracing of the project plan.

3. The Refined Section Networks

Since the specialist groups or subcontractors, are responsible for the execution of their particular phases, and the manager cannot be expected to know all the technical aspects of the specialties, each department, section or division head is to create his detailed section (called an inter-face drawing) of the overall plan.

Again, this can be performed by people down to the foreman level. The network planning technique is a tool for the professional, experienced man. The planning can be delegated to an assistant, if he is sufficiently experienced in the work. The superior always has the final perogative of review and acceptance.

This step also involves the collection of activity duration estimates. One objective usually voiced is that it is difficult, even impossible to obtain the necessary information from subcontractors or suppliers. It will be found that a direct discussion with the people involved, with a minimum of orientation with those who are ignorant of the technique, will produce adequate information for the first schedule generation. Conservative estimates will show up on the Critical Path.

4. The Master Plan

After a co-ordination meeting, where the interconnections of the sectional nets are resolved and agreed upon, they are all meshed into the completely detailed Master Plan. This is the overall project working plan, from which information is to be generated. However, it is not consonant with the "management by exception" concept to expect the manager to work with such a large, unwieldy drawing. On large projects, the next step is recommended.

5. The Summary Drawing

From the detailed Master Plan, a Summary drawing is prepared for the Project manager's use in controlling his project. It should not exceed three hundred arrows, have the critical jobs identified, and the slack paths grossed, usually showing the long duration activities that may be in the domain of a particular department. For example, the engineering department may be required to generate some fifteen specifications, all of which are parallel activities; but only one or two of these, such as pre-amplifier and power supply specifications are the longest duration. There could be one or two arrows identified as "Eng. Dept. - Pre-Amp" and "Eng. Dept. - Pow. Supply" on the summary drawing. The Summary Drawing is consonant with the reporting system, in that the reports received by the manager have been abstracted by lower level personnel until they match his Summary Drawing. Figure 15.0 indicates the evolution of these drawings.

The 5 steps listed above are obviously general, slanted towards the big project in a big company. In a small firm, all of the steps would be combined by the one planner, or outside consultant, to obtain the initial plan. The whole concept is to generate information for evaluation as to feasibility in time, costs and resources.

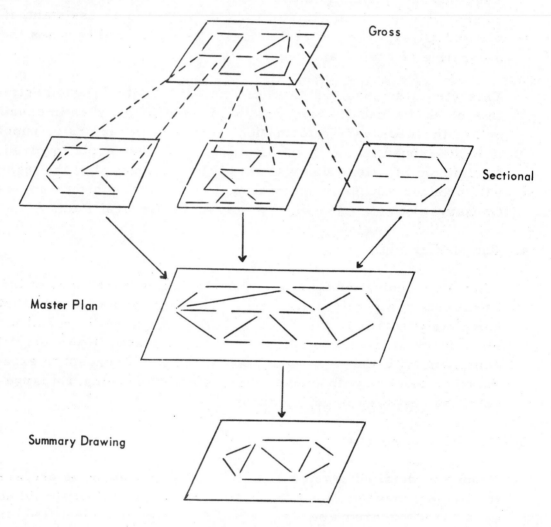

The Evolution of Network Drawings

Fig. 15.0

CHAPTER 16

COMPUTER ORIENTATION

There is no question that in a very short period of time after inaugurating Network Planning, the user will find himself in need of electronic data processing machines - the computer and its peripheral equipment. This is <u>not to be feared</u>! The use of computers is widespread, and anyone can be trained to use the necessary equipment (key punch, sorter, and off-line printers) in a very short period of time. As an indication of simplicity, this technique of Network Planning, and the use of digital computers, is now being taught to junior college students in quite a few colleges throughout the country. The computer is to be regarded as a mechanical abacus and sorter of information; a tool to perform the "Forward" and "Backward" Passes and calculate the schedule, in order to relieve a human from the tedium and possibility of error in all the event summations. The computer only needs three numbers for it to perform the Network calculations - the "i", the "j", and the duration of that "i-j" activity. Once the master plan has been drawn, reviewed and agreed upon. the next step is computation. For those projects that are of such a magnitude (100 activities or more) that a computer is required, the next step is the filling out of data sheets that eventually will enter the information into the machine. This is the weakest link of the technique—human errors occur here more than in any other phase. The machine acts only on numbers; and it has been instructed (programmed) to recognize certain anomalies, such as an open link in a path. But as long as the computer sees a closed network, it acts on the numbers inserted into it. Any errors in the numbers inserted are carried right through the program, and the schedules produced will reflect them.

Each specific computer program has specific locations for the input data. Since the majority of computers have some form of a card input, this shall be the basis of description for a computer input. Fig. 16.0 is a typical 80 column IBM card.

Figure 16.0 has been key punched and shows the field locations (column grouping) for the input information for a particular program. In this program, the "i" numbers are inserted in column 1 to 4. Fields are right justified; i.e. the units are in the far right column, tens in the next column to the left, hundreds in the next column to the left, etc. Thus an "i" number of 322 would have a 3 digit in column #2, a 2 digit in column number 3, and a 2 digit in column #4. The "j" field is columns 5 to 8; "duration" field columns 9 to 12; "normal cost" field, columns 13 to 17; and Description field, columns 18 to 50. The description field normally will take only alphabetical and numerical characters. It must be remembered that each program (computer instructions) will

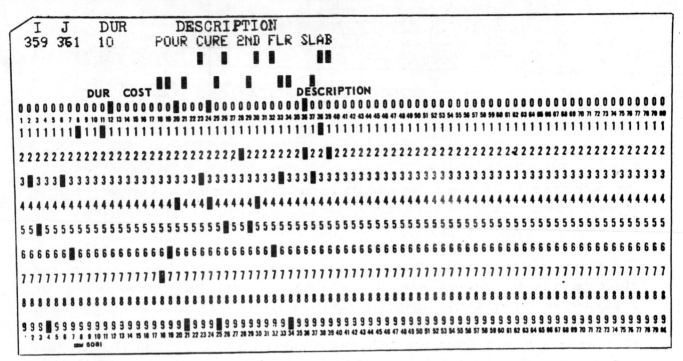

Fig. 16.0

have its own field locations. The instructions received with the program deck will detail the specific locations on the input card for the input data.

The cards are key punched from a data sheet. A data sheet is nothing more than a row - column sheet, with the 80 total columns shown. Each row is an entry for a specific activity arrow on the project network diagram. Fig. 16.1 indicates a typical data sheet format for a key punch operator.

Note that the information must be placed column by column by whomever fills out the data sheet. Key punching will be an exact one to one transfer of information, column by column.

The system described herein is basic. Some of the computer programs will have additional input. On just the time scheduling alone, some of the larger computers will also take a start date for a calendar dated output, transaction codes for updating information (a specific column location will receive a C for a completed activity, a D for Deleted, N for a new activity, etc.) This text will discuss the fundamentals of updating in Chapter 18, and show how this is done manually. The automatic computer updating will then be understood more readily.

In the basic system described above, with just an "i", "j", "duration" and "description" information as input on a simple program, expansion of this to include additional information that may be helpful to the line manager is

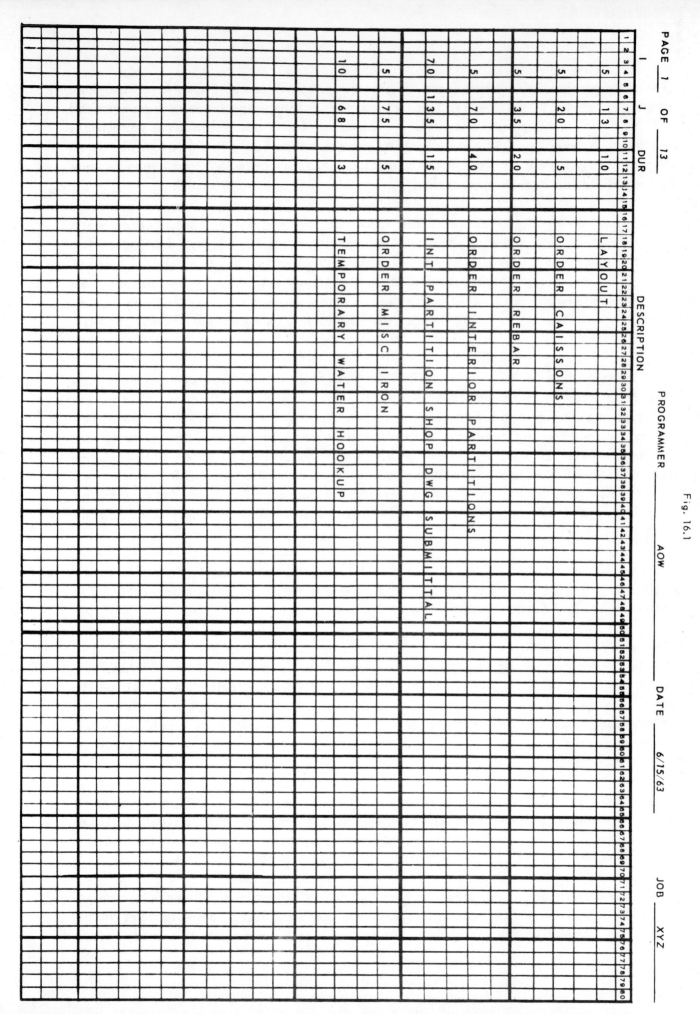

Fig. 16.1

easily accomplished without expensive re-programming costs or time. For example, part of the description field in the basic program may be reserved for cost codes, responsibility codes (sub-contractor code numbers, suppliers, sections, areas. etc. all can be given a code number). In the example of an input in Figs. 16.0 and 16.1, columns 18 to 21, or cols. 47 to 50 may be reserved for such a coding system.

Note also from Fig. 16.1 that the take off of information from the network plan on to the data sheet need not be in any sequence. Either the computer, or a separate machine called a Sorter, may be used to sort the input information. It is strongly recommended that the input data always be sorted in an "I" major sequence, so that when information on a particular activity is desired, a minimum of searching time through a sequenced list will result. This applies to a random Event numbered Network also.

The information from the data sheet is punched on business cards. This Key Punching step is then verified on a machine called a Verifier, which repeats the typographic method of punching cards and ascertains that there has been no error in the original key punched card. A card is punched for each arrow on the network. In addition, there are "header" cards that enter the date, the description of the project, and any other two lines of editing information desired. For some programs that require a sequential "i-j" entry into the computer, the cards go through a Sorter, which delivers the pack of cards with the proper "i-j" sort.

This phase is called "off-line" and has the lowest rental cost. The data deck, consisting of all activity cards and the header cards, can be generated for approximately 5¢ a card, at a rate of 250 per hour.

There are essentially two models of a key punch machine, one which also prints the key punch information at the top of the card while it is punching the holes in the particular column (this is called "interpreting" a card), while other model doesn't print, just key punches. The advantage to a printed card is obvious, it allows quick searching for any error sources. A machine called an Interpreter will take a pack of unprinted, punched cards, and reproduce a printed pack of key punched cards.

In addition to the data deck, there is a program deck of cards and a pinboard for the computer. These are the internal instructions for the computer. A set of instructions, for switch positions, and sequence of switch action on the computer console, is also furnished.

After the insertion of the program deck of cards, some computers will accept the data deck and immediately print out the schedules. This is part of the "on-line" phase of the computer. Others will deliver a pack of cards, called the "output deck". These must be separately inserted into the off-line printer. There are many advantages to an output deck. They can be sorted or collated, and print outs obtained with the information in any sequence or distribution desired, such as critical jobs by department, jobs by criticality and early activity start time, by total float, etc. The larger computers, the magnetic type kind, have built in sort routines in their program, so that the information will come out, and then duplicated in the various sorts as mentioned above. One disadvantage of this on a small project (small here means under 2000 arrows. This "small" project could represent a $20 million dollar project, but the number of arrows may only total 1500 or so.) is that quite often you receive all this redundant information whether you want it or not. You will be paying for it at the higher rate. The large machines run up to $700 per hour rental rate, but can rapidly perform the Critical Path or PERT schedule.

Another strong recommendation: have the network plan with you at the initial run. Every Network program has built in error analyses; the computer will tell you if an arrow has been left out of the input (it must see a completely closed network) and identify the open event numbers; it will spot loops in the logic; and it will tell you if you have exceeded its capacity (too much information). The formulae for determining capacity are described below. When you receive error stops, you must either stop the computer run, find and correct the errors, and then arrange to get back on the computer. Sometimes it is urgent to obtain the schedule, and the error correction is made while retaining, and paying for the computer time. Thus, the presence of the network plan will benefit the search for errors.

Each time an up-dated schedule is required, a revised data deck must be inserted, and a complete schedule is obtained. The revision will consist of the replacement of completed activity cards by an "update" card (see Chapter 18). The larger machines have the "updating" routine already in the program, and the executed activity cards contain "transaction" code symbols to inform the computer.

Tape Input

First, input cards are key punched and verified, as above. Then a card-to-tape converter is employed to store the input information on magnetic tape for insertion into the computer. The instructions for the tape input computer are contained in a program magnetic tape, and sometimes there is an additional tape called a systems tape. These both contain instructions for the computations and several methods of sorting output information. An output tape is created, which is then placed on either an off-line printer or a smaller computer that acts as an off-line printer. Here, the information comes out with fixed sorts. The tape programs, however, offer the advantage of accepting only change information, not a complete repeat of the problem deck.

As the project progresses, the reports get smaller. Once an activity is completed, it is not reported on again.

The tape input machine is generally a higher capacity machine, for large network diagram input information. There are "canned" programs available from computer manufacturers at little or no cost. However, a difficulty may be encountered, in that you must accept the output information in that programs format, and you must enter information in accordance with those particular machine instructions. You may find yourself bending and fitting your systems and procedures to suit the machine—this is contrary to the logical condition of having the machine work for you.

<u>Computer Capacity</u>

There are two general rules of thumb for ascertaining the size (the memory capacity in terms of storage "bits") of a computer necessary to handle the particular project plan.

1. In general, there are 1.5 times as many arrows, including dummies, as there are events in the network plan. Thus, if a sequential, non-consecutive numbering system in steps of 5 was used, and the final or objective event (O.E.) has a "j" of 950, then it can be safely assumed that there are approximately 300 arrows $\left[\dfrac{950 \times 1.5}{5}\right]$

2. Each computer with a specific memory capacity, has a limiting number of arrows it can handle. The general computer capacity formula is

$$J_{O.E.} + n = k$$

or the j number of the final or objective event, plus the number of arrows, equals some constant. Note that the $j_{O.E.}$ is the last assigned "j" number, not the actual number of events. The computer program will also have a numerical limit as to the digit numbers in the objective event j number (999, 1999, 5999, etc.).

As an example, the capacity constant (K) of the IBM model 1620 computer, with a 20k (20,000 bits) memory capacity is 1614, a G E 225 is 3099.

Table 16.0 below lists some of the currently available computer programs. These are free; by contacting the manufacturer, or the local salesman, they are readily available. Instructions will be included with the copy of the program deck. The column labeled "Equipment" is primarily an indication of the memory capacity of the machine. In the parlance of computers, a 20 K capacity means 20,000 "bits", or storage bins for a digit.

SOME TYPICAL AVAILABLE COMPUTER PROGRAMS

Computer	Equipment	Technique	Capacity	Event Numbering	Remarks
Autonetics-Recomp II		PERT	704 Act.	Random Max. 99,999	Final activity must be Dummy. Max. 192 arrows if calendar dated. No Probability.
Bendix G-15		CPM	108 events or 400 activ.	Random, but final j's all i's	No Activ. Desc.
Bendix G-20	LP-10 Crd. Rdr. 1MT-10 8K, cc-10	PERT (Activ. Orntd.)	200 Activ. (w/core) 1800 Activ. (w/1 tape) 2500 Activ. (w/2 tapes)	Random	Time-1000 Activ. 8 min. Multiple start & finish events. 37 col. desc. Probability.
Burroughs-205		CPM			
Burroughs-220		PERT	500 Activ. or 300 events	Random	Can handle up to 10 interface networks. Probability. Resource forecasting. Multi-initial & terminal events.
Burroughs-220	10K	CPM	400 Activ.	I < J	
Control Data-1604		PERT			
GE-225	8K words 4 tapes	CPM & Cost expediting	1000 events or 2100 Activ.	Random Max. 999	By assigning arbitrary weight factor to activity will allocate float. 52 col. description
H-400 Minneapolis-Honeywell	2.1 K wds. 4 tapes	PERT event or Activ. oriented	20,000 Activ.		Multiple start-end events. Network summarization.

SOME TYPICAL AVAILABLE COMPUTER PROGRAMS

Computer	Equipment	Technique	Capacity	Event Numbering	Remarks
IBM-360		CPM Precedence	10,000 Events	Random	Has Cost Control, Multiple Nodes
IBM-1130	8K Disc	CPM Precedence	999 Max J 1400 Act.	Random	Manpower Distrib. Multiple Start, ends, bar chart
IBM-650 (Bendix)	RAMAC tape drive	PERT	999 events	$I < J$	Probability. 20 col. description Multiple begin & end events.
IBM-704	8K 16K 32K	PERT	99,999 events	$I < J$ Max. 99,999	Probability. Updating. Needs 2 cards for activity desc. 4 report sorts
IBM-704	32K w/tapes	CPM & Cost expediting	2400 events or 3500 Activ.		
IBM-1401 (Bendix)	4K w/1405 RAMAC	PERT	1000 events or 15,000 Act.	Random	Allows 200 initial & 100 terminal events. Input format as Lockheed 7090. Probability.
IBM-1401 (IBM-Dallas)	16K w/4 tapes, Multi-Divide, Hi-Lo-Eq. compare. Advanced Programming.	CPM & PERT Cost	999 events	Random Max. 999	Time & cost updating. Tab &/or Bar output. 16 output sorts. Scheduled dating.
IBM-1401 (IBM-St. Louis)	4K Adv. Prog 8K Modify Addr. 12K Hi-Lo-Eq. Comp. 16K 3 tapes Sense Sws.	CPM	8 x Max. Ev. No. 10 x Activ. 3552, 7552, 11,552, 15,552	Random	Std. less format Card and/or on line Printout
IBM-1401 (IBM)	4K	CPM	575 events	$I < J$ Consec.	25 col. description

SOME TYPICAL AVAILABLE COMPUTER PROGRAMS

Computer	Equipment	Technique	Capacity	Event Numbering	Remarks
IBM-1401 (IBM-Seattle)	4K	CPM	99,999 events	$I < J$ $J-1 \leq 249$	25 col. desc.
IBM-1440	8K two-1311 Disk Storage Drums #1442 Card Rd. Punch #1443 Ptr.	CPM or Precedence Network	2,000 Activ.	Random Max. $j = 1,999$	Output includes status reports, Bar Chart generation, Cost Sum. Scheduled Event Times may be inserted.
IBM-1620 (GE)	20K	PERT	999 events	Random	Multiple terminal events.
IBM-1620 (GE)	Auxiliary	PERT	500 events or 750 Activ.	Random Max. 750	Two output sort options
IBM 1620 (IBM) Less	60K 40K	CPM & Resource Allocation	1400 Acts. 740 Acts.	$I < J$ Max. 999	Can limit resources No activity desc. 10 resources, 10/ Activity. Uses standard less Time output as input.
IBM-1620 (IBM-Beaumont Texas)	60K 40K	Resource & Allocation	600 Activ. w/2 crfts. 240 w/12 230 w/2 90 w/12	$I < J$	Written in Fortran Also available on IBM 7070 & 7090
IBM-1620 (IBM) Miss-Less	20K 40K 60K	CPM	Max. Ev. No. + No Act. 1654 3654 5654	Random Max. 999	Includes cost sum. 33 col. description Loop finding routine
IBM-1620 (IBM)	20K 40K up	PERT	695 Activ. 999 Activ.	Random	Updating routine Multiple begin & end events Req's. auto-divide for probability 31 col. description

SOME TYPICAL AVAILABLE COMPUTER PROGRAMS

Computer	Equipment	Technique	Capacity	Event Numbering	Remarks
IBM-7070 (Collins Radio)	w/1401	PERT	999 events	Random	Event No. sort output only
IBM-7070	5K	CPM	1350 Activ. or 700 events	Random Max. J 999 Do not use "0"	Time-600 activ., 2 min. Last event No. must be 999 Cost expediting.
IBM-7070	1401-8K 2 tapes 7070-10K 7 tapes	Act. oriented PERT	1500 activ.	I < J consecutive 5 digit max.	One end event. 46 col. description 4 report sorts Probability
IBM-7070/ 7074 (by IBM) Taxis	10K 7501 Crd. Rdr. 8 tapes	PERT & CPM	2000 Activ. or 1000 events	Random 10 char.	25 col. description Has updating routine. 5 output sorts, 1 terminal event; scheduled dating. Time-- 500 act. 3 min. including 2 reports Probability.
IBM-709/ 7090	714 & 720 32768 words 6 tapes	PERT	5120 Activ.	Random 7 digit	No loop detection Probability, scheduled dates 4 output sorts
IBM-7090	32K 12 tapes	PERT & PERT cost	20,000 Activ.	Random	Schedule Dating Updating Routines Probability Will handle 99 networks
IBM-NORC (Navy-Dahlgren)		PERT	Unlimited	Random	Updating routines
LPG-30 (Royal Precision)	4.1 K words	PERT	511 events or 2048 activ.	Random Max. 511	Probability. Multiple output sorts
NCR-304 & 315 (National Cash)	6 tapes or 2 CRAM units	PERT Activity Oriented	5000 Activ.	Random	Updating routines. 30 col. description Probability.

SOME TYPICAL AVAILABLE COMPUTER PROGRAMS

Computer	Equipment	Technique	Capacity	Event Numbering	Remarks
Philco-2000 (Philco)		PERT	750 Activ.	Random	Names "WCC PERT" 7 output sorts
RPC-4000 (General Precision)		PERT Activity Oriented	1023 events or 2048 Activ.	Random Max. 99,999	Computes ES, EF, TF. Schedule dating. Updating Routine.
Rem-Rand S.S. 80 (ORI)		PERT II	Unlimited	Random	Scheduled dates. Output sort options.
Rem-Rand S.S. 80/90 (Rem-Rand)		Resource Allocation	3000 events/ project, unlim. projects.	I < J Max. 2,999	Uses CPM output as input. Inter or intra project scheduling, levelling up to 30 resources. Each project max. 200 days duration, splits.
Rem-Rand, UNIVAC I (Rem-Rand)		CPM & Cost expediting	400 activ.	I < J	Activities, not crews
Rem-Rand, UNIVAC I, II 1103A & 1105 (Rem-Rand)		CPM & Cost expediting	1. 11-239 events 03A- 512 events 05- 1000 events	I < J Consec.	
Rem-Rand, UNIVAC II (Rem-Rand)		PERT	1399 events/ network Multi-networks	Random	Multi-output sorts Probability
RCA 501		PERT & Resource Forecasting	2000 Activ. or 1000 events		9 resources, 9/ activity Output in graph form. Gives Hi, Low & Avg. points for rates of expenditure

It is advisable to check with the local representative of the particular computer company, to determine availability and latest revisions to their program.

General Recommendations for the Initial Computer Run

1. Obtain a low rental source of computer time. The best source of this information are the computer salesmen in your area. Contact the various manufacturers, and inquire as to whether any of their users will rent time to an outside firm. The salesmen generally are happy to find additional customers for their users. It is surprising the amount of overselling of data processing and computation equipment that has occurred. There are many firms that are almost desparately trying to justify, even amortize a computer installation. Once the source is located, the same salesman will obtain the "canned" program for the Critical Path or Pert schedule desired.

2. Have the input data cards "interpreted" (printed as well as punched) if possible.

3. Have the input data cards sorted in an "I" major (I in ascending numerical order) sequence.

4. Have a print of the network plan at the computer site when making the run.

5. Check the first output schedule, by checking the durations of the Critical Activities on the print out, to be sure that they match the duration on the plan. A carelessly written zero on the input data sheet may look like a 6 to a key punch operator. Remember, the computer only sees "i", "j" and durations, and it cannot tell the difference between a dummy and an activity.

PROBLEM

16.1 A network, numbered sequentially in steps of three's (3) has an objective event "j" number of 936. Does the IBM 1620 20K computer have sufficient capacity to handle the arrow diagram?

CHAPTER 17

CONTROL OF SIMULTANEOUS PROJECTS

Up to this point, we have been discussing intra-project scheduling, dealing with one project only. The aspects of the techniques are applicable to the situation where an organization has more than one project in the house at the same point in time. There are two specific cases that will be considered in inter-project scheduling:

Case 1

This covers the situation where there are interrelations of equipment or activities that affect some or all of the projects. As an example, in the construction of four identical dormitories, each building can be considered a separate project in itself. However, the contractor has elected to sub-contract a certain number of concrete forms, thus there is a restraint between the four projects due to availability of the forms. A form of project priority has been set, in that the sequence of building erection has been determined by the topology of the site, and by the presence of buildings on one half of the site that have to be demolished before grading and building site preparation can start. In this type of case, the individual networks are tied together by the dummy restraints; and the four separate networks become one large overall network, as indicated in Figure 17.0. Once the sequence of projects has been set, the Master Network configuration is easily established. The Master Project Plan is then

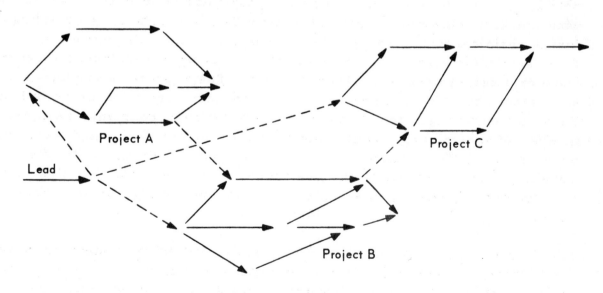

Combining Separate Networks Into
A Master Network Plan

Fig. 17.0

handled as before, as one project. This is the most common practice today. As contracts are signed, that project's arrow diagram is tied into the company's Master Plan. This is a dynamic method, the Master Plan ever changing as project's are completed, and added to the master diagram.

Another approach in multi-project scheduling is the conversion of the output schedule information to a time scale format, as discussed in Chapter 11. Two projects, "A" and "B", of Fig. 17.1 have an overlapping schedule—Project "B" is received 12 time units after the start of Project "A". They are laid out on a time scale Master Plan, Fig. 17.2 with the Early Start and Late Finish times shown as bracketed time boundaries. It will be remembered that the time scale does have limitations: (1) The schedule information comes after the Forward and Backward Pass, and (2) while an excellent "first analysis" tool, may become onerous in the continual control of one or more of the projects, since the time scale could have to be redrawn to accommodate consequential effects of changes. There are mechanical devices on the market, utilizing peg boards and strings, or metal bars on an aluminum extruded frame, that can serve as a Master Schedule Board.

The time scale board does serve as an excellent source of quick managerial review, for allocation of resources, and of visual status of any project in the house.

Case II

This is the situation where the organization has several totally unrelated projects at various phases of progress at any particular point in time. An example would be an Engineering-Construction firm that is designing and building different chemical and power plants in several countries throughout the world; or an electronic systems manufacturer that is designing and building separate and unrelated systems for several clients. These projects are unrelated to each other, but the organization has a fixed number of resources to handle these projects; i.e., the Engineering Construction company has a certain number of piping designers that must perform all the design and layout work on each plant. This situation is handled in a manner similar to the Resource Levelling and Allocation Method described in Chapter 11. It is first necessary to create three priorities: (1) A Project Priority, (2) Activity priorities on each project and (3) A time priority.

The Project priority is established by the organization on whatever of several bases it chooses, or a combination thereof. It may be on the magnitude of the capital cost, the date of the project contract, or the weight given to old and repetitive clients, or to new ones. The nature of any project may create its own priority, such as a vital project involving national defense.

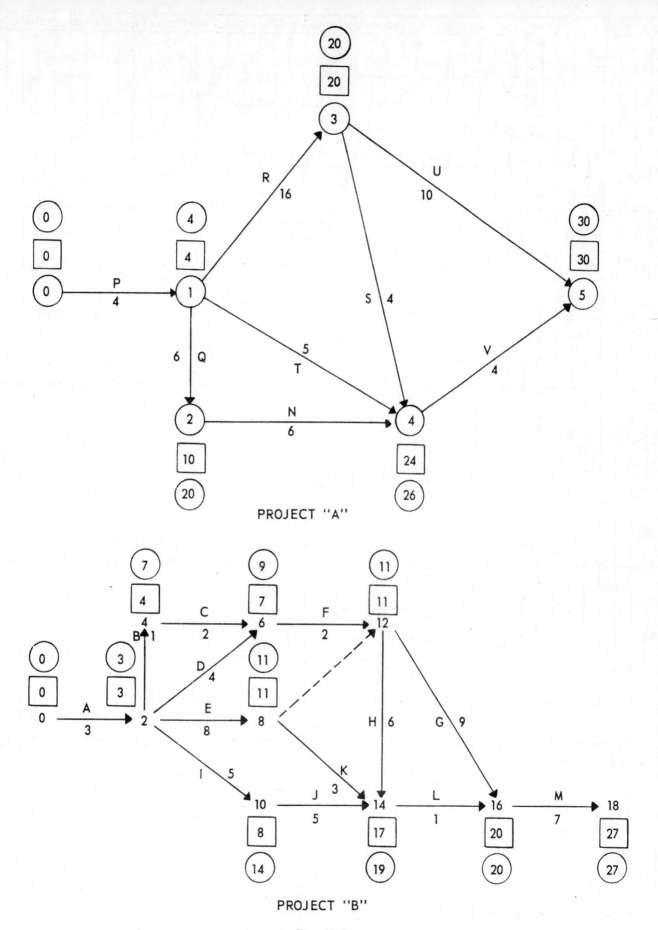

PROJECT "A"

PROJECT "B"

Fig. 17.1

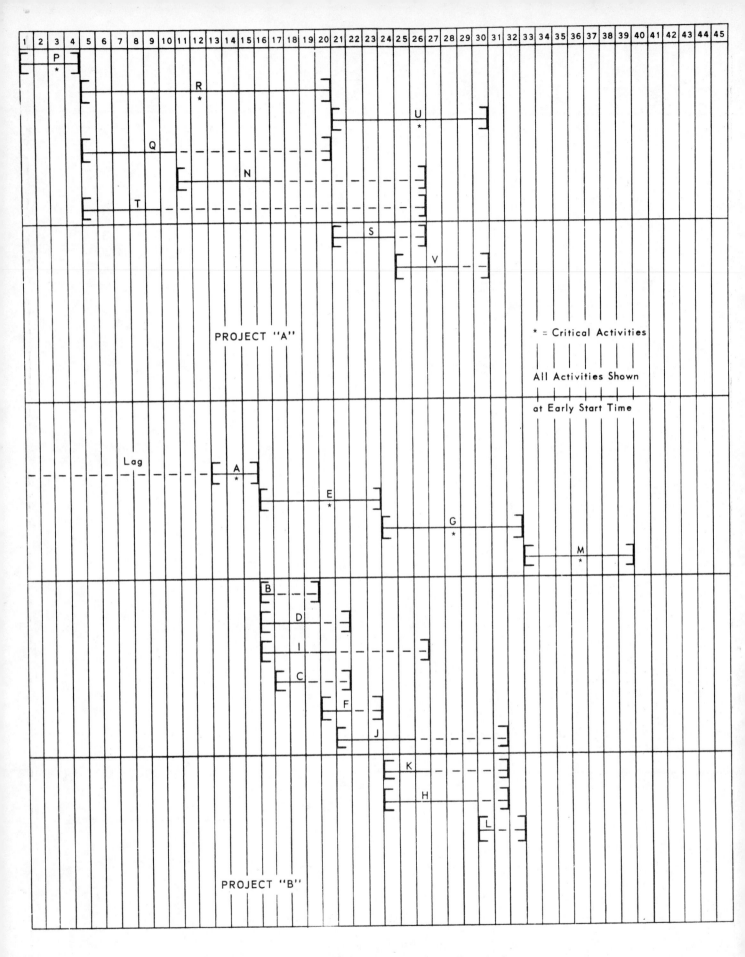

Fig. 17.2   Multi-Project Schedule on Time Scale

240

Activity priorities: Each operation may be assigned a weighed factor, a judgment value representing the importance of an activity, or the uncontrol-ability of an activity. This is the Distributed Float concept described in Chapter 12. The amount of Distributed Float can be used directly as an activity priority number, or another classification based on Distributed Float may be used. One such classification is that 0 priority is given all critical jobs; #1 priority to jobs of Distributed Float that is 10% or less than the normal duration of the project; #2 prority to Distributed Float from 11 to 29% of the Project Duration, and so on.

The time priority most used is either the Normal Early Activity Start Time, or a Distributed Early Activity Start Time (based on the allocation of Distrib-uted Float to each activity) of each particular activity. Then a simple analogy of having all the arrows on all the projects fall off the network, and lay at random on the floor, gives an idea of how this method works. The activities are then sorted in order of their (1) time priority, (2) activity priority, and (3) project priority. They are then queued up in ascending order of this pri-ority number, and each day they may be scheduled against the specific re-sources available. Each day there is a reshuffling, so that as operations are delayed in their starting date, their priority changes; and all activities must be requeued, highest priority first. Since this is strictly a numbers informa-tion technique, some practical considerations must be utilized by the Project Manager to effectively control his project, as well as the Squad Leader or Shop Superintendent who has to execute the diverse activities within the limits of his resources. Thus, the output may say that the Piping Design group should stop work on the layouts on Project #8 and start work on the Preliminary Layouts of Project #12. But, if the layouts of Project #8 are 80% complete, the Squad Leader should have the authority to decide to complete Project #8 and further delay Project #12 a few days. The information derived from Net-work techniques is not always favorable, but it does present the Manager with sufficient information to make efficient managerial decisions, such as author-izing overtime, or hiring additional temporary help.

As an example of this "queueing" method, we shall return to the network plans of Fig. 17.1. The Project Priority will be assigned in sequence of re-ceipt of the project contract, thus Project "A" will be Project Priority #1, "B" Project Priority #2. In order to establish Activity Priorities, the Dis-tributed Float will be used. Distributed Float will be calculated by the method of Chapter #12, using the activity weights listed below in Table 17.0. The Distributed Event Times have been rounded off, and are shown in Table 17.1.

The Distributed Event Times are shown independently for each plan; the 12 time unit lag for Project B is not shown. We are interested in establishing the Distributed Float for each activity, which will serve as the Activity Pri-ority Number. The third priority number, the Time Priority, will be the actual

ACTIVITY WEIGHTS

Project A Activity	Weight	Project B Activity	Weight
P	3	A	8
Q	2	B	2
R	1	C	2
S	3	D	10
T	4	E	12
U	2	F	6
V	1	G	12
N	3	H	18
		I	10
		J	14
		K	12
		L	0
		M	8

Table 17.0

DISTRIBUTED EVENT TIMES

Project A Event No.	T_D	Project B Event No.	T_D
0	0	0	0
1	4	2	3
2	13	4	5
3	20	6	8
4	26	8	11
5	30	10	11
		12	11
		14	18
		16	20
		18	27

Table 17.1

Early Start Time for each activity. This priority will reflect the actual starting times, as shown on Fig. 17.2, accounting for the lag of 12 time units in Project B.

From the above information (the reader may calculate the Distributed Float for each activity as a review of the material in Chapter 12), we are now ready to establish the sequence, or "queue" of the activities in the two projects. Note the ascending numerical order priority of all the activities in Table 17.2.

Once the sequence of priority numbers is established, then the matrix of available resources per project time unit, versus this sequenced list, is established, and scheduled exactly as the problem in Chapter 11, Fig. 11.14 and Fig. 11.15 for the three craft problem, by entering remainders in the time unit columns for subsequent assignment. The only additional rule to be observed is the effect of delaying an activity on its priority number. As the float is consumed by delay, the priority number reduces, and the delayed activity may be re-queued ahead of previous sequenced activities.

MULTI-PROJECT ACTIVITY SEQUENCE

Activity	Time Priority (Early Start)	Activity Priority (Distributed Float)	Project Priority
P	00	00	01
R	04	00	01
Q	04	03	01
T	04	17	01
N	10	07	01
A	12	00	02
E	15	00	02
B	16	01	02
D	16	01	02
C	18	01	02
I	18	03	02
U	20	00	01
F	20	01	02
S	20	02	01
G	23	00	02
H	24	01	02
V	24	02	01
J	25	02	02
K	27	04	02
L	30	01	02
M	32	00	02

Table 17.2

PROBLEM

17.1 The contractor who is building the New House of Problem 12.1 on page 109, and the Car Wash Station of Problem 12.3, page 110 (which starts 12 working days after the start of the house) decides to set up the Simultaneous Project Scheduling procedure of this chapter. His craft assignments for the House are the same as those of Problem 11.3, page 93. He is concerned about the availability of carpenters (a maximum of 4) and plumbers (a maximum of 3). On the Car Wash Project, the following activities require these crafts:

Activity	Carpenters	Plumbers
11. Form and rebar Wash Pit Walls, East column footings	2	
12. Form office-locker room section slab, West column footings	2	
15. Mechanical Rough In (Underground)		3
25. Install windows and doors	2	
26. Install plumbing in Wash Pit		2
27. Install Hot Water Tank, Spray Headers, plumbing fixtures		3
33. Install millwork, office and locker rm.	3	
34. Final plumbing connections, test		2
40. Install interior partitions	2	

Set this up as a two craft problem. Since the capital cost of the Car Wash Station is higher than that of the House, the Car Wash Station is given Project Priority #1, the House Project Priority #2. Establish the schedule, based on the above limits of 4 carpenters available, 3 plumbers. The threshold for each craft is one man, and all activities may be split. (Refer to problem 11.3.b, page 93).

17.2 What would the schedule be for Problem 17.1 above, if the Project Priorities were reversed.

CHAPTER 18

PROJECT REPORTING AND UPDATING

General

Naturally, the essence of a "good" management control system is the timely, accurate and explicit flow of information. This is to be done as simply and economically as possible, in order to have the system work for the manager; not the manager working for the system. The philosophy herein is to minimize the system requirements--simplicity is the key.

This section will necessarily be of a general nature. Each organization will evolve its own system and procedures; starting with a transition from past practices, and gradually taking advantage of the network technique to arrive at the evolved optimum system of information flow, then analysis: it is certain that no organization will abruptly discard present systems and procedures. Whatever existing method of collecting time against a project; comparing costs against a cost code; and projecting time and cost trends for that project will become the basis of evolution. One other fact is certain; the existing system will not be fully compatible with the network method of planning and scheduling. Cost Codes will either be too gross, or too fine, to suit the specific activity arrows on the first networks used by an organization. Thus, a transition will have to be made by that company to suit their initial networks.

The Flow of Information

It will be assumed that a periodic time sheet will be signed by all personnel working, assigning consumed time to a project. For a contractor, it will be a Foreman's Daily Report. Here this information is entered; the name, badge number, classification of the men under the foreman's jurisdiction. Their time, in hourly units, is assigned to specifically described activities, and verified by the foreman.

An obvious comparable situation in a manufacturing operation is a section head's collection of time by his sub-ordinates on his operations of a project. The collected time sheets are then sent to a higher echelon manager; in the construction industry to the superintendent. Each noted activity to which time has been charged must be identified with a suitable "i - j" reference number. This may be done by the foreman, or a clerk in the superintendent's office. One way or another, charged time must be accumulated against the activities on the arrow diagram of the project. The next step in the flow of information is the accumulation of time charges for a given period of time, and the forwarding of the information to either a home office from the field, or to the next higher echelon of management. In industry this next echelon may be represented by a staff group whose function is the gathering of data, preparing of requisite input information, and the issuance of an updated schedule and an analytical report. This group is a service group, serving to coordinate the information flow upward.

Activity Action Priorities

In order to maintain the efficiency of the network planning method, a range of "action" priorities should be assigned to all of the operations or activities on the project. The highest priority items deserve management attention, with delegation of surveillance responsibility down the lower echelons of supervision.

It was noted in Chapter 6 that "Near Critical Activities", those whose Float Time is very near the Critical Path, deserve as much attention as the Critical Activities. On a two year project, those activities with an initial Total Float of two weeks (10 days) or less can easily lose that small amount of Float, and rapidly become Critical at any time during the life of the project.

A basic priority system structured on the Total Float of an activity is as follows:

Action Priority #1: 0 Total Float - top management attention.

Action Priority #2: Total Float equal to 10% or less of the project duration in time units - middle management attention.

Action Priority #3: Total Float between 11% and 40% of the project duration in time units.

Action Priority #4: Total Float between 41% and 60% of the project duration in time units.

Action Priority #5: Total Float above 61% of the project duration in time units.

The Reporting Period

Almost any organization has a periodic reporting period; weekly, bi-monthly, quarterly, etc. This has been derived from traditional practice. While it is almost certain that this periodicity will not be changed, it may be pertinent to inject a new thought here. Why is a periodic report necessary? The most common answer is: "Because that's the way management wants it." But an organization using network techniques of Critical Path, will find that an examination of the initial schedule will reveal not an even distribution of activity finish times over the entire project schedule. Dependent on the configuration of the network, and the original duration estimates, there will be an uneven distribution of activity finish times. It might be more logical to arrange reporting periods by the schedule distribution of Early Activity Finish Times-- the first report may not be required for 6 weeks; the second two weeks later;

the third four weeks later; the fourth two weeks later, etc. It goes without saying, this concept of aperiodic (irregular) reporting periods based on the distribution of Early Activity Finish Times will be difficult to inaugurate, but the concept is in keeping with the basis of efficiency of the network technique.

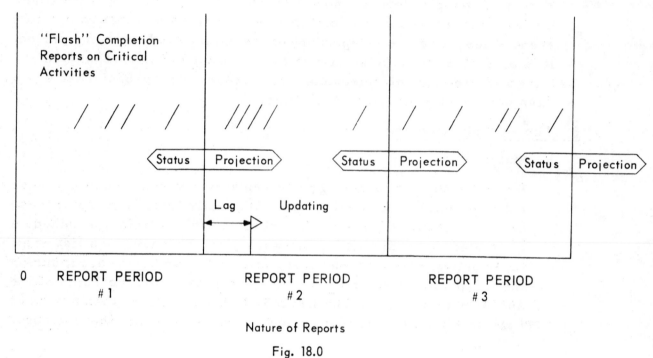

Nature of Reports

Fig. 18.0

The schematic of Fig 18.0 encompasses a complete spectrum of reports. Not all phases of this report system may be put into immediate effect. Some aspects are more important to one particular type of industry than to another. The nature and size of the project will also have a bearing on the number of aspects put into the Project Reporting and Control System.

1. "Flash" Reports

This is a simple report, indicating the completion of a critical activity by an operating group. These are issued when that completion event occurs, and are independent of the established reporting period. It is a "finger on the pulse" technique. This type of report may be obtained informally—by telephone call. Requests may be issued a few days before the scheduled finish time.

2. The Periodic Report

a. Status

Here, the field or operating groups submit upward to the next higher echelon, status information on certain activities. In general, these certain activities have been pre-established from the schedule generated from the project network plan. This will be a list of all those activities whose Earliest Activity Finish Times occur during the reporting period. The list is originated in the home office and sent to the field. The list may be nothing more than a pack of IBM machine cards, duplicated by machine from the computer output.

The field, or operating group, examines this list and submits the following status information.

1. Activities completed in this reporting period.

    a. The total number of direct man-hours, by craft category, assigned to the completed activity.

    b. Date completed.

2. Activities in progress in this reporting period.

    a. Total number of direct man-hours, by craft category, charged to the partially completed activity as of report date.

    b. The estimated remaining working time units (days, weeks, hours, etc.—the consistant time unit used in the estimates of the project network plan) to complete as of report date.

    c. The estimated crew size, or craft man-hours, required to complete.

3. Activities that could have started in the period but were delayed

   a. Any pertinent reasons for the delay.

4. Blank Information (DELIVERY VERIFICATION)

   If no report is received, it can only be assumed that no action was taken. For a physical activity involving human labor, the initial estimate is still assumed valid, and still remaining. For delivery dates--no report, no change from the original date. The most difficult aspect of updating by collecting STATUS information is the neglect of equipment and material delivery dates. After the original schedule is obtained, the human tendency is to assume that the original delivery date is static, or will be maintained. Quite often this is not the case, resulting in false updated schedules. DELIVERY DATES should be checked again at each Update Status Collection period. It is suggested that Delivery activities be separately coded (use a few columns of the Description Field, if no Code Field is standard with the computer program being used), and sorted and listed for each report, to particularly flag these items Most extensions of schedules are caused by slippage of Delivery Dates, and most initial delivery dates are assumed valid without subsequent checking. Thus, a dangerous schedule trend of slippage is not detected in advance, defeating the purpose of this technique.

5. Re-evaluation

   It is perfectly valid to re-estimate durations of activities during the Status Update Collection Phase, since more pertinent knowledge of project conditions, performance records, and other project conditions are now known. Remember, this is a numerical information system, and the numbers (estimates) can change at any time. Estimates tend to become more realistic as the project progresses in time.

b. Projection (PROJECT LOOK AHEAD)

A second list, or pack of IBM cards, can also be sent to the field. This list would comprise those activities whose Early Activity Start Time, from the initial network schedule, could occur in the next reporting period. The field would indicate probable starting dates of these activities.

This acts as both a check on the progress of the project, and a situation analysis from the site of work. Some factors in the project may have been overlooked in the initial plans, and an indication, in advance, of a delay that would consume float allows management to anticipate and prepare in advance for such situations.

The concept is not to spotlight any particular individual, department, sub-contractor, etc., but to anticipate troublesome delays that occur in the field or plant that could not be or were not anticipated in the original planning. The Projection report is made by a visual comparison of the previous Latest Start Date (LS) of the reported activity with the Anticipated Start Date of the activity. Any activity trending towards its Latest Start Date over the life of the project is trending towards schedule trouble.

Updating a Critical Path Schedule

I. General

In order to obtain the maximum value from CPM, it is necessary to establish a simple, flexible and timely method of collecting and transmitting information as to the status of the various operations shown on the CPM Arrow Diagram. The information as to the status of the project is that of a particular day, e.g., the last day of the month, Friday of the particular week, etc. The purpose of the status information is to insert the project progress, or lack of it, into the project plan, and generate a new updated schedule that may be analyzed for slippages or improvement.

Information is required as of the particular day of Update for:

(1) The completion of physical activities.
(2) Those physical activities that are in actual progress as of the Update date
    (a) either report the number of working days left
        to complete a particular operation or activity,
        or report the percent complete of that activity
        (in duration time units) as of the Update date.
(3) Delivery dates for material and equipment that are as firm as possible.
(4) Any changes in the sequence of operations from the original plan.
(5) Any "Holds" or basic design changes that affect operational sequences.
(6) Any additions or deletions to the original Scope of Work.

II. Methods of Reporting

There are several methods of reporting the status of the project. These will be exemplified by the attached network plan for, "Building A New

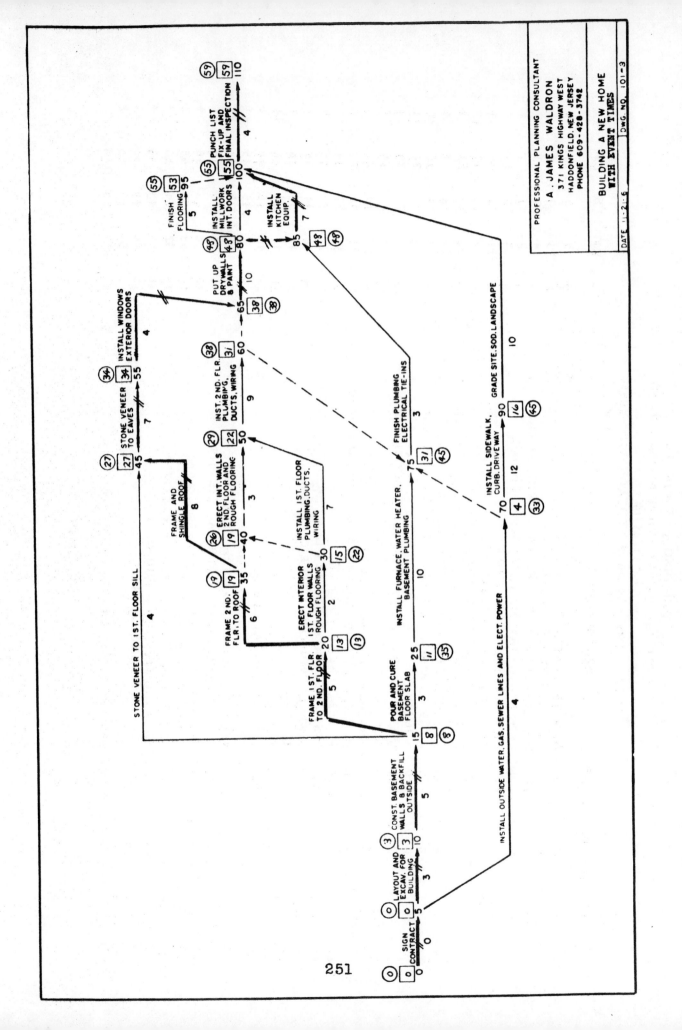

BUILDING A NEW HOME

CRITICAL PATH METHOD NORMAL TIME

AJW 11/21/68

TOTAL TIME - 59 DAYS   TOTAL COST - 20640

I	J	DURATION	COST	DESCRIPTION	ES	EF	LS	LF	TF
0	5	0	0	SIGN CONTRACT	0	0	0	0	*
5	10	3	700	LAYOUT AND EXCAVATE	0	3	0	3	*
5	70	4	950	INSTALL OUTSIDE WATER GAS SEWER POWER	0	4	29	33	29
10	15	5	1320	CONSTRUCT BASEMENT WALLS BACKFILL	3	8	3	8	*
15	20	5	850	FRAME FIRST TO SECOND FLOOR	8	13	8	13	*
15	25	3	470	POUR AND CURE BASEMENT	8	11	32	35	24
15	45	4	2200	STONE VENEER TO FIRST FLOOR SILL	8	12	23	27	15
20	30	2	430	INSTALL FIRST FLOOR WALLS ROUGH FLOOR	13	15	20	22	7
20	35	6	1200	FRAME SECOND FLOOR TO ROOF	13	19	13	19	*
25	75	10	350	INSTALL FURNACE WATER HEATER BASEMENT PLUMBING	11	21	35	45	24
30	40	0	0	DUMMY	15	15	26	26	11
30	50	7	650	INSTALL FIRST FLOOR PLUMBING DUCTS WIRE	15	22	22	29	7
35	40	0	0	DUMMY	19	19	26	26	7
35	45	8	1130	FRAME AND SHINGLE ROOF	19	27	19	27	*
40	50	3	490	INSTALL SECOND FLOOR WALLS ROUGH FLOOR	19	22	26	29	7
45	55	7	2060	STONE VENEER TO EAVES	27	34	27	34	*
50	60	9	480	INSTALL SECOND FLOOR PLUMBING DUCTS WIRE	22	31	29	38	7
55	65	4	370	INSTALL WINDOWS EXTERIOR DOORS	34	38	34	38	*
60	65	0	0	DUMMY	31	31	38	38	7
60	75	0	0	DUMMY	31	31	45	45	14
65	80	10	840	PUT UP DRYWALLS PAINT	38	48	38	48	*
70	75	0	0	DUMMY	4	4	45	45	41
70	90	12	1320	INSTALL SIDEWALKS CURB DRIVEWAY	4	16	33	45	29
75	85	3	490	FINISH PLUMBING ELECTRICAL TIE-INS	31	34	45	48	14
80	85	0	0	DUMMY	48	48	48	48	*
80	95	5	730	FINISH FLOORING	48	53	50	55	2
80	100	4	490	INSTALL MILLWORK INTERIOR DOORS	48	52	51	55	3
85	100	7	1870	INSTALL KITCHEN EQUIPMENT	48	55	48	55	*
90	100	10	750	GRADE SITE SOD LANDSCAPE	16	26	45	55	29
95	100	0	0	DUMMY	53	53	55	55	2
100	110	4	500	PUNCH LIST INSPECTION	55	59	55	59	*

FIG. 18.0

Home," Dwg. #101-3, the CPM Computer schedule thereof, and the Project Control Chart, Dwg. #101-1. The reporting system described is based on the Project Start date of February 3, 1969, and the status being reported by February 28, 1969. The information is forwarded to the consultant, or data processor or planner by someone in the user organization who is aware of project status. The information can be obtained by a clerk, secretary, draftsman or junior engineer from personnel cognizant of actual project progress. The Project Manager, or senior Line Executive will designate frequency of reporting, who is to collect data, who will furnish data, and who will process it. In large organizations this could mean separate departments to be co-ordinated by the Planning Group. In a small company, one person may perform all three functions.

1. Narrative Report

The status of Building a New Home might be reported thusly:

PROJECT: BUILDING A NEW HOME #101-3
STATUS: February 28, 1969

Layout, excavation, basement walls and floor slab complete. Framing to the roof is in progress with three days remaining. Stone veneering to eaves is 70% complete. Interior framing and rough flooring on 2nd floor is 60% complete. Interior walls and rough flooring on 1st floor is complete. 1st floor plumbing-ducts-wiring is 85% complete. Furnace and basement plumbing is 60% complete.

Changes in sequence of operations is commonly reported narratively, but a marked up print of the current Arrow Diagram will serve as effectively.

BUILDING A NEW HOME
CRITICAL PATH METHOD NORMAL TIME
AJW 11/21/68

Para. II.3   TOTAL TIME - 59 DAYS   TOTAL COST - 20640   **Para. II.2**

Actual Work in Progress →

% Complete →

I	J	DURATION	COST	DESCRIPTION	% Complete	ES	EF	LS	LF	TF
C	5	0	0	C SIGN CONTRACT	100%	0	0	0	0	*
5	10	3	700	C LAYOUT AND EXCAVATE	100%	0	3	0	3	*
5	70	4	950	R-2 INSTALL OUTSIDE WATER GAS SEWER POWER	50%	0	4	29	33	29
10	15	5	1320	C CONSTRUCT BASEMENT WALLS BACKFILL	100%	3	8	3	8	*
10	20	5	850	C FRAME FIRST TO SECOND FLOOR	100%	8	13	8	13	*
15	25	3	470	C POUR AND CURE BASEMENT	100%	8	11	32	35	24
15	45	4	2200	C STONE VENEER TO FIRST FLOOR SILL	100%	8	12	23	27	15
15	30	2	430	C INSTALL FIRST FLOOR WALLS ROUGH FLOOR	100%	13	15	20	22	7
20	35	6	1200	C FRAME SECOND FLOOR TO ROOF	100%	13	19	13	19	*
25	75	10	350	R-4 INSTALL FURNACE WATER HEATER BASEMENT PLUMBING	60%	11	21	35	45	24
30	40	0	0	DUMMY		15	15	26	26	11
30	50	7	650	R-1 INSTALL FIRST FLOOR PLUMBING DUCTS WIRE	85%	15	22	22	29	7
35	40	0	0	DUMMY		19	19	26	26	7
35	45	8	1130	R-3 FRAME AND SHINGLE ROOF	60%	19	27	19	27	*
40	50	3	490	R-1 INSTALL SECOND FLOOR WALLS ROUGH FLOOR	66%	19	22	26	29	7
45	55	7	2060	R-2 STONE VENEER TO EAVES	70%	27	34	27	34	*
50	60	9	480	INSTALL SECOND FLOOR PLUMBING DUCTS WIRE		22	31	29	38	7
55	65	4	370	INSTALL WINDOWS EXTERIOR DOORS		34	38	34	38	*
60	65	0	0	DUMMY		31	31	38	38	7
60	75	0	0	DUMMY		31	31	45	45	14
65	80	10	840	PUT UP DRYWALLS PAINT		38	48	38	48	*
70	75	0	0	DUMMY		4	4	45	45	41
70	90	12	1320	INSTALL SIDEWALKS CURB DRIVEWAY		4	16	33	45	29
75	85	3	490	FINISH PLUMBING ELECTRICAL TIE-INS		31	34	45	48	14
80	85	0	0	DUMMY		48	48	48	48	*
80	95	5	730	FINISH FLOORING		48	53	50	55	2
80	100	4	490	INSTALL MILLWORK INTERIOR DOORS		48	52	51	55	3
85	100	7	1870	INSTALL KITCHEN EQUIPMENT		48	55	48	55	*
90	100	10	750	GRADE SITE SOD LANDSCAPE		16	26	45	55	29
95	100	0	0	DUMMY		53	53	55	55	2
100	110	4	500	PUNCH LIST INSPECTION		55	59	55	59	*

Fig. 18.1

254

Fig. 18. 2

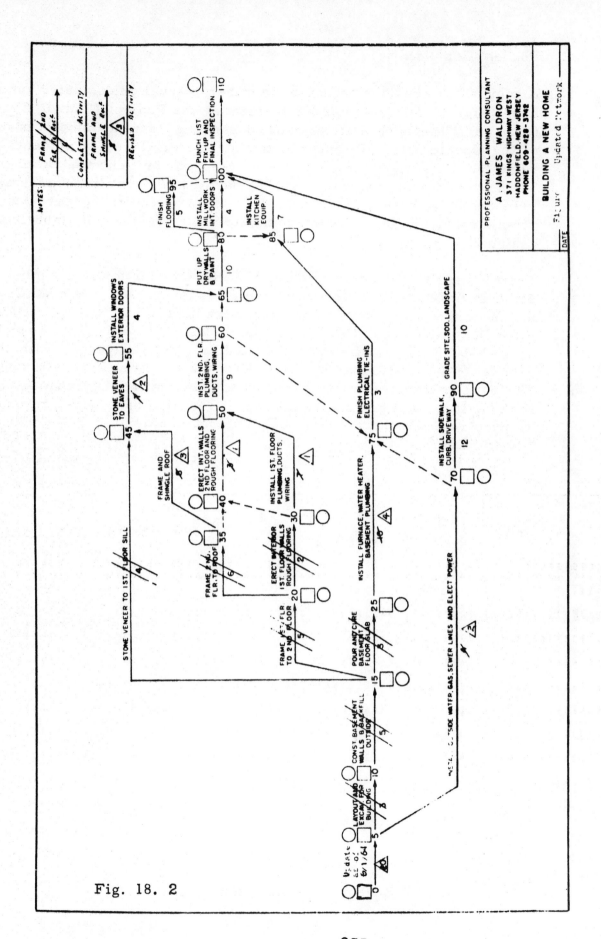

255

2. Percent Complete

A copy of the previous schedule, in either manual or computer output form (Fig. 18.1), or a marked up print of the Project Control Chart (Fig. 18.2) has each activity marked showing Percent Completion as of the Update date. Delivery dates for material and equipment, if reliably established, may also be marked on the schedule or chart. The Arrow Diagram may also be so marked. Blanks are always assumed to be unstarted activities. If costs have been assigned, its percentage remaining (100% less percent complete) becomes the new input cost assigned.

Another method of transmission of STATUS information is a duplicate set of Input Data Cards when a computer is being used. An IBM Card is sent to the proper cognizant party, with the I, J, DURATION, DES-CRIPTION, COST OR CONTRACTOR CODE NUMBER, etc., printed on its top. This is marked in the field or plant, by pencil, with the completion date, or Remaining Days, as of the end of the Report Period. The card is returned to the processor, This card is a transmission vehicle, it is not a new input data card. The new input data card must be punched from the received STATUS information. Fig. 18.3 is a typical STATUS Card.

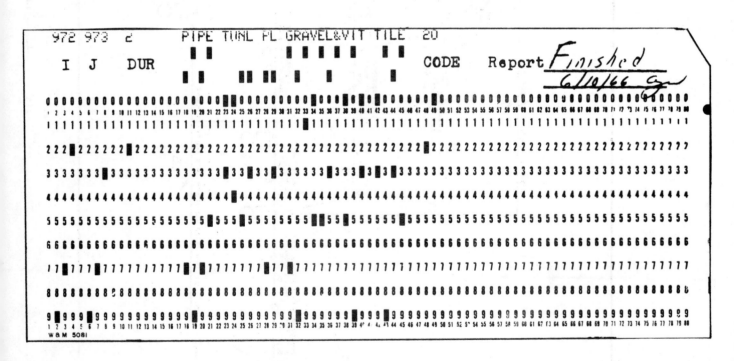

Fig. 18.3 Project Activity Status Card

3. Actual Completion and Work In Progress

As in II, 2, above, the CPM schedule (as in Fig. 18.1) or the Project Control Chart (Fig. 18.2) is marked with either a C for a completed activity (The actual completion date is not necessary here, since the purpose of updating is to obtain the projected schedule of remaining work and costs. Actual completion dates are recorded elsewhere, as in job diaries or time sheets, and are always available.), or R-( ) Days for activities in progress at the time of update, with the number of remaining days (or basic work time units, such as weeks, shifts, hours, etc.) marked against the particular activity on the schedule. If costs have been assigned they will be prorated to the ratio of remaining days/original days for the remaining activity cost at time of update.

Of these three methods, the one most definitive is II, 3, above, with the previous schedule (manual or computer) being the best vehicle for transmission of status information.

III. Structuring the Arrow Diagram For Time Schedule Updating

The most effective method of Updating is to have a constant calendar reference, e.g., Day #31 will always be **Mar. 17, 1969**, the example project. The simplest way to accomplish this is to have one lead arrow on the diagram, which goes into a start baseline event, from which the initial activities emanate. Common practice is to initially label this arrow thusly: "0-1 (0) Duration, Start Baseline **Feb. 3, 1969** "

Updating the Schedule

Let us assume the initial Project Schedule of the Project Plan is as shown in Fig. 18.4 below. The chain X, Y, Z, AA was the original Critical Path; the chain A, B, C, D was a 10 day Float chain; chain M, N, O, P is a 12 day Float chain; E, F, G, H, is a 20 day Float chain.

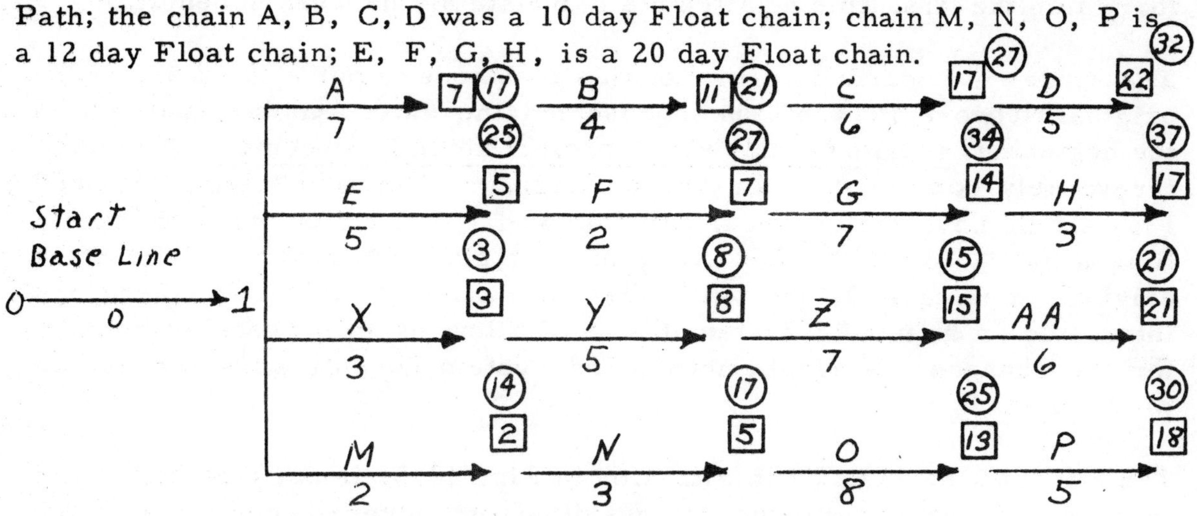

Fig. 18.4 Initial Project Schedule

257

Assume also that the Project is to be updated bi-weekly, or every 10 working days (based on a five day week). The STATUS information is received, by whatever means selected. The received information is translated on the diagram of Fig. 18.5. Activities in progress have their remaining durations shown. The lead arrow now reflects the elapsed working days of the first update period.

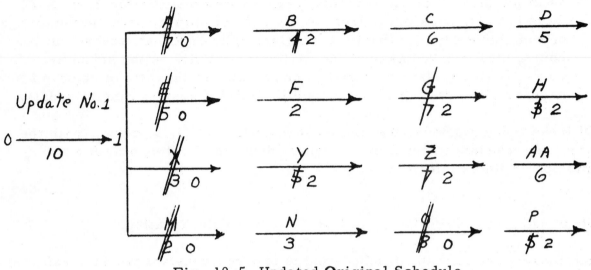

Fig. 18.5 Updated Original Schedule

The most probable reaction to the reader, as he examines the chart in Fig. 18.5, is that of one watching a house of cards tumble to the ground. How can operation G be partially completed before its predecessor F is not even started; is it possible that O can be finished before N starts?

Let the reader be assured that not only can the sequences of operations be radically changed from that of the initial Arrow Diagram, but it can be almost guaranteed that this is the way he is going to receive STATUS information the first time he attempts to update his network schedule.

The reader is reminded again that this technique is not a rigid one. The original Network Plan, which represented the most feasible approach in the beginning, is not the only way a project could be executed! No one is irrevocably commited to an Arrow Diagram. The shibboleth, "A Good Plan should never be changed, just followed," is a ridiculous artificiality. It is almost a certainty that every project plan, no matter how good the original thinking, will see valid changes during its life. Thus, we are interested in a flexible system that will allow us to reasonably account for the changes. A rigid, unchanging system doesn't work for us; we work for it.

The affect of the STATUS information in Fig. 18.5, is not remarkable, it is an every day occurrence. It results from several conditions at the project:

1. The initial sequence was based on assumptions, entirely feasible during the planning phase. However, unexpected conditions arise during the execution of the project that were not foreseen in the beginning. A case in point:

Dig #1 → Dig #2 → Dig #3 → Dig #4 →

Fig. 18.6

In the above initial plan, a trench was to be dug, from North to South, in sections numbered #1, #2, #3, #4. The project started on schedule, and Section #2 is even completed ahead of schedule. However, when the first shovel full is turned in Section #3, it is discovered that there are process pipes in the ground at this Section that were not shown on the original diagram. Does the project stop because we have a chart that says we cannot start digging Section #4 until we've finished digging Section #3? It is to be hoped that the answer is negative, that a field decision will be made by the Superintendent to jump to Section #4, continue the digging of the trench until the unindicated piping problem is resolved. The Superintendent is expected to make the hour-by-hour field decisions, and should do something as shown in Fig. 18.7.

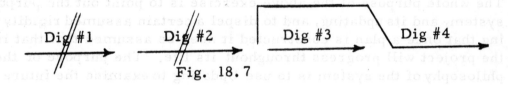

Fig. 18.7

A second case in point:

In the foundation plan below, Fig. 18.8, the initial plan was based on starting at the most accessible corner, the Northeast corner, and the work would flow from A4 to A1, then to B1 to B4 to C4 to C1, and then to D4. Each row of footings was represented by an arrow on the initial plan.

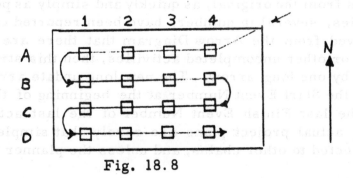

Fig. 18.8

In the actual excavation, work did start at Footing A4, and went ahead of schedule until Footing A1 was started. Then, unexpectedly, a spring was encountered at A1 during excavation. Should the remaining footing excavation be held up until the A1 area is dammed and dewatered because we have a chart that states that B1 to B4 cannot start until A4 to A1 is finished? Or should a rational decision be made in the field to jump around the problem and continue the other excavation activities.

2. No matter how detailed a description of an activity, it still is relatively gross in relation to the hour-by-hour actual work on the project. The hour-by-hour, or day-by-day "work package", under lower level supervision, has every chance of being interrupted for many reasons. A design activity, "Power Supply Design," originally estimated at 5 days, may be interrupted after the second day due to a more urgent need for that designer on another project. A normally successor activity, "Chassis Design," might be started by another designer before its normal predecessor is completed. No chart is going to prevent this from happening.

   It is also true that certain sequences cannot change. The author would become very nervous if he discovered the Structural Steel was being erected on a construction project if the foundations were not complete.

The whole purpose of the above exercise is to point out the purpose of the system, and its updating, and to dispel a certain assumed rigidity in thinking that once a plan is constructed it need be assumed that that is the way the project will progress throughout its life. The purpose of the flexible philosophy of the system is to use Updating to examine the future schedule of the remaining activities, regardless of what field changes took place. We are really concerned with what is the new project schedule of the project in Fig. 18.5, not with the fact that O was finished before N started.

Updating, therefore, is a one to one comparison with the activities, the actual performance is related to the overall schedule.

The purpose is to find the remaining schedule, even with radically changed sequences from the original, as quickly and simply as possible. If a string of activities, several in number, have been reported complete, and it can be observed from the Arrow Diagram that there are no interconnecting dummies or other uncompleted activities, then this string or chain can be replaced by one long arrow. The new long update arrow will have as its I number the Start Event Number at the beginning of that chain, and its J number the last Finish Event Number of the last activity on that chain. However, actual project plans are rarely that simple, most chains are interconnected to other chains, and unless the planner is extremely careful, the replacement of a chain of completed activities by one long update activity may omit some interconnecting dummies or uncompleted activities. This will give "opens" in the Forward and Backward Pass in the Update Schedule calculation, because there will remain Event Numbers that no longer have a connection to the diagram.

On a large project, the "one to one" comparison can be easily handled by a Collator, which replaces the original activity input card by a zero duration "Completed" activity card (with the same I-J identification as the original) for completed activities, and a new card with a reduced duration for those activities that are in progress at the end of the report period.

The Update Forward Pass starts with the initial load arrow being replaced with an Update Arrow (and its card for a computer run) which has the elapsed time as its duration. The reason for the elapsed duration is to maintain the constant calendar reference date. If Day #15 is May 22nd, then every time a schedule time appears at Day #15 it will always be May 22nd.

The Update Forward Pass for Fig. 18.5 is shown in Fig. 18.9 below.

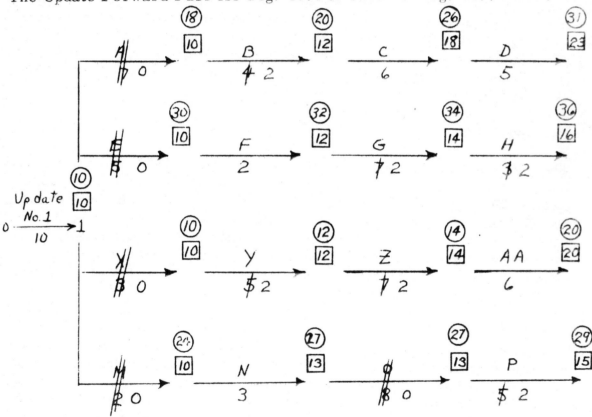

Fig. 18.9

Note that the resultant Updated Schedule gives us the information we want, the remaining future schedule. The original Critical Path is still on schedule (compare the Event Time at the head of **Z** with the original Event Time at that event), as are all the other chains of activities. The author would examine the status of the chain E, F, G, H; for some reason a large Float chain is being pushed to its Early Finish Time. There may be inefficiencies in the use of resources here, or a misunderstanding, or poor estimates.

IT CANNOT BE STATED TOO STRONGLY OR TOO OFTEN: THIS IS A FLEXIBLE TECHNIQUE: THE DIAGRAM IS NOT A RIGID PLAN. IT MUST ALSO BE NOTED THAT THE ARROW DIAGRAM NEED NOT BE REDRAWN EVERY TIME A CHANGE OCCURS. The one to one comparison of the activities, regardless of changes in actual sequential performance gives us the necessary information--the remaining schedule, how much ahead or behind we are in the schedule, and what is the new Critical Path.

It is unfortunate that several computer programs used in Updating a Network Schedule have been written in such a manner that they will reject the input of Fig. 18.5. The programmer, a human, decided that once the computer was told the O follows N, it will always follow N, or if a predecessor is partially completed, then the successor activity cannot start. This is a fault of the human, not the system, and this computer program forces the manager to work for it, rather than it working for the manager. A change routine must first be entered into the computer before the new STATUS information is inserted. This unfortunate situation is particularly true of magnetic tape machines, where the information has been stored sequentially.

IV. Negative Float or Slack - The Fallacy

There are computer programs today that allow the insertion of the Actual Start and Finish dates of an activity; then it will compare the original Late Start and Finish dates from the initial schedule; and if the activity was actually started later than the original LS time, the computer will print a negative number in the Total Float or Slack column, ostensibly to indicate how many days or weeks behind schedule that activity is. In essence, the program accepts these actual starts and finishes as a T_S time, as described in Chapter 9.

There are several fallacies with this approach which result in ficticious schedules that are confusing and require an inordinate amount of time in explanation that the project is really not behind schedule. The basic fallacy is the Rigidity Abuse, described in Chapter .. Once the original schedule is calculated, accepted, then stored in the memory of the computer, it is automatically assumed that the activity sequences can never deviate from the original sequence. Unfortunately, life is not that neat and simple. In fact, the author can almost guarantee that the project will never be executed in exactly the sequence originally established.

One typical problem is the situation where a simple chain starts with a negative float number and ends up with a positive float number. This can be seen in Figs. 18.4 and 19.9. In Fig. 18.9, the program assumes that the Critical Path of operations X, then Y, then Z, then AA has to be done in only that sequence. The updated information, as plugged in in Fig. 18.9, will produce a negative float of -7 days for activity X (The original schedule, Fig. 18.4, had an original Late Finish Date of Day #3, yet the update runs produce a Finish Date of Day #10), the second activity in the chain, Y, will get -4 days of float or slack, and operations Z and AA both get -1 day of float or slack. Thus, a simple straight chain of activities will start with a negative float of 7 days, and end with a positive float of 1 day.

To illustrate how ridiculous this approach is, an actual case history will be described. In the design and construction of a chemical plant, the main pipe runs from unit to unit were layed out around the periphery of the plant on pipe supports. These were simple stanchions inbedded in concrete footings, containing a Tee Bar for pipe support. They were spaced 10 feet apart. Instrument tubing runs, for the pneumatic instruments, were supported parasitically from the stanchion. Fig. 18.10 is an elevation of the typical pipe support.

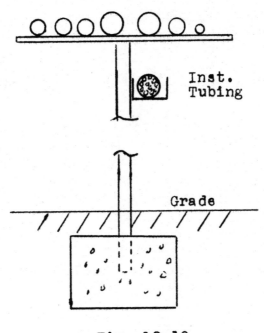

Fig. 18.10

From the Network Plan for the project, the section of the network relating to the pipe supports, pipe runs, and instrument tubing runs were as in Fig. 18.11 below, with the initial results of the Forward Pass shown.

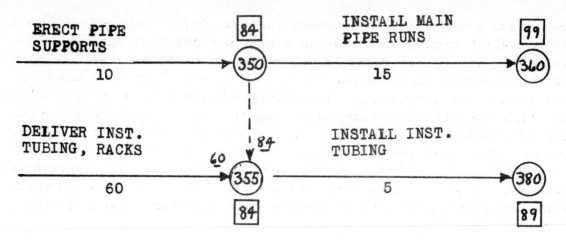

Fig. 18.11

What actually occurred at the project was this: by Day #72, all of the footings and stanchions were erected, but the Tee Bars were not, as indicated in Fig. 18.12.

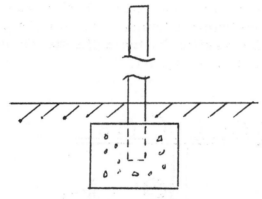

Fig. 18.12

At Day #72, the Instrument Sub-Contractor visited the site and noted that the stanchions were all erected and he could proceed to install his instrument tubing runs. Receiving permission to do so, he started on Day #74. The project schedule was updated on Day #75. In making the update computer run, the scheduler inserted the actual start date of Day #74 for the activity, "Install Instrument Tubing." In essence, the computer made a T_S run, as described in Chapter 9. The computer "saw" a network section of Fig. 18.13.

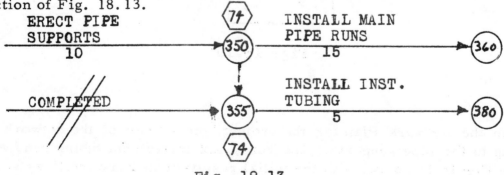

Fig. 18.13

In effect, the updated run said that the activity, "Erect Pipe Supports" is 10 days late, since the activity, "Install Instrument Tubing," from that network, could not start until the shown predecessor was finished, so it assigned -10 days of float time to the "Erect Pipe Supports" activity. What was even worse, the program then made a new Forward Pass, and gives the other successor activity, "Install Main Pipe Runs" 10 additional days of float over the original schedule on the premise that it could also start on Day #74.

The most common reaction of data processing people when this case is discussed is to respond by saying, "You should remove the dummy before making the update computer run." As soon as you do that, you are working for the system. There is nothing wrong with the technique, the limitation is in the computer program, and the programmer's philosophy. Again, updating is a one to one comparison of activities. The purpose is to obtain the remaining schedule, based on the major milestones in the project, regardless of how sequences change.

Remember, no matter how much thought and realistic preparation went into the initial network plan, you can be assured that it will never be executed in that exact literal manner. Many of the original sequences were optional; contingencies arise daily in the field; not all conditions are known at the inception of a project. Again, the milestone attainment is the major goal of the technique, not that the project is literally done X, then Y, then Z and then AA, as in Fig. 18.4. The goal is to get that chain finished on time, regardless of how the individual sequences change.

V. Input Format

The Network lead arrow will be given a duration equal to the elapsed time (in working time units) as of the date of the particular update. It will be noted in the House example that the lead arrow reads, "0-5, 20 days, Update as of February 28, 1969." Completed activities are given a zero (or blank) time. Activities in progress are given revised times, representing the remaining working time units for completion. The concept is simplicity, and the system being described herein is applicable to manual computations, and for use on small computers. The network plan is always closed, in order to obtain a "Forward and Backward Pass" for schedule computation. Fig. 18.10 is an input data sheet for a small computer, the format is for the IBM-MISS-LESS program for the IBM Model 1620 computer. Each computer has its own specific format and basically accomplishes what is described here; replacing reported activities with completions or revised activity times. In the 1620 program described herein, the original input key punched cards are replaced by the Update cards. This is done either manually, or by data processing techniques (a Collator is used to automatically replace cards). Unstarted activities have their original cards in the input data deck.

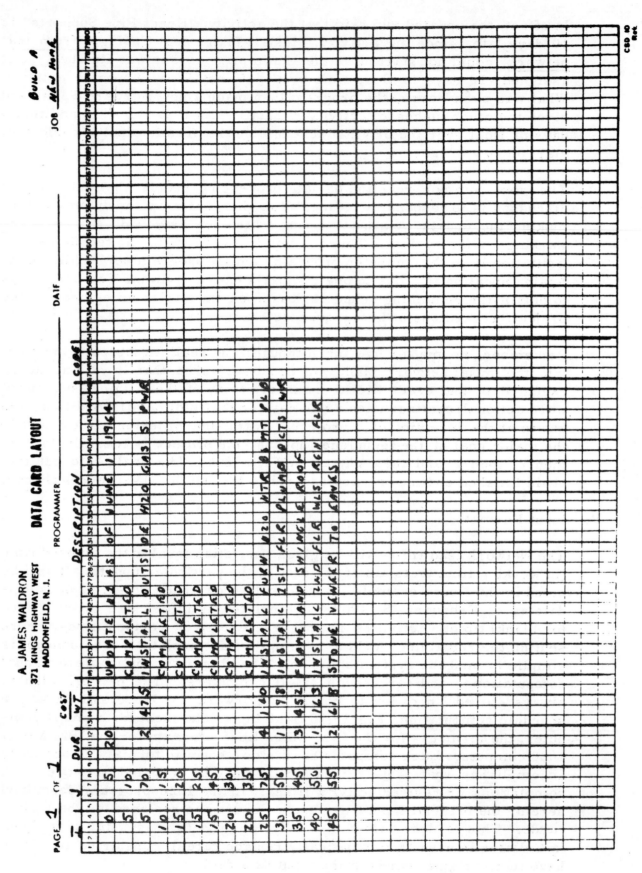

Fig. 18.14

BUILDING A NEW HOME
CRITICAL PATH METHOD
UPDATE #1 AS OF FEB.28, 1969
COMPLETION 51 DAYS   REMAINING COST $9786

I	J	DURATION	COST	DESCRIPTION	ES	EF	LS	LF	TF
0	5	20	0	UPDATE AS OF FEB. 28. 1969	0	20	0	20	*
5	10	0	0	COMPLETED	20	20	20	20	*
5	70	2	475	INSTALL OUTSIDE WATER GAS SEWER POWER	20	22	23	25	3
10	15	0	0	COMPLETED	20	20	20	20	*
15	20	0	0	COMPLETED	20	20	20	20	*
15	25	0	0	COMPLETED	20	20	43	43	23
15	45	0	0	COMPLETED	20	20	24	24	4
20	30	0	0	COMPLETED	20	20	20	20	*
20	35	0	0	COMPLETED	20	20	21	21	1
25	75	4	140	INSTALL FURNACE WATER HEATER BASEMENT PLUMBING	20	24	43	47	23
30	40	0	0	DUMMY	20	20	20	20	*
30	50	1	98	INSTALL FIRST FLOOR PLUMBING DUCTS WIRE	20	21	20	21	1
35	40	0	0	DUMMY	20	20	21	21	1
35	45	3	452	FRAME AND SHINGLE ROOF	20	23	21	24	1
40	50	1	163	INSTALL SECOND FLOOR WALLS ROUGH FLOOR	20	21	20	21	*
45	55	2	618	STONE VENEER TO EAVES	23	25	24	26	1
50	60	9	480	INSTALL SECOND FLOOR PLUMBING DUCTS WIRE	21	30	21	30	*
55	65	4	370	INSTALL WINDOWS EXTERIOR DOORS	25	29	26	30	1
60	65	0	0	DUMMY	30	30	30	30	*
60	75	0	0	DUMMY	30	30	47	47	17
65	80	10	840	PUT UP DRYWALLS PAINT	30	40	30	40	*
70	75	0	0	DUMMY	22	22	37	37	15
70	90	12	1320	INSTALL SIDEWALKS CURB DRIVEWAY	22	34	25	37	3
75	85	3	490	FINISH PLUMBING ELECTRICAL TIE-INS	30	33	37	40	7
80	85	0	0	DUMMY	40	40	40	40	*
80	95	5	730	FINISH FLOORING	40	45	42	47	2
80	100	4	490	INSTALL MILLWORK INTERIOR DOORS	40	44	43	47	3
85	100	7	1870	INSTALL KITCHEN EQUIPMENT	40	47	40	47	*
90	100	10	750	GRADE SITE SOD LANDSCAPE	34	44	37	47	3
95	100	0	0	DUMMY	45	45	47	47	2
100	110	4	500	PUNCH LIST INSPECTION	47	51	47	51	*

Fig. 18.15

BUILDING A NEW HOME
CRITICAL PATH METHOD DIRECTED SCHEDULE
UPDATE #1 AS OF FEB.28, 1969
COMPLETION 51 DAYS

I	J	DURATION	DESCRIPTION	EXPECTED START	REQUIRED FINISH
0	5	20	UPDATE AS OF FEB 28, 1969	0	20
5	10	0	COMPLETED	20	20
5	70	2	INSTALL OUTSIDE WATER GAS SEWER POWER	20	23
10	15	0	COMPLETED	20	20
15	20	0	COMPLETED	20	20
15	25	0	COMPLETED	20	20
15	45	0	COMPLETED	20	20
20	30	0	COMPLETED	20	20
20	35	0	COMPLETED	20	20
25	75	4	INSTALL FURNACE WATER HEATER BASEMENT PLUMBING	20	28
30	50	1	INSTALL FIRST FLOOR PLUMBING DUCTS WIRE	20	21
35	45	3	FRAME AND SHINGLE ROOF	20	23
40	50	1	INSTALL SECOND FLOOR WALLS ROUGH FLOOR	20	21
45	55	2	STONE VENEER TO EAVES	23	25
50	60	9	INSTALL SECOND FLOOR PLUMBING DUCTS WIRE	21	30
55	65	4	INSTALL WINDOWS EXTERIOR DOORS	25	29
65	80	10	PUT UP DRYWALLS PAINT	30	40
70	90	12	INSTALL SIDEWALKS CURB DRIVEWAY	22	35
75	85	3	FINISH PLUMBING ELECTRICAL TIE-INS	30	35
80	95	5	FINISH FLOORING	40	45
80	100	4	INSTALL MILLWORK INTERIOR DOORS	40	45
85	100	7	INSTALL KITCHEN EQUIPMENT	40	47
90	100	10	GRADE SITE SOD LANDSCAPE	34	45
100	110	4	PUNCH LIST INSPECTION	47	51

Fig. 18.16

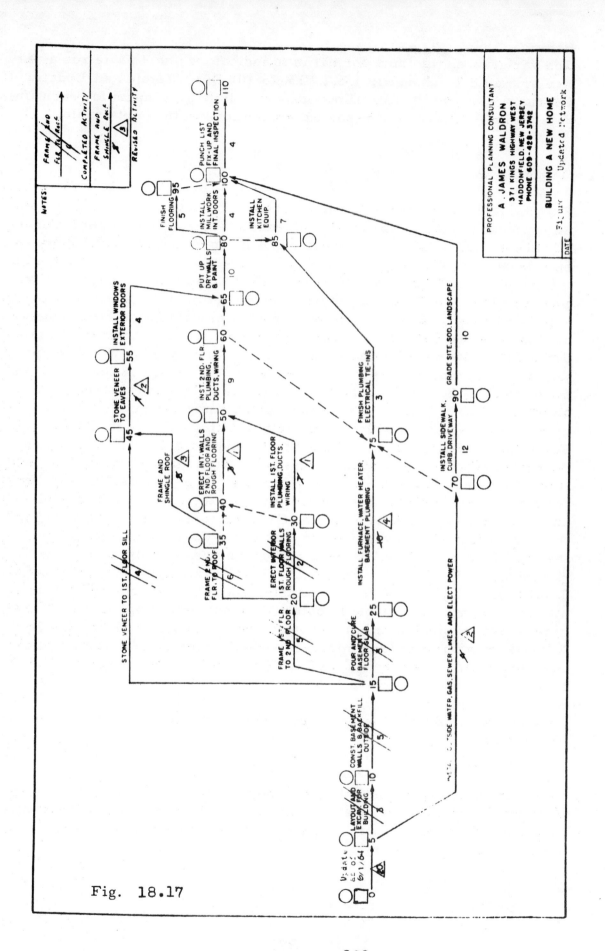

Fig. 18.17

269

One variation on the input format is to indicate which Update run a particular card is changed, i.e., "15-25 (0) Dur. Completed Update #1 February 28, 1969." This allows separate listing of inputs for each Update period. A record of reported completions is then obtainable.

VI. Output Format

Fig. 18.15 represents the Updated Schedule output, ready for dissemination and analysis. In this particular example, the project is ahead of its initial schedule. Make note of the fact that if sub-contractors and suppliers have been given the original schedule, or starting date information from the initial schedule, they are to be notified of the improved schedule. Projects are known to lose the progress made because a sub-contractor or supplier was not notified of a moved up starting date from the original on an Updated schedule. The sub-contractor had accepted the initial starting date, and in the meantime had made other commitments. When the project was ready for him he was not ready for the project. The output information may be distributed in many forms, and by several methods, from an informal phone call to a separate sorted computer output. One technique is to take the complete CPM output with its four activity times of Early Start, Early Finish, Late Start, and Late Finish, and issue a "Directed Schedule" of just two times for a particular activity. The first time is labeled Expected Start, and is the same as the Early Start Time from the CPM schedule. The second time is labeled, "Required Finish," and is the Early Finish Time of the CPM schedule, plus a percentage of the Total Float Time available for that activity. If management decides that one third of the Float Time can be given to sub-contractors, then the "Required Finish Time" would be the Early Finish plus one third of the Total Float. Fig. 18.16 is an example of a Directed Schedule. It is an attempt to control a "People Problem," by not revealing all the Float to groups who are not familiar, or who have a superficial knowledge of Network Planning. Dummies are usually eliminated. Note that the Directed Schedule is in essence a Bar Chart, with one Start and one Finish time.

This new schedule will serve as the vehicle for the next status report (Paragraph II, 3.).

VII. Management will determine the frequency of Updated reports, in order to maintain control. Control also depends on the timeliness of the information, a month old report may be worse than useless. Suggested reasonable periods would be as follows:

Project Overall Duration	Report Periods	Allowable Lag
6 months or less	Weekly	24 hours
7 to 12 months	Bi-Weekly	72 hours
Above 12 months	Monthly	1 week

VIII Notes

1. If a revised network is required within the Updated Schedule, a common practice is to mark off completed activities, and indicate revised times as in Fig. 18.17

2. Note that the Status Report, and the revised Plan actually changed a logical sequence, in that there is still some remaining time on Activity #35-45, "Frame and Shingle Roof," and some work completed on a predecessor, #45-55, "Stone Veneer to Eaves." This indicates that work originally planned sequentially was executed concurrently. This occurs quite frequently and is well within the prerogative of the Line Manager (Superintendent) and should not be curtailed. It is not necessary to redraw the network for such changes. The value of the technique shown herein, that of the changing lead arrows, and zero duration on completed activities, is that it allows a flexible information system to be utilized without a lot of unnecessary drafting revisions.

3. Fig. 18.18 is the Updated Time Scale Chart of the Project.

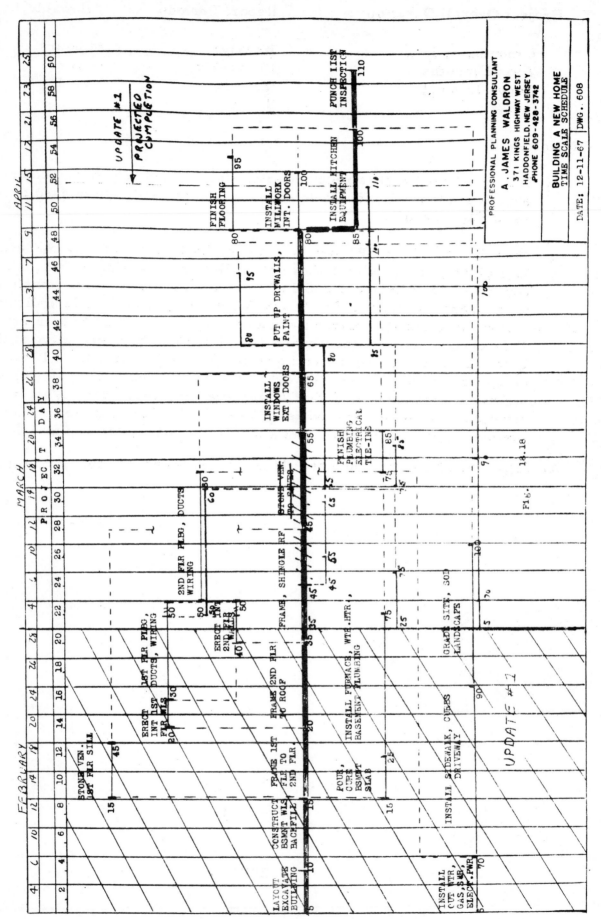

BUILDING A NEW HOME
TIME SCALE SCHEDULE

PROFESSIONAL PLANNING CONSULTANT

A. JAMES WALDRON
371 KINGS HIGHWAY WEST
HADDONFIELD, NEW JERSEY
PHONE 609-428-3742

DATE: 12-11-67   DWG. 608

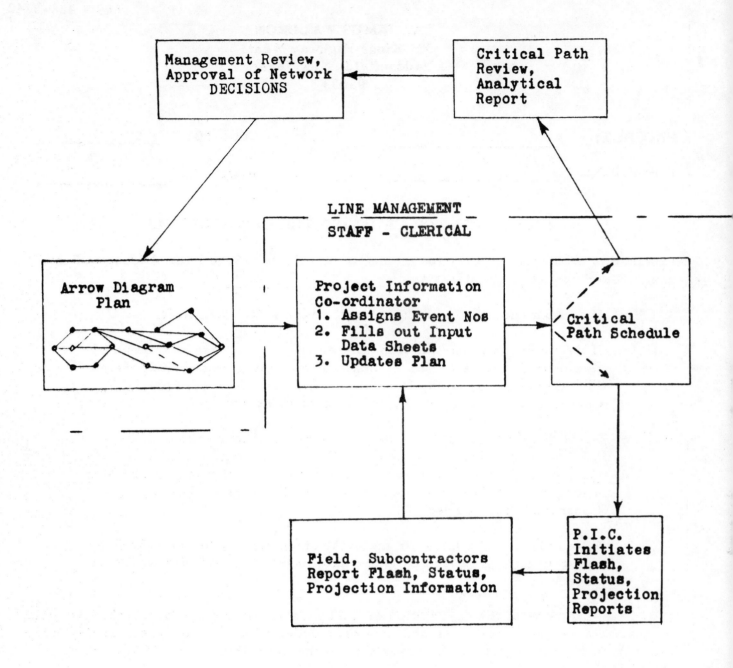

Management Review, Approval of Network DECISIONS	Critical Path Review, Analytical Report

LINE MANAGEMENT
STAFF - CLERICAL

Arrow Diagram Plan	Project Information Co-ordinator 1. Assigns Event Nos 2. Fills out Input Data Sheets 3. Updates Plan	Critical Path Schedule

Field, Subcontractors Report Flash, Status, Projection Information	P.I.C. Initiates Flash, Status, Projection Reports

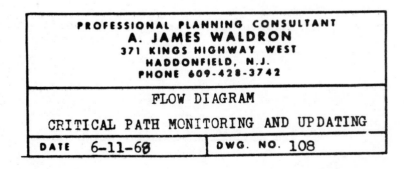

PROFESSIONAL PLANNING CONSULTANT
A. JAMES WALDRON
371 KINGS HIGHWAY WEST
HADDONFIELD, N.J.
PHONE 609-428-3742

FLOW DIAGRAM

CRITICAL PATH MONITORING AND UPDATING

DATE 6-11-68	DWG. NO. 108

Fig. 18.19

A. JAMES WALDRON
371 Kings Highway West
Haddonfield, New Jersey
08033

PROJECT: _____   CODE NO: _____

LOCATION: _____   NAME: _____

INPUT DATA FOR UPDATING THE CPM SCHEDULE

Attached is a copy of your section of the overall Project Schedule. Please indicate next to each operation its status as of _____.

    a. If an operation has been completed, mark a "C" next to it.

    b. If an operation is partially completed, or in progress at the end of this period, indicate your estimate of the remaining working days for completion by noting it with a prefix "R". For example, if an operation has two (2) days left, it would be noted "R-2".

    c. Blank (no information) will be assumed not to have started.

    d. Correct any original activity estimates if new information indicates a longer or shorter duration. Note if this change is due to a Contract Change Order.

    e. Note any delivery dates for material or equipment on the schedule, by writing its promised calendar date against the activity on the schedule.

If you are unsure what the description of the operation on the enclosed schedule means, then write a narrative report of what has been done, what is in progress at the end of this period. Your report will be translated to the activities on the CPM Arrow Diagram.

In accordance with the referenced contract specifications, these Update Reports are requested at this office no later than _____.

Thank you.

Signed: _____

Date: _____

Fig. 18.20 Update Transmission Form Letter

The basic updating system is simple, and clerical in nature. Under the principle of "management by exception" the system and procedure is to be set up that the bulk of data collection, handling, and distribution is by a clerical person. Only the initial plan, and the estimates are to be generated at the management level. Assigning event numbers, calculations or filling out data sheets, obtaining the computer run, distributing the schedule information, setting up and collecting the Flash, Status, and Projection information for making the update run is primarily clerical in nature. The analysis of the information and decisions based on that information is at the management level. Fig. 18.19 is an Information Flow Diagram, delineating the management and clerical levels. The "Project Information Co-ordinator" may be a junior engineer, draftsman, or clerk. Several companies are using a woman secretary in this category.

Fig. 18.20 is a typical Form Letter used in the transmission of the previous schedule to cognizant parties for their update information

Naturally, as projects increase in size and complexity, more sophisticated systems and procedures will be established, from the above fundamentals. A staff function, in the data handling aspect, may appear. This is quite necessary at times. The line manager is to use the staff groups every chance he gets, since it is their function, not his, to collect and distribute the data. It is not a staff function to analyze the data, unless an experienced line person is serving with the staff group. The biggest pitfall here is the ease with which a line manager may wind up working for a staff group in order to obtain the vitally needed schedule information. This generally comes under the heading of "people problems," rather than inadequacy of a powerful yet simple technique. Parkinson's Law and its corollaries seems to be in effect in these cases:

(1) The work will fill up the time allowed available.
(2) Available physical space will be rapidly filled up
    with men and machines.
(3) Paper work will fill up the system evolved.
(4) The initial authorized budget will rarely be underrun.

When a line manager finds himself working for the staff people; when he is led down to a level of minute detail, and abandoned there to grope his way out of a quagmire of minutae and trivia, heaving with continually changing milestones and mushrooming clerical and overhead costs; he is suffering from an attack of business' Parkinson's Disease.

One easily detected symptom of the above sickness is the "paper short circuit" readily seen in Fig. 18.19 This is the lower right hand loop. It has happened in many cases, that a tremendous circulation of paper was created on projects, but no information came out for management distillation and decision making. The solution to this problem is outside the scope of the subject matter of this text. The handling of people problems quite often falls into the realm of child psychology.

A Color Code System of Monitoring

In order to gain maximum efficiency in the utilization of network techniques, a black and white print of the project arrow diagram may be marked with various colors (crayons or thin strips of colored acetate tape). This will graphically indicate the progress status as of the last reporting period.

The date of the last report is inserted in an inverted triangle. The triangle is located on the print at the pertinent event, or in the middle of an arrow, indicating work on that particular activity was in progress at the time of the last report.

Color Code

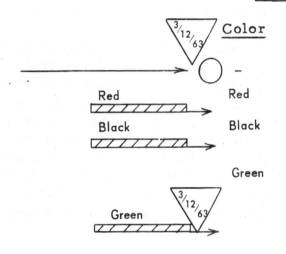

Color	Significance
–	Report Date
Red	Original Critical Path
Black	Completed Activities
Green	Work in progress (length of green bar over arrow sometimes approximates % complete of that activity). If there has been a slippage, the green bar is followed by the following colors:
Orange	Indicates that float has been used, and the activity is now Action Priority #2 (remaining float is 10% or less of the remaining duration of the project.

The orange color is extended to all the future activities on that path, or parallel paths that should be executed in the next reporting period. Project completion will be met, but this chain of activities deserves more attention. |

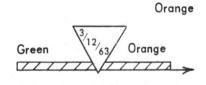

Indicates some slippage, and activity is now in Priority #3 (Remaining float is 40% or less of remaining project duration).

Yellow color bars are extended for all future subsequent activities that should be executed in next reporting period.

Red, with red slant bars thru arrow.

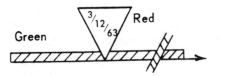

Indicates slippage that may extend project. The subsequent activities for execution in the next reporting period are now critical. If project completion date will not be met, the number of days behind schedule is marked in red after the inverted triangle of the last report date.

A color coded project "map" is rapidly read by anyone with the slightest familiarization of the network technique. This map should be confidential, and its location and access controlled.

Dissemination of Information

As a matter of practical psychology, it is suggested that consideration be given to the amount of information from the network schedules that will be distributed within and without an organization. This is particularly pertinent to the amount and type of information that flows downward to the lower echelons in a company. Human nature being what it is, there will be a tendency towards waiting until close to the Latest Activity Start Time for a particular activity or chain of activities to be started by operating groups. Since this is a management system, it is recommended that management take full advantage of the system. Management may withhold some of the float, as safety factors, by issuing "directed" schedules. The directed schedule is one where activity start times are selected somewhere between the Early and Late Activity Start Times. The duration may or may not be the originally estimated one, but a more or less conservative one, dependent on the many human factors involved. A "directed" bar chart, with start and finish times derived from the network schedule (with appropriate retention of float by the manager) is an excellent vehicle for controlled dissemination of schedules. The Distributed, or "Ideal" Schedule of Chapter 12 states this purpose well.

Chapter Summary

The philosophy of a project control system has been described in this chapter, based on the logical and efficient attributes of network planning techniques. It should be sufficient for the reader to start a truly accurate, timely and explicit Project Control system, evolving from present company systems and procedures. Ultimate development into a comprehensive and practical system depends only on the imagination of the user.

CHAPTER 19

COST CONTROL

The previous chapters have been devoted to the concepts of Planning, Scheduling and Controlling a project based on logic (or, preferably, experienced common sense), and the utilization of information on time, resource allocation and optimization of costs to make supervisory and managerial decisions.

However, in the day-to-day management of a project, control of costs is urgent--this is where profit or bankruptcy materializes. Control, in its general definition, is (1) comparison of an established standard with actual conditions, and (2) the application of necessary corrective action when a deviation from the standard occurs. A room thermostat in a building's heating system measures the actual condition by its sensing element, a bi-metallic thermal element, and compares for any deviation from the desired setting on the thermostat dial. Actual conditions are also indicated by a thermometer. Corrective action is taken by sending a signal to the final control element (the valves or stoker on the furnace) for more or less heat (energy) to the system being controlled. So, too, in a project control system a standard is first established. This is the project estimate (direct material, labor, sub-contract costs, indirect costs, overhead and burden costs, and profit). Actual costs are then accumulated from time sheets, invoices and contracts. Control is exerted (by the application of energy--human in this case). A foreman's production, a poor estimator, an inefficient buyer are subject to corrective action in many forms by the manager. The analogy to the thermostat is the same--corrective action must be taken! In project control this means decisions based on the information received. The basis for the source information is the project plan and its contingent estimates.

An immediate problem for the Project Manager (or Controller, for want of a better name) is the fact that the Project Network Diagram Plan will not be compatible with existing Cost Codes in his organization. Most Cost Account or Code systems are based on categories and unit costs. The technique developed herein is a "Work Package" oriented system--it is activity oriented. A company may estimate today a cost of $19 per cubic yard of concrete placed for slabs on grade. However, this could well be shown on the Arrow Diagram Plan by three (3) consecutive arrows; "Form Slab on Grade," "Reinforce Slab on Grade," and "Pour and Cure Slab on Grade." These three activities could well encompass six or seven Cost Code Numbers or Accounts on a typical construction company Cost Code system.

Another problem is the relation of material to labor, then this relation to the Activity Arrow. Material is generally estimated by a survey, a physical counting up from drawings and specifications. It is ordered in toto, for price advantages. Thus, all the reinforcing steel on a construction project will be ordered on one Purchase Order and will probably have one Cost Code Number.

On the diagram there are bound to be many "Work Packages," activities that will use this particular material. In general, the activity arrow will cover more than one Cost Code number in typical Cost Code or Account Systems. The Network Planner will have to develop an evolutionary translation--previous coding systems are not going to be arbitrarily discarded once the first Arrow Diagram is drawn, estimated, and scheduled.

Since material is generally a survey take-off, it is more accurately estimated than the human labor necessary to put it into place. Project Control is basically human labor control, which continues throughout the life of the project. In general, material and labor are two separate entities, and should be treated as such.

In order to develop the concept of Cost Control, this text will somewhat simplify the material and labor direct costs into the estimated cost, and the actual cost, of an activity work package. The "Building A New Home" Project, Problem 1.1 of Chapter 1, and the Network Plan of Chapter 8, will be the basis of Cost Control development. Fig. 19.0 is a typical construction company Cost Account Numbering System. Fig. 19.1 is the "Work Package" Estimate (Direct Material and Labor) for the home building project. It has been based on the activity list. To simplify, Indirect Costs, Overhead and Burden are included pro-rata. They will be manipulated in the same way the Direct Costs are handled when separately listed. Fig. 19.2 is the summary of these costs by the Account Number. This is the format that higher levels of management usually want.

Fig. 19.3 is a fundamental Cost Control form. The Basis Estimate initially is The Original Estimate. The basis estimates are compared with the actual accumulated and committed costs (an awarded sub-contract, even though unliquidated or performed, is entered as an actual cost as of the date of award). Where the original estimate is revised to suit a Change Order, it is so indicated by a Code Number in this instance. The codes in Column 6 of Fig. 19.3 indicate whether the estimate is the original, a revised and reimbursible estimate, or a revised and non-reimbursible estimate. In this latter case, the projections are made with the original reimbursible estimate, since the revised and non-reimbursible estimate includes that original plus the amount that is not reimbursible. The Actual Costs To Date are accumulated from Time Sheets, invoices, or committed sub-contracts. The Actual Code, relating to the previous Actual Cost To Date field, indicates re-worked activities where an overrun will not be reimbursible above the original estimate; extra work which will ultimately be substantiated by a Change Order, Change Memo, Field Work Order, or supplementary Purchase Order; this code is also used to indicate completed activities. Completed activities may be dropped out of subsequent Cost Control reports, if desired. The Estimated Cost to Complete is a projection made (manually or by the computer if a machine Cost Control system is being used) by subtracting Actual Costs from Original Estimated Costs. This field may also contain a new entry which is in essence a re-esti-

00. UNDISTRIBUTED COST
01. Temporary Office, Buildings
2. Temporary Toilet
3. Temporary Utilities
4. Temporary Finance
5. Architectural Fee
6. Building Permit
7. Licenses
8. Taxes and Insurance
9. Sales Tax
10. Bond
11. Hazard Insurance
12. Barricade, Temporary Fence
13. Truck and Haul
14. Travel Expense
15. Ice, Snow Removal
16. Hoist and Chutes
17. Equipment
18. Tools
19. Penalties
20. Temporary Heat
21. Temporary Roads, Parking
22. Photographs, Tests
23. Field Office Expense

10. PREPARATION OF SITE
01. Grading
    01. Fill outside building
    02. Fill under building
    03. Cut
2. Walks and drives
3. Demolition
4. Layout
5. Clean and Miscellaneous
6. Landscape
7. Fence
8. Drains
9. Curb and Gutter
10. Topsoil
11. Septic Tank
12. Railroad
13. Dewatering
14. Text Borings

20. CONCRETE
01. Footing Excavation
2. Backfill
3. Footings
4. Foundation Walls
5. Grade Beams
6. Beams
7. Columns
8. Pedestals
9. Slabs on Grade
    01. Fine Grade
    02. Screeds
    03. Mesh
    04. Pour Concrete
    05. Finish
10. Slabs off Grade
    01. Screeds
    02. Mesh
    03. Pour Concrete
    04. Finish
11. Topping
12. Membrane W. P.
13. Slag under slab
14. Hardner and Curing
15. Roof Slabs
16. Steps
17. Forms
    01. Footings
20.17.02. Foundation Walls
    03. Beams
    04. Columns
    05. Pedestals
    06. Grade Beams
    07. Slab off grade
    08. Sonotubes
    09. Steel Pans
    10. Steel Tax
    11. Remove Forms
    12. Tiers and Form Oil
18. Window Sills
19. Lintels
20. Miscellaneous Concrete

21. Reinforcing Steel
22. Text
23. Precast Concrete
24. Splans Blocks
25. Expansion Joint
26. Concrete Joist
27. Corruform
28. Dovetail Slot
29. Inserts
30. Rub Concrete
31. Fill In Blocks
32. Grade Beam Excavation

30. MASONRY
01. Concrete Block
2. Face Brick
3. Common Brick
4. Concrete Brick
5. Structural Tile
6. Cut Stone
7. Marble and Limestone
8. Wall Coping
9. Mortar
    01. Mixer
10. Scaffolds
11. Anchors and Ties
12. Clean Masonry
    01. Acid
    02. Brushes
    03. Point Up
    04. Rub Block
13. Set Lintels
14. Masonry Reinforcing
15. Thru-wall Flashing
16. Parging
17. Waterproofing
18. Expansion Joints

40. STEEL
01. Structural Steel
2. Miscellaneous Steel
3. Ornamental
4. Bar Joist
5. Steel Roof Deck
6. Metal Toilet Partitions
7. Chain Link Fence
8. Movable Metal Partitions
9. Counter Door
10. Steel Siding

50. CARPENTRY
01. Framing
    01. Treated Cants
    02. Treated Nailers
    03. Treated Plates
    04. Treated Facia
    05. Blocking
2. Sheathing
3. Decking
4. Siding
5. Finish Lumber
6. Millwork
7. Casework
    01. Wood
    02. Metal
8. Paneling
9. Wall and Ceiling Stripping
10. Sheetrock
11. Furring and Grounds
12. Windows and Screens
13. Doors
14. Chalk and Tackboard
15. Rough Hardware
16. Felt
17. Finish Hardware
18. Overhead Doors
19. H. M. Doors and Frames
20. Insulation
21. Weather Stripping
22. Flooring
23. Counter Tops
24. Clean Windows
25. Folding Doors
26. Cabinet Hardware

27. Caulking
28. Ram Sets
29. Breeze Sash
30. Bird Screen
31. Wood Bumper

60. SUB-CONTRACTORS
00. General
01. Plumbing
2. Heating
3. Ventilation
4. Air Conditioning
5. Electrical
6. Electrical Fixtures
7. Roofing and Sheet Metal
8. Plastering
9. Painting
10. Special Wall Finishes
    01. Vitricon
    02. Fabrican
    03. Lorrain
    04. Others
11. Ceramic Tile
12. Asphalt Tile
13. Marble
14. Floor Finishing
15. Sprinkler System
16. Elevator
17. Hoist and Monorail
18. Acoustical Ceiling
19. Glass and Glazing
20. Venetian Blinds
21. Refrigeration
22. Walk-in Cooler
23. Stainless Steel Kitchen
    Equipment
24. Roof Deck
    01. Insulrock
    02. Porex
    03. Tectum
    04. Flexicore
    05. Light Weight Concrete
25. Termite Control
26. Paving

70. SUPERVISION
01. Superintendent
2. Timekeeper, Clerks
3. Room and Board
4. Watchman

80. MISCELLANEOUS
01. Medicine Cabinets
2. Mirrors
3. Letters and Signs
4. Special Equipment
5. Vaults
6. Toilet Accessories
7. Food Service Equipment
8. Punch List, General

Fig. 19.0

BUILDING A NEW HOME
COST ACCOUNT AND WORK PACKAGE ESTIMATE

I	J	DURATION	TOTAL COST	DESCRIPTION	ACCOUNT NUMBER	ESTIMATE DIRECT CHARGES LABOR	MATERIAL
0	5	0	0	SIGN CONTRACT	---	---	---
5	10	3	700	LAYOUT AND EXCAVATE	10.04	700	---
5	70	4	950	INSTALL OUTSIDE WATER GAS SEWER POWER	60.00	600	350
10	15	5	1320	CONSTRUCT BASEMENT WALLS BACKFILL	30.01	840	480
15	20	5	850	FRAME FIRST TO SECOND FLOOR	50.01	570	280
15	25	3	470	POUR AND CURE BASEMENT	20.09	300	170
15	45	4	2200	STONE VENEER TO FIRST FLOOR SILL	30.06	600	1600
20	30	2	430	INSTALL FIRST FLOOR WALLS ROUGH FLOOR	50.22	210	220
20	35	6	1200	FRAME SECOND FLOOR TO ROOF	50.01	790	410
25	75	10	350	INSTALL FURNACE WATER HEATER BASEMENT PLUMBING	60.01	100	250
30	40	0	0	DUMMY	---	---	---
30	50	7	650	INSTALL FIRST FLOOR PLUMBING DUCTS WIRE	60.00	300	350
35	40	0	0	DUMMY	---	---	---
35	45	8	1130	FRAME AND SHINGLE ROOF	60.07	600	530
40	50	3	490	INSTALL SECOND FLOOR WALLS ROUGH FLOOR	50.22	270	220
45	55	7	2060	STONE VENEER TO EAVES	30.06	1760	300
50	60	9	480	INSTALL SECOND FLOOR PLUMBING DUCTS WIRE	60.00	320	160
55	65	4	370	INSTALL WINDOWS EXTERIOR DOORS	50.12	120	250
60	65	0	0	DUMMY	---	---	---
60	75	0	0	DUMMY	---	---	---
65	80	10	840	PUT UP DRYWALLS PAINT	60.09	610	230
70	75	0	0	DUMMY	---	---	---
70	90	12	1320	INSTALL SIDEWALKS CURB DRIVEWAY	10.02	700	620
75	85	3	490	FINISH PLUMBING ELECTRICAL TIE-INS	60.00	380	110
80	85	0	0	DUMMY	---	---	---
80	95	5	730	FINISH FLOORING	50.22	500	230
80	100	4	490	INSTALL MILLWORK INTERIOR DOORS	50.06	130	360
85	100	7	1870	INSTALL KITCHEN EQUIPMENT	80.07	620	1250
90	100	10	750	GRADE SITE SOD LANDSCAPE	10.06	640	110
95	100	0	0	DUMMY	---	---	---
100	110	4	500	PUNCH LIST INSPECTION	80.08	500	---
					SUB TOTAL	12,160	8,480

Fig. 19.1

ACCOUNT NO.	DESCRIPTION	LABOR	MATERIAL	TOTAL
10.02	WALKS & DRIVES	$ 700	$ 620	$1320
10.04	LAYOUT, EXCAVATE	$ 700	---	$ 700
10.06	LANDSCAPE	$ 640	$ 110	$ 750
20.09	SLAB ON GRADE	$ 300	$ 170	$ 470
30.01	CONCRETE BLOCK	$ 840	$ 480	$1320
30.06	STONE VENEER	$2360	$1900	$4260
50.01	CARPENTRY FRAMING	$1360	$ 690	$2050
50.06	MILLWORK, INT. DOORS	$ 130	$ 360	$ 490
50.12	WINDOWS, EXT. DOORS	$ 120	$ 250	$ 370
50.22	FLOORING	$ 980	$ 670	$1650
60.00	SUB CONTRACTS	$1600	$ 970	$2570
60.01	PLUMBING	$ 100	$ 250	$ 350
60.07	ROOFING	$ 600	$ 530	$1130
60.09	PAINTING	$ 610	$ 230	$ 840
80.07	KITCHEN EQUIP.	$ 620	$1250	$1870
80.08	PUNCH LIST. GEN	$ 500	---	$ 500

Fig. 19.2 COST SUMMARY BY ACCOUNT NUMBER

COST SUMMARY REPORT

PROJECT: BUILDING A NEW HOME

DATE: 2-28-69

ACCOUNT NO.	DESCRIPTION	BASIS ESTIMATE	EST. CODE	ACT. COST TO DATE	ACT CODE	ESTIMATE TO COMPLETE	DOLLAR DIFFERENCE
							(Overrun)
10.02	WALKS AND DRIVES	1320	1			1320	
10.04	LAYOUT, EXCAVATION	700	1	600	3		100 -
10.06	LANDSCAPE	750	1			750	
20.09	SLAB ON GRADE	470	1	470	3		
30.01	CONCRETE BLOCK	1320	1	1410	3		(90)
30.06	STONE VENEER	4260	1	3612		618	30 -
50.01	CARPENTRY FRAMING	2050	1	2130	3		(80)
50.06	MILLWORK, INT. DOORS	490	1			490	
50.12	WINDOWS, EXT. DOORS	370	1			370	
50.22	FLOORING	1650	1	735		893	22 -
60.00	SUBCONTRACTS	2570	1	1202	1	1543	(175)
60.01	PLUMBING	1350	2	1250	2	140	(40)
60.07	ROOFING	1130	*3	1330	3	452	(652)
60.09	PAINTING	840	1			840	
80.07	KITCHEN EQUIPMENT	1870	1			1870	
80.08	PUNCH LIST GEN	500	1			500	
TOTAL		$21,640		12,739		9786	(885)
PROJECTED COMPL.		$22,525					
DOLLAR DIFFERENCE		*(885)		*Overrun			*(885)

EST. CODES

1. ORIGINAL ESTIMATE
2. REVISED ESTIMATE – REIMBURSIBLE
3. REVISED ESTIMATE – NON-REIMBURSIBLE

ACT. CODE

1. REWORK – NON-REIMBURSIBLE
2. EXTRA – REIMBURSIBLE
3. ALL ACTIVITIES IN THIS ACCOUNT ARE COMPLETE

FIG. 19-3

Fig. 19.3

COST CONTROL REPORT

PROJECT: BUILDING A NEW HOME

DATE THIS REPORT Day 25, 3/07/69
REPORT COVERS TO Day 20, 2/28/69

ORIGINAL ESTIMATE $20,640
REVISED ESTIMATE 21,640

DOLLAR DIFF. $(885)

ACTUAL COST TO DATE $12,739
EST. COST TO COMPLETE 9,786
PROJECTED COMPLETION COST 22,525

6 EST. CODE:
1. Original Estimate
2. Revised Estimate – Reimbursible
3. Revised Estimate – Non-Reimbursible

10 ACT. CODE
1. Rework – Non-Reimbursible Addition
2. Extra – Reimbursible
3. Completed

Activity I-J 1	Cost Code 2	Basis Estimate Labor 3	Basis Estimate Material 4	Basis Estimate Total 5	Est. Code 6	Actual Cost to Date Labor 7	Actual Cost to Date Material 8	Actual Cost to Date Total 9	Act. Code 10	Est. Cost to Complete Labor 11	Est. Cost to Complete Material 12	Est. Cost to Complete Total 13	Dollar Diff. (Over Run) 14	Comments 15
5-10	10.04	700	–	700	1	600	–	600	3	–	–	–	100–	Completed –
5-70	60.00	600	350	950	1	400	300	700	1	350	125	475	(225)	Reject by Inspector
10-15	30.01	840	480	1320	1	960	450	1410	3	–	–	–	(90)	Completed –
15-20	50.01	570	280	850	1	620	280	900	3	–	–	–	(50)	Completed –
15-25	20.09	300	170	470	1	300	170	470	3	–	–	–		Completed –
15-45	30.06	600	1600	2200	1	570	1800	2370	3	–	–	–	(170)	Completed –
20-30	50.22	210	220	430	1	180	210	390	3	–	–	–	40–	Completed –
20-35	50.01	790	410	1200	1	830	400	1230	3	–	–	–	(30)	Completed –
25-75	60.01	600	750	1350	2	500	750	1250	2	140	–	140	(40)	Furnace Changed C.O. #1
30-50	60.00	300	350	650	1	200	302	502		80	18	98	50–	
35-45	60.07	600	750	1350	3*	580	750	1330		452	–	452	(652)	Under Estimate. Refer to Original Est.
40-50	50.22	270	229	490	1	200	145	345		90	73	163	(18)	
45-55	30.06	1760	300	2060	1	942	300	1242		618	–	618	200–	
50-60	60.00	320	160	480	1					320	160	480		
55-65	50.12	120	250	370	1					120	250	370		
65-80	60.09	610	230	840	1					610	230	840		
70-90	10.02	700	620	1320	1					700	620	1320		
75-85	60.00	380	110	490	1					380	110	490		
80-95	50.22	500	230	730	1					500	230	730		
80-100	50.06	130	360	490	1					130	360	490		
85-100	80.07	620	1250	1870	1					620	1250	1870		
90-100	10.06	640	110	750	1					640	110	750		
100-110	80.08	500	–		1					500		500		

mate. The new entry will override the "automatic" projection mentioned previously. The Dollar Difference (or Deviation) is listed to indicate under and over runs. The sum of all the Actual Costs To Date, plus Estimated Costs To Complete is compared to the Basis Total Estimate to indicate the Project Dollar Difference for a project under or over run.

Fig. 19.3 indicates the Cost Control Report for the sample project as of Day #20, or Feb.28, 1969. Please note that this is the same date that the Time Schedule Updating was performed, and is described in Chapter 18. Note from the Original Cost Account and Work Package Estimate, Fig. 19.1, two changes have been made. Activity 25-75 has had a Revised Reimbursible Estimate, and Activity 35-45 has had a Revised Non-Reimbursible Estimate due to an initial under-estimate on the material associated with this activity. Some organizations prefer to show an additional field between Columns 5 and 6 of Fig. 19.3, and that field would be called "REVISED ESTIMATE" and would have the sub-columns "LABOR", "MATERIAL", "TOTAL". In Fig. 19.3, Activity 35-45, Dollar Difference of $652 is obtained by comparing the sum of Actual Cost To Date and Estimated Cost to Complete ($1330 plus $452) to the Original Estimate of $1130 (refer to Fig. 19.1), not the Revised Non-Reimbursible Estimate of $1350.

The Cost Control Report of Project Day #20 (Feb.28, 1969) indicates a Project Overrun of $885 at the present rate. If the profit on this project was taken at 5%, or $1,082 (which is 5% of the Estimated Cost of $20,640), then the project is rapidly approaching a loss situation. Fig. 19.3 reveals that Activities 5-70, 10-15, 15-20, 20-35, have gone over on the labor estimates, which may indicate performance of a foreman or supervisor, or performance by whoever is estimating labor. Activities 15-45 and 5-70 have gone over on material estimates. Activities 5-70, 35-45 and 40-50 are projecting overages on labor.

Another significant fact herein will be revealed by a comparison of the Updated Schedule as indicated in Chapter 18. The project has greatly improved its Completion Date (now projected at Day #51, or Apr.14, 1969 as compared to the original Completion Date of Day #59, or Apr.24, 1969. Generally, there is not much future for the contractor in finishing ahead of schedule, but over his budget.

Fig. 19.3 is the Cost Control Report of the contractor, and is for his internal control. It is not intended for his client. Fig. 19.4 is an Updated Cost Summary by Cost Account Number.

As a variation on the Cost Control philosophy, and Fig. 19.3 above, comparisons are sometimes made of "Actual Cost To Date" with "Estimated Cost To Date", and the Dollar Difference generated. This means a judgement is made as to what percentage of the original Work Package should have been committed, or consumed, by the report date. The "Estimated Cost To Date" is subtracted from the original estimated cost, and the remainder is "Estimated Cost To Complete."

COST SUMMARY REPORT

PROJECT: BUILDING A NEW HOME   DATE: 2/28/69

ACCOUNT NO.	DESCRIPTION	BASIS EST.	EST. CODE	ACT.COST TO DATE	ACT CODE	EST. TO COMPLETE	DOLLAR DIFF. (OVERRUN)
10.02	WALKS AND DRIVES	1320	1			1320	
10.04	LAYOUT, EXCAVATE	700	1	600	3		100-
10.06	LANDSCAPE	750	1			750	
20.09	SLAB ON GRADE	470	1	470	3		(90)
30.01	CONC. BLOCK	1320	1	1410	3		
30.06	STONE VENEER	4260	1	3612		618	30-
50.01	CARPENTRY FRAMING	2050	1	2130	3		(80)
50.06	MILLWORK,INT.DOORS	490	1			490	
50.12.	WINDOWS, EXT. DOORS	370	1			370	
50.22	FLOORING	1650	1	735		893	22-
60.00	SUB CONTRACTS	2570	1	1202	1	1543	(175)
60.01	PLUMBING	1350	2	1250	2	140	(40)
60 07	ROOFING	1130	*3	1330	3	452	(652)
60.09	PAINTING	840	1			840	
80.07	KITCHEN EQUIP.	1870	1			1870	
80.08	PUNCH LIST GEN	500	1			500	

TOTAL $21,640   12,739   9786   (885)

PROJECTED COMPL. $22,525·

DOLLAR DIFFERENCE: *(885)   *OVERRUN   *(885)

EST. CODES
1. ORIGINAL ESTIMATE
2. REVISED ESTIMATE-REIMBURSIBLE
3. REVISED ESTIMATE-NON-REIMBURSIBLE

ACT.CODE
1. REWORK-NON-REIMBURSIBLE
2. EXTRA-REIMBURSIBLE
3. ALL ACTIVITIES IN THIS ACCOUNT ARE COMPLETE

Fig. 19.4

Again, "Estimated Cost To Complete" can be a re-estimate rather than a sub-tractive projection. In practice, "Estimated Cost To Complete" should be a re-estimate. Control depends on surveillance; a bookkeeping system and/or a data processing machine will survey numbers, not conditions. Human judge-ment (Estimated Cost To Complete) can never be replaced by a system or a machine.

The frequency of cost collection, and re-estimating, depends on the duration and magnitude of the project. The same rules that were developed for updating time schedules apply to Cost Control.

Fig. 19.5 is a typical invoice, based on the Network Diagram Plan, and sub-mitted by the contractor to the client. Note for his Activity 35-45, where a big overage has occurred, he has essentially guessed at the invoice amount of $1,000. This would be hard to substantiate by time sheets and suppliers in-voices. The Cost Control report indicates $452 projected to be spent, and he has already spent $1330 on this activity. Fig. 19.5 indicates the initial in-voice. Each subsequent invoice would show accumulated sub-totals.

The simplified example here does not show a separate breakout for Overhead and Profit. It can be assumed that it is included proportionately in the costs shown in the above figures. If Overhead and Profit are shown on a Cost Con-trol Report as separate entries, it is suggested that Profit be considered as the first Cost entry. It will lead the cost items, and gives an immediate pic-ture as to the status of Profit on the progressing project.

The Percent Complete Fallacy

It is important to note here the problem of "Percent Complete" reporting. A common fallacy is the assumption that the Work Package Time (Activity Dur-ation) and the Work Package Cost (Labor & Material) are directly proportional. This is not so, since cost is comprised of Labor and Material. Labor and dur-ation may be proportional, but material, the quantity survey item, is not spread over the duration of the activity. In the Cost Projection Curves that follow (Fig. 19.6), Material costs are assigned to the first day of the particular ac-tivity, where Labor is distributed over its entire duration. Thus, if an activity has been estimated at 10 days duration, with a total Labor and Material cost of $1,000, and is reported 50% complete in time (5 days left to execute), it cannot be assumed that $500 of the cost has been expended or commited. It could well be that $700, representing all of the material, and 50% of the labor, has already been expended by the end of the 5th day. Time and Cost are not proportionately related, except on a pure labor activity. A Project Control System must separate Material and Labor costs for effective control. The ef-fect of commiting Material costs on the first day of the activity (less, of course, any specified retainage) produces a more realistic cost picture. Extended re-tainage or hold-back will be accumulated until the specified completion, then assigned.

287

X Y Z CONSTRUCTION CO.

INVOICE NO.: 1   Date: **Mar.10,1969**
PROJECT: Building a New Home
PERIOD: **Feb.3 to Feb.28, 1969**

Activities Completed This Date	Value	Previously Invoiced	This Invoice
5-10 Layout, Excavate	$ 700	--	$ 700
10-15 Basement Walls, Backfill	$1320	--	$ 1,320
15-20 Frame 1st to 2nd Flr	$ 850	--	$ 850
15-25 Pour Cure Basement Slab	$ 470	--	$ 470
15-45 Stone Veneer to Sill	$2200	--	$ 2,200
20-30 1st Flr Walls, Rough Flr	$ 430	--	$ 430
20-35 Frame 2nd Flr to Roof	$1200	--	$ 1,200
Activities Started This Period			
5-70 Install Outside Services	$ 950	--	$ 700
25-75 Install Furnace, Htr	** $1350	--	$ 1,250
30-50 1st Flr Plumb. Ducts Wiring	$ 650	--	$ 502
35-45 Frame & Shingle Roof	$1130	--	$ 1,000
40-50 2nd Flr Walls, Rough Flr	$ 490	--	$ 345
45-55 Stone Veneer to Roof	$2060	--	$ 1,242

** Revised, Change Order #1

This Invoice	$12,209
Retainage, 10%	$ 1,221
Amount Due	$10,988

Invoiced To Date: $ 10,988
Total Contract: $ 21,640

Percent Invoiced
   To Date   50.7%

Attested: _J.M. SHIFTY_

TITLE: President

Fig. 19.5

The second cause of "Percent Complete" problems is the fact that a small percentage of activities in any plan control the time status of the project; the position on the Critical Path (10% or less of all the activities) determines if the project is on schedule. The other 90%, with Float time, cause a Time-Cost Domain relationship, not a proportional curve relationship. If all the activities are started at their Early Start time, costs would accumulate as indicated in Fig. 19.6. If all the activities are started at their Latest Start time, the second boundary of the Time-Cost Domain is established. The position of the project in Time is the position on the abscisa, the horizontal time ordinate, and this position is controlled by the execution of the Critical Activities.

Fig. 19.6 is the COST PROJECTION Chart for "Building A New Home." The original estimated costs have been shown first accumulating if all activities started at their Early Start time. Material costs are assigned to the first day of each activity; Labor costs are pro-rated over the duration of the activity. Note that 50% of the project could be expended by the 21st day if all activities started early. If all activities are started at their Latest Start time (a maximum risk condition, since all activities are Critical in this case), costs will accumulate (or be commited) on a later schedule. Thus, 50% of the costs in this case do not occur until the 30th day.

Therefore, the Percent Complete reporting must give two reports to be meaningful; the percent of time left to complete, and the percent of remaining costs. Time and costs cannot be lumped into a single Percent Complete Report.

From the Cost Control report of Fig. 19.3, the accumulated costs have also been indicated on the Cost Projection. It is definitely indicated as penetrating the Early Start-Cost boundary, indicating the good possibility of finishing ahead of schedule, but over the estimate or budget. Conversely, if accumulated actual costs trend towards penetrating the Late Start-Cost boundary, the project will probably be within the estimate or budget, but will probably be behind schedule. Somewhere between the two boundaries lies the ideal schedule. Refer to Chapter 12 for one method of establishing the ideal schedule based on Distributed Float. An approximation may be made by roughly splitting the area in the Time-Cost domain. This may be used for Cost Forecasting. If the executor of the project is financing the project by borrowing, he need borrow only that amount necessary to keep operating until sufficient cash from his invoices is derived to support the balance of the project. Generally speaking, the executor (contractor) prefers to move along the Early Start boundary, so he is in essence "playing" with the owner's money. The unbalanced estimate, that of overweighing initial activities with money, for early inflow, is perhaps somewhat easier to see on the Network Cost Project, as compared with the Bar Chart (Gantt Chart) approach. The cost milestones easily noticed on Fig. 19.6, will enable the executor (contractor) to borrow less. If he is working on a credit commitment, a series of these cost Projections, overlaid on a common calendar, will project the number of concurrent projects the executor can handle financially.

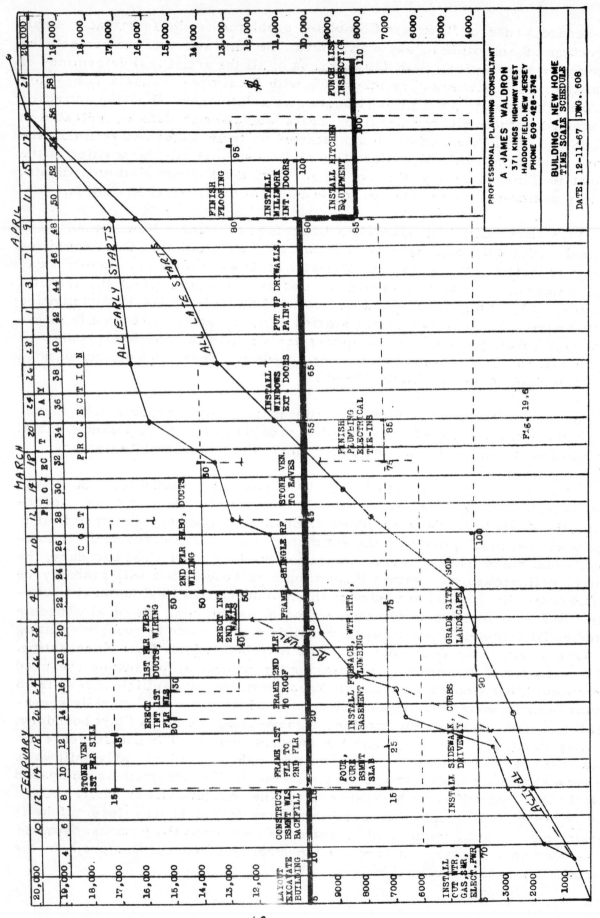

Fig. 19.6

A. JAMES WALDRON
PROFESSIONAL PLANNING CONSULTANT
371 KINGS HIGHWAY WEST
HADDONFIELD, NEW JERSEY
PHONE 609-428-3742

BUILDING A NEW HOME
TIME SCALE SCHEDULE

DATE: 12-11-67   DWG. 608

The owner's savings from the Cost Projections are even more graphic. If the project is being financed from assets and securities, these can be liquidated at the latest feasible time, protecting whatever income is being generated from the assets and securities for the maximum period of time, keeping the principal as high as possible for as long as possible. If the owner receives the funds for his project at the beginning (such as a bond issue for a new school, or funds raised for a church), then a greater portion of this money can be placed in higher-return long term investments, handling the balance in lower return short term investments. The longer the overall project duration, the stronger this incentive becomes.

PERT COST

The philosophy of Cost Control developed above has been essentially adopted by the government. PERT Cost uses the common framework of the Network Plan to integrate costs (and resources) into an overall project. Since most government programs are of enormous size and costs, the PERT Cost system is designed to break the costs down to a reasonable level, then allow upward summarization of accumulated costs in a pyramidal fashion. The highest level report will be a summary of all the lower levels, with time and cost projections accordingly lumped. After the project is defined, broken into end items, then into end item subdivisions, the project is broken into further detailed units.

The subdivision of the work breakdown structure continues to successively lower levels, reducing the dollar value and complexity of the units at each level, until it reaches the level where the end item subdivisions finally become manageable units for planning and control purposes. The end item subdivisions appearing at this last level in the work breakdown structure are then divided into major work packages (e. g., engineering, manufacturing, testing). At this point, also, responsibility for the work packages will be assigned to corresponding operating units in the contractor's organization.

The configuration and content of the work breakdown structure and the specific work packages to be identified will vary from project to project and will depend on several considerations: the size and complexity of the project, the structure of the organizations concerned and the manager's judgement concerning the way he wishes to assign responsibility for the work. These considerations will also determine the number of end item subdivisions that will be created on the work breakdown structure before the major work packages are identified and responsibility is assigned to operating units in a contractor's organization.

Further Functional or Organizational Subdivisions. An organization unit will usually identify smaller work packages within the major work packages assigned to it. This division of work may take the form of more detailed functional (e. g., engineering) identification, such as systems engineering, electrical engineering, mechanical engineering, or it may take the form of a more detailed end item identification within engineering, such as instrumentation engineering, power cable engineering, missile section assembly engineering, and so forth. The form chosen for more detailed identification will depend, again, on the structure of the performing organization and the manager's judgement as to the way he wishes to assign responsibility for the work. The number of these smaller subdivisions will naturally depend on the dollar value of the major work packages and the amount of detail needed by the manager to plan and control his work. Normally, the lowest level work packages will represent a value of no more than $100,000 in cost and no more than three months in elapsed time.

THE WORK PACKAGES FORMED AT THE LOWEST LEVEL OF BREAKDOWN THEN, CONSTITUTE THE BASIC UNITS IN THE PERT/COST SYSTEM BY WHICH ACTUAL COSTS ARE (1) COLLECTED, AND (2) COMPARED WITH ESTIMATES FOR PURPOSES OF COST CONTROL.

A "TREE", starting with the overall project, and then breaking down into lower level end items (an "end item" represents hardware, services, equipment or facilities that are to be delivered to the government, or that constitute a commitment on the part of the contractor), and finally down to "WORK PACKAGES."

A "Work Package" as defined in the PERT Cost Manuals is regarded as a unit of work required to complete a specific job, such as a report, a design, a piece of hardware, or a service, which is within the responsibility of a single operating unit in an organization. Up to this point, a "Work Package" has been considered an activity. In PERT Cost, due to the enormity of the projects and, therefore, the complete Project Master Plan, a Work Package consists usually of groups of activities performed by one unit in a company or organization.

Fig. 19.7 shows the PERT COST TREE for our reference project, "Building A New Home." The House is broken into two major end items at the second level, those of Site Work and Structure. The "Structure" level is further broken down to "Foundation", "Structural Shell", and "Interior Work." The fourth level shown for the "Structure" parent category has both the Structural Shell and Interior Work parents broken down further. In this simplified example, many of the activities become "Work Packages" by themselves. A more typical example of a PERT Cost Work Package is item 10B022. The two stone veneering activities on the Network Plan become one "Work Package."

The Summary Numbers, usually assigned by the Contracting Office of the particular government agency, are used to identify the work packages to be grouped together in summarizing manhour (labor) and cost information for an item appearing on a higher level of the work breakdown structure. Present PERT Cost programs allow up to 16 levels to be established.

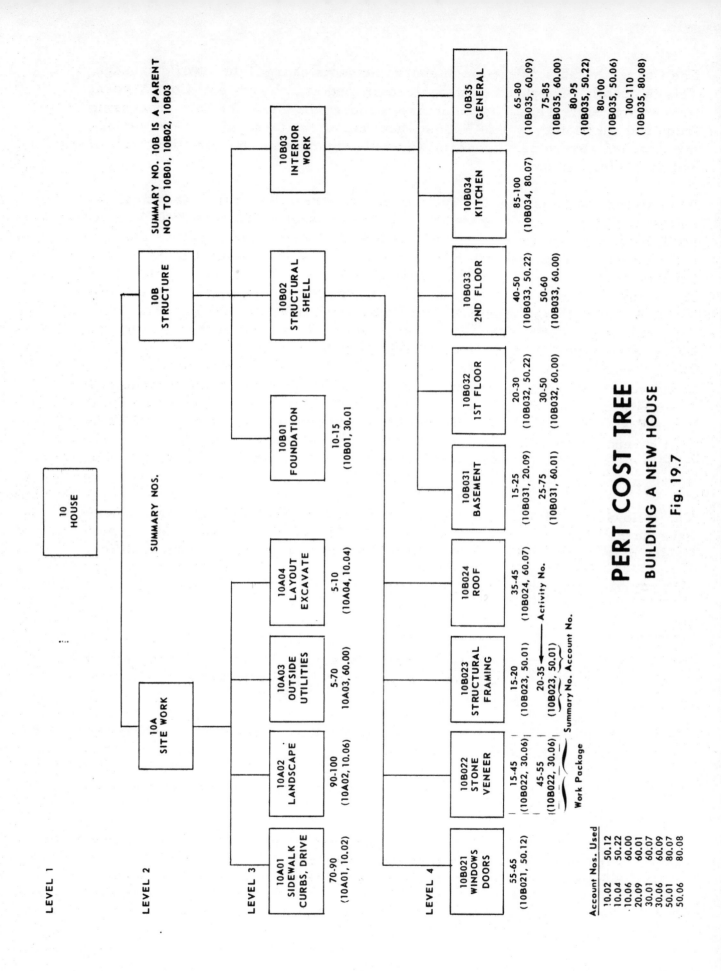

PERT COST TREE

BUILDING A NEW HOUSE

Fig. 19.7

The Charge Number is used to identify the costs charged to a work package. This is usually furnished by the contractor, and is to be his own Cost Account Number system. Thus, the Summary number will allow the data processing (required inevitably by PERT Cost, because of the mass of information desired on the government program) to easily summarize and sort the manhour and cost information.

In an attempt to minimize the Parkinson's Law reaction to the statistical result of the three time estimate (the t_e), the present PERT technique has expanded its nomenclature to include the one time estimate, either as an estimate or as an assigned scheduled duration. This is now identified by t_S - a scheduled duration. When the Forward and Backward Passes are performed, the results are called S_E - The Earliest Completion Time. This is the same as the Earliest Finish Time (EF) developed in Chapter 8. The Backward Pass produces the S_L, the Latest Completion Date. This corresponds with the Latest Finish Time (LF) developed in Chapter 8.

Note two things: this is primarily an exercise in semantics, in order to differentiate between the three time estimate and the one time estimate (be it estimates or "assigned" it is still an initial estimate), and secondly, PERT is still a report oriented rather than a control oriented system. PERT is primarily designed for the Planner to report his status to a client government agency.

The following are typical output reports from the PERT Cost system. These have been taken from the "PERT COST SYSTEMS DESIGN", a DOD-NASA Guide, dated June, 1962, and available from the Government Printing Office.

PERT/COST Management Reports

Figure

19.8 Management Summary Report -- A summary for each management level, showing the current date and projected time and cost status of a project.

19.9 Manpower Loading Report -- A projection of monthly manpower requirements by skill categories.

19.10 Manpower Loading Display -- A graphic representation of the Manpower Loading Report.

19.11 Cost of Work Report -- A graphic means of comparing the "Actual expenditures and commitments to date" and the "Budgeted rate of expenditures" with the "Contract estimate for the work performed to date."

PERT/COST MANAGEMENT SUMMARY REPORT — PROGRAM: MWS — SUMMARY LEVEL: 4 - CONTROLS — REPORT COVERS THE PERIOD: 1 July 1961 - 31 March 1962

PERT/COST MANAGEMENT SUMMARY REPORT — PROGRAM: MWS — SUMMARY LEVEL: 4 - CASE — REPORT COVERS THE PERIOD: 1 July 1961 - 31 March 1962 — PROJECT: A10 VEHICLE — CONTRACTOR: — CONTRACT NO. 6593 — DATE THIS

PERT/COST MANAGEMENT SUMMARY REPORT	PROGRAM: MWS	SUMMARY LEVEL: 3 - PROPULSION	REPORT COVERS THE PERIOD: 1 July 1961 - 31 March 1962
	PROJECT: A10 VEHICLE — CONTRACTOR: MISSILE SYSTEMS CO.	CONTRACT NUMBER 659	DATE THIS REPORT: 3/31/62

ITEM	COST OF WORK: WORK PERFORMED TO DATE $ — Original Estimate	Actual Costs	Overrun (Underrun)	TOTALS AT COMPLETION $ — Contract Estimate	Latest Revised Estimate	Projected Overrun (Underrun)	SCHEDULE (Δ - Scheduled completion date of total item; E - Earliest completion date, L - Latest completion date of most critical element within item; 1961 J A S O N D, 1962 J F M A M J J A S O N D, 1963 J F M A M J)	DAY	SLACK STATUS (Weeks)	REMARKS (S = Schedule Slippage, O = Cost Overrun, U = Cost Underrun)
TOTAL PROPULSION	850,000	1,050,000	200,000	2,500,000	2,850,000	350,000	E L Δ	28 31	-8.0	1. Case (S,O) 2. Controls (O)
Case	72,000	186,000	114,000	596,000	814,000	218,000	E L Δ	12 17	-8.0	1. Mounting (S,O)
Controls	392,000	411,000	19,000	704,000	793,000	89,000	E L Δ	27 05	10.0	1. Staging Transducer (O)
Servo	126,000	174,000	48,000	387,000	447,000	60,000	E L Δ	18 02	2.0	
Nozzles	67,000	64,000	(3,000)	378,000	346,000	(32,000)	E L Δ	28 31	-8.0	1. Cone (S,U)
Ignition	114,000	141,000	27,000	262,000	279,000	17,000	E L Δ	08 12	5.0	
Aux. Power Units	79,000	74,000	(5,000)	173,000	171,000	(2,000)	E L Δ	31 31	4.0	
							TIME NOW			

Figure 19.8

MANPOWER LOADING REPORT				SKILL: 16
PROGRAM: MWS		REPORT DATE: 3/31/62		
PROJECT: A10 VEHICLE		CONTRACT NUMBER:		98-7865
LEVEL: (3) PROPULSION				
Month	Performing Unit	Charge No.	Estimated Man-hours	Activity Slack (weeks)
6/62	6821	39786340	1000	-4.0
	6821	39782191	2000	8.0
	5211	39784213	4000	12.0
			7000*	
7/62	6821	39782315	800	1.0
	6821	39782191	1000	8.0
	5211	39784213	200	12.0
			2000*	

Figure 19.9

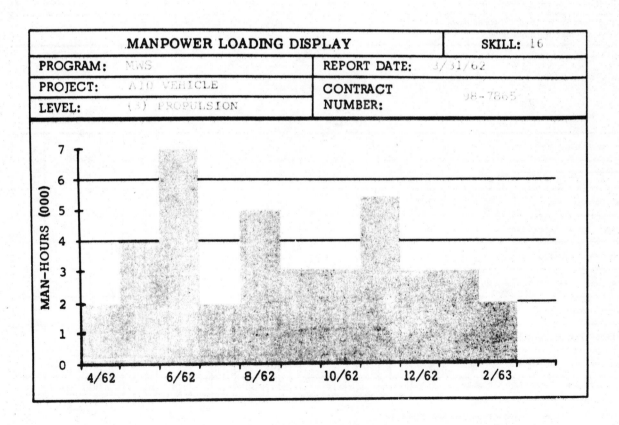

Figure 19.10

Reports Control Symbol

2AFSC-R32

As of Date

30 September 1961

CURRENT YEAR MILESTONE SCHEDULE

LINE	MILESTONES	FY 1961 (J F M A M J) / Cy 1962 (J A S O N D) / FY 1962 / Cy 1961 (J F M A M J)
1	Fuel Quantity Gaging	
2		
3	Engine PERT	
4		
5	Complete R&D DEI	
6		
7	Prototype Delivered	
8		
9	Letter Contract Award	
10		
11	Material Support Plan	
12		
13	PERT Implementation	
14		
15		
16	Deliver 1st Production Article	
17		
18	R&D Testing Ends	
19–33		

Typed Name and Title of Authenticator

FRANK J. NORTON, CAPT USAF

Signature of Authenticator

Frank J Norton

Figure 19.11

PERT MILESTONE REPORT

DATE 1/24/61   WEEK 107.9

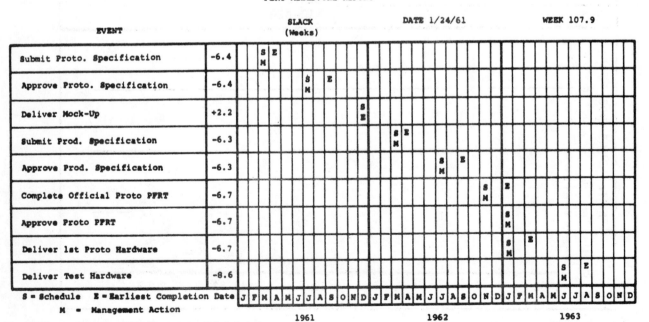

EVENT	SLACK (Weeks)
Submit Proto. Specification	-6.4
Approve Proto. Specification	-6.4
Deliver Mock-Up	+2.2
Submit Prod. Specification	-6.3
Approve Prod. Specification	-6.3
Complete Official Proto PFRT	-6.7
Approve Proto PFRT	-6.7
Deliver 1st Proto Hardware	-6.7
Deliver Test Hardware	-8.6

S = Schedule   E = Earliest Completion Date

M = Management Action

Slack = Degree of Constraint to End Schedule

1961   1962   1963

Figure 19.12

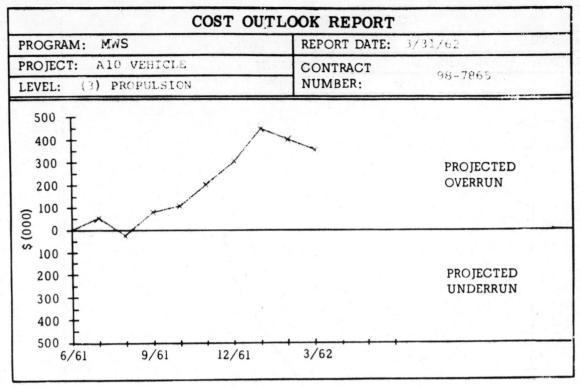

Figure 19.13

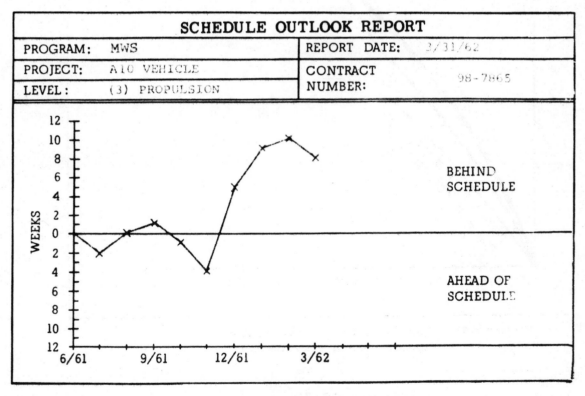

Figure 19.14

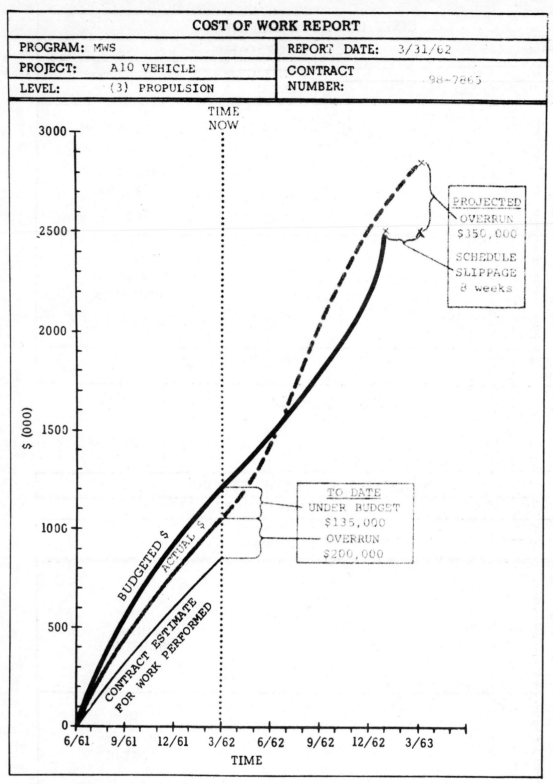

Budgeted $ = Planned rate of expenditure
Actual $ = Expenditures & commitments made to date

Figure 19.15

PERT/COST Management Summary Report

The PERT/COST Management Summary Report (Fig. 19.8) shows the overall schedule and cost status of both the project as a whole and of each of the major component items. It also indicates the problem areas that require management attention.

The report provides each manager with the following information relative to his area of responsibility.

- the cost overrun or underrun to date (a measure of cost performance), through a comparison of the estimated costs with actual costs for the work performed;

- the projection of cost overrun or underrun for the total project, which is obtained by comparing the original cost estimate for the project with the actual costs plus the estimated costs to complete the project (latest revised estimate);

- the amount of schedule slippage as indicated by the difference between the established schedule for project completion and the present expected date for project completion.

- the identification of trouble spots -- that is identification of those areas of the project where the cost or time status requires management attention.

Each Management Summary Report will normally be accompanied by a brief written analysis.

In the example shown, the manager responsible for the Propulsion effort would see that:

- the Propulsion System Development Effort, scheduled for completion on 31 January 1963, is now expected to be completed on 28 March 1963, a slippage of eight weeks;

- there is a cost overrun to date of $200,000 and a projected $350,000 overrun at project completion;

- the major contributor to this cost overrun is the Case, which is also critical in terms of completing the project on schedule;

- the Nozzles, which are also critical to the total effort,

show both a cost underrun and a time slippage that may require more resources to get back on schedule;

. the Controls effort is ahead of schedule, but shows a cost overrun, which might indicate that extra costs are being incurred unnecessarily.

PERT/COST Management Summary Reports are prepared for managers at each level or echelon of the project or program structure. One report is prepared, for example, for the entire Missile Weapon System (level 1). At level 2, a similar report is prepared for each major element of the program, such as ships, missiles and fire control. At the next lower level, level 3, the major elements of the program are subdivided again and a PERT/COST Management Summary Report is prepared for each manager to whom responsibility is assigned. The missile, for example, is divided into elements such as propulsion, re-entry body, and ballistic shell. The report illustrated here is for the Propulsion Manager, and as can be seen from the illustration, further subdivisions and management reports are prepared at such lower levels of the project as are considered necessary by the program manager.

In analyzing the status of a project, the responsible manager would examine the reports for those end items where trouble is indicated. He would then refer to the lower level reports as required to isolate the trouble.

In the situation illustrated, the Propulsion Manager would examine the Management Summary Report for the Case, since the Case is a major contributor to the cost overrun and the schedule slippage.

Manpower Loading Reports and Displays[1]

The two manpower reports showing the time-phased planning for manpower assignments are usually prepared by skills within a project. These reports are used by management to plan the application of manpower and to determine the need for overtime, additional hiring, or rescheduling of activities.

The Manpower Loading Display (Fig. 19.10) shows the overall requirements for drafting manpower (skill 16) while the Manpower Loading Report (Fig. 19.14) indicates the allocation of man-hours among the various work packages (activities or groups of activities) as identified by the Charge Number. This report allows the responsible manager to shift resources to alleviate the problems of idle manpower or excessive manloading requirements.

---

[1] Manpower loading reports are optional. They are not essential to the operation of the PERT/COST System but are useful in the internal planning and subsequent re-allocation of resources.

Referring to the Manpower Loading Display and Report examples, the manager can determine that:

- the irregular loading pattern may make overtime necessary in some months, even though preceding or succeeding months have unused capacity;

- the heavy loading in June, 1962, is tentatively planned largely for a work package with positive slack (Charge Number 39786340), while, in the same time period, there is a work package with negative slack (39786340) which might benefit from those resources;

- the heavy loading in June, 1962, could be reduced by shifting resources within some activities from June to July or by rescheduling slack activities for July, thereby lowering the man-hour requirement in June.

Cost of Work Report

The Cost of Work Report (Fig. 19.11) shows the project manager:

- the budgeted costs to perform the work;

- the actual costs (expended and commited) to date;

- the contract estimate for the work performed (progress) to date;

- the projection of costs to project completion, based on actual costs to date and estimates-to-complete for work not yet performed.

A comparison of the actual costs accumulated to date and the contract estimate for the work being performed to date will show whether the work is being performed at a cost which is greater or less than planned.

In the example shown in Fig. 19.11, the manager can quickly determine that:

- the project is $135,000 under budget in terms of the time-phased cost plan through March 1962;

- the project is $200,000 over the cost estimate for the work that has now been performed;

- a $350,000 overrun is anticipated at the completion of the project;

.   a schedule slippage of eight weeks is predicted for the
    project.

Rate of expenditure information, similar to DD Form 1097 data, can be shown on the Cost of Work Report, or can be displayed separately.

Schedule & Cost Outlook Reports

The Cost Outlook Report (Fig. 19.12) and the Schedule Outlook Report (Fig. 19.13) show the trend of successive monthly projections of the time and cost to complete the work. Each month new projections are obtained from the project Status Report and these projections provide new entries for the Cost and Schedule Outlook Reports.

For the Propulsion System, the manager could determine that:

.   as of the end of October 1961, the anticipated completion
    of the System was one week ahead of schedule, with a
    projected overrun of $100,000;

.   as of the end of November, the outlook was for comple-
    tion of the System four weeks ahead of schedule but at a
    projected cost overrun of $200,000;

.   by the end of January, both the cost and schedule pro-
    jections were significantly poorer;

.   from the end of January to the end of March (the cur-
    rent time), the cost projection improved considerably,
    while the schedule projection deteriorated further and
    then showed slight improvement.

By relating the trend of these projections to previous management decisions, the manager can observe the effects of these decisions on the cost and schedule for the project. He can determine, on a month to month basis, whether or not the actions taken to control schedules and costs are producing the desired results.

Milestone Reports

Selected PERT/COST network events will represent major milestones of accomplishment toward the completion of a program. It is often useful to give additional program status guidance to management by reporting on these identifiable milestones. This may be accomplished either by written comments in the "remarks" section of the PERT/COST Management Summary Report in which major milestones and their status are shown for each level of management. The report forms presented in Fig. 19.14 and 19.15 are included as

ing use of the described technique, but even then (1974) a small percentage of those organizations who could benefit by these approaches will be only just aware of them, much less using them. If the reader noticed the dates, then assumed that the material is "old" and has been superceded, it is suggested that his thinking be re-evaluated. These charts are examples to explain a technique--the dates are irrelevant.

Secondly, it has been the subtle intention to point out that the technique is similar to a Science Fiction Time Machine. Its normal function is to project into the future, as a predictive model and tool. The technique can also go backward in time to reconstruct history in an organized and coherent form from available records and documents. The historical manipulation is the method of presenting information in claims and litigation applications. The Time Scale Charts of Chapter 11 are the basic vehicle of presenting original versus actual schedules.

INVOICELESS COST CONTROL

A simplified cost control procedure designed to minimize paper work, yet allowing the project manager to maintain adequate fiscal control over his project, is based on the assignment of costs to each activity on the Time Scale Project Control Chart that was developed in Chapter 11A. Fig. 19.16 is such an example, utilizing our common example of the BUILDING A NEW HOME project. Each time-scaled activity is assigned a dollar value representing materials, labor, overhead burden and profit for that activity. It is determined by the sub-contractor, department, or agency performing that work.

Current practice is for each sub-contractor, department, or agency performing that activity to submit a requisition periodically (usually monthly) during the life of the project. The requisition or invoice represents the value of work or services performed during that period. Payment is made, once the work or service has been verified, with a retainage of a percentage of that invoice in order to maintain control.

Thus, if at the end of a month or any other contract designated period, the Mason Sub-contractor has finished 90% of a wall, he will submit an invoice for that amount of work. Yet 90% of a wall does no one any good. If a farmer retains a contractor to fence in a pasture for his prize thoroughbred, and at the end of a month 90% of that work is complete, it is obvious that the work completed is of no value to the farmer.

The concept of Invoiceless Cost Control not only minimizes paper work, it eliminates the abuse of Periodicity (see Chapter 23) and properly assigns the costs to the "work package" activities on the Project Control Chart. The system is extremely simple. Each sub-contractor, department, or agency receives a reproducible copy of the Time Scaled Project Control Chart. They had participated in the construction of the arrow diagram that generated this because they had furnished the logic and the time duration estimates.

Naturally each sub-contractor or other party insists that his costs be kept confidential. Each one, therefore, notes on his Project Control Chart his assigned costs per activity. These are all accumulated on a Master Project Control Chart that is held confidential by the project manager. Each Subcontractor or other participant receives a copy of his cost annotated Project Control Chart. Time progress is maintained on the basic Time Scale Project Control Chart. This is used at the periodic (usually weekly) Project Progress Meeting where a line is drawn down the day of the meeting on the Project Control Chart, and completed activities are colored in. If marked off activities are to the right of the data line, that activity is ahead of schedule. If the marked activities are to the left of the data line, and do not meet the date line, than that activity, or chain of activities is behind schedule. A Project "Look Ahead" study is readily available--if a delinquent sub-contractor is behind schedule, the amount can be readily seen, and a near future event can be determined so that in the next period (usually weekly) he is instructed to reach that event, finish that amount of work by whatever means he has to, and come back on schedule.

Once an activity is completed and accepted by the usual inspection and verification procedures (quality control is still maintained by this concept), there is a payment authorized for the already priced activity. One client of the author's pays within 48 hours after acceptance of the completed activity. Delivery of equipment and material is handled the same way. Any retainage of costs is also maintained. Any sub-contract work that is not represented by an activity, such as rental of cranes or other equipment, or an activity that cannot be estimated, such as dewatering, which may be on a per diem basis, can be represented by a series of bars or a continuous time scaled bar that are not connected to the logic of the project plan. These costs are pro-rated over milestone periods on the project plan.

While the original concept was to primarily reduce paper work, another benefit accrued on an actual project where this approach was used, and which was not anticipated. There was continuous co-ordination and expediting; the sub-contractors who did finish work at the end of the project were continually expediting those sub-contractors who performed initial excavation and structural work. Each knew when his activity was finished and accepted, payment within 48 hours would be received.

Either the total cost assigned to an activity is considered due at the completion and acceptance of that activity (scheduling of inspectors becomes paramount to all involved), or the costs can be broken down per time unit for that activity, with the material costs assumed accumulated at the first time unit of that activity, with the other labor, burden and profit factors essentially pro-rated over the life of that activity. For example, a "Form and Reinforce" activity may assume that the reinforcing steel material costs payable the first day of that activity. This approach allows the project manager to establish control of those activities that are partially completed, and subsequent activities partially started.

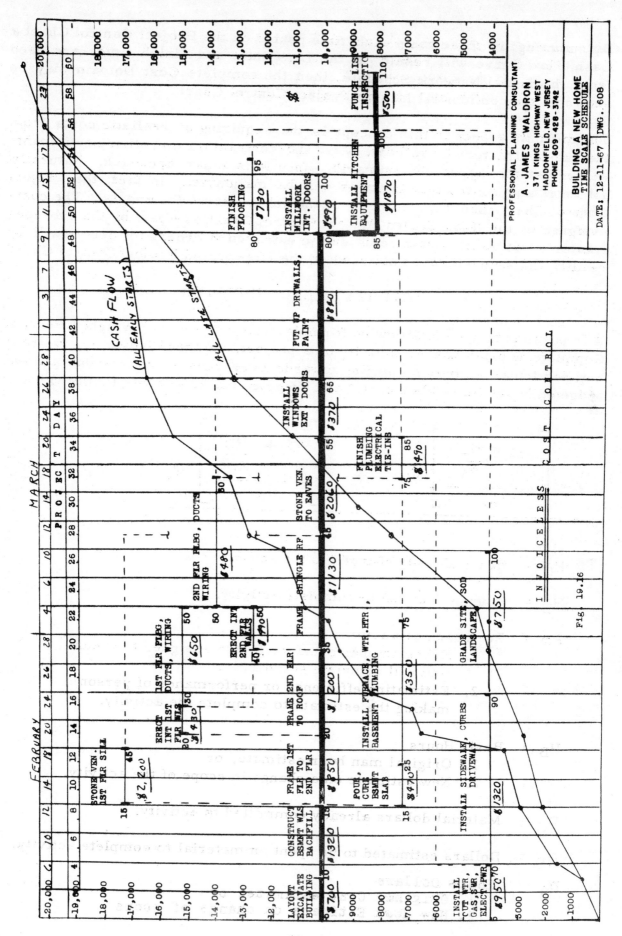

Fig. 19.16

By summing up these costs each time period on the Project Control Chart a Cash Flow Curve will result. If the Late Start and Finish times are used from the initial Network Schedule, then the complete Cost Domain may be shown on the confidential Master Project Control Chart.

The difficult aspect of this concept is the acquiring of realistic costs from the sub-contractors by activity. Most sub-contractors and other participants will resist this approach, primarily since it is a new approach. Naturally, all legal and contractual controls will be maintained; in fact, the Project Control Chart should become part of the sub-contract documents. The costs assigned to the Time Scaled Project Control Chart are part of the sub-contractor's contract. Extra costs due to extra work orders or delays will be readily and sequentially established on the Time Scaled Project Control Chart.

ACTIVITY TREND INDICATOR

It is possible to use the following formula, which will give a Weighed Percent Complete number to each activity on a project, as trend indicator, to see if any particular activity is tending towards over-runs or under-runs against budgeted hours (thus labor costs) and budgeted material costs (dollars).

$$Z = \frac{\left[\dfrac{H_C + \dfrac{H_E}{P_F}}{H_B}\right] W_1 + \left[\dfrac{D_C + \dfrac{D_E}{E_F}}{D_B}\right] W_2}{2}$$

H_C = Hours already charged to the activity.

H_E = Hours estimated to complete activity.

P_F = Performance Factor
    1. Efficiency performance factor assigned to group who will complete activity, or
    2. Estimating efficiency or performance of person making the estimate to complete the activity.

H_B = Basis Hours
    1. Original man hour estimate, or
    2. New estimate on a change in scope of the activity.

D_C = Material dollars already committed to activity.

D_E = Dollars estimated to be spent on material to complete activity.

D_B = Basis Dollars
    1. Original Cost estimate, or
    2. New Cost Estimate on change of scope

E_F = Estimating Factor
1. Allowance of escalation of material costs, or
2. Estimating efficiency factor of person making estimate to complete activity.

W_1 = Weight Factor on hours, to balance time aspect of activity. Will equal 2 if no material costs are involved on activity.

W_2 = Weight Factor on material, to balance material costs of activity. Will equal 2 if no labor hours are charged to activity.

P_{C_W} = Weighed Percent Complete of activity.

P_{C_E} = Estimated Percent Complete of activity.

Z = Activity Trend Indicator

$$P_{C_W} = \frac{P_{C_E}}{Z} \qquad W_1 + W_2 = 2$$

if $0.97 < Z > 1.03$, activity within budget

if $Z > 1.03$ activity over run indicated

if $Z < 0.97$ activity under run indicated

CHAPTER 20

PRECEDENCE DIAGRAMMING

A variation of the Arrow Diagramming technique, which derives from Industrial Engineering Flow Chart techniques, is one wherein the activity or "work package" is shown in a box or circle, and interconnected by arrows for the logical relationship. Another sub-variation is the construction of the activity "boxes" on a vertically gridded chart, with each vertical line a sequential step in the project. Horizontal lines show the logical sequential (or precedential) relationship of the various activities. A fairly common term for this system is "Activity On Nodes." This technique eliminates the "dummy" or logic transfer agent, since the interconnections show the precedential interrelationships.

A comparison of the development of the "Activity On Nodes" diagram with the Arrow Diagram is shown in Fig. 20.0.

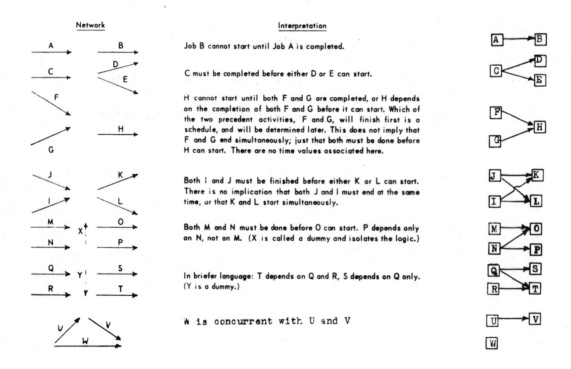

Fig. 20.0 Precedence Diagramming

Fig. 20.1 indicates the most common method of Precedence Diagramming, with the activity boxes interconnected by arrows. The Network Plan for the fabrication of rocket engine sections, Fig. 2.13 of Chapter 2, is shown in Fig. 20.1.

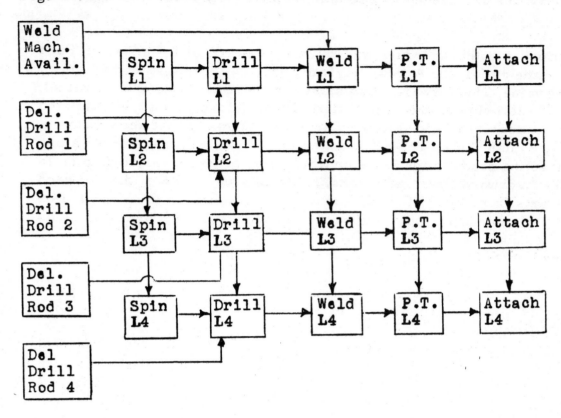

Fig. 20.1 Precedence Diagram, Manufacture of Rocket Engine Sections

Any logic restriction can be shown on the Precedence Diagram without the dummy. However, there are times when an isolated activity, or chain of activities, which have no other predecessor or successors, such as operation W on the last line of Fig. 20.0 above, and these would be shown by themselves. When the Forward and Backward Pass are made to establish the schedule, in the exact manner of the Arrow Diagram, as described in Chapter 5, these isolated activities could be overlooked. So a zero duration activity is added at the beginning and the end of the diagram to close the network, as shown in Fig. 20.2 below.

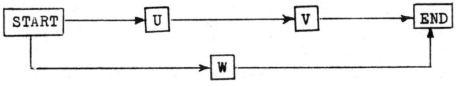

Fig. 20.2 Start and End Activities

The proponents of Precedence Diagramming claim that it is easier to define the sequence of work; the diagram can be drawn more quickly (less erasing to insert the Logic Rectifying Dummy); more easily understood by the uninitiated; and fewer activities are needed to define a project. More about these claims later.

The calculations for the schedule are by the Forward and Backward Pass exactly the same as for the Arrow Diagram, with the interconnecting arrow having a zero value. The Forward Pass starts at the START box (a zero value), proceeds to the left side of the next Activity Box or Boxes, and the estimated duration of that activity is added to the Early Start Time of that activity to obtain the Early Finish Time. The EF is noted in the box, and transferred along the interconnecting arrow to the left side of the next predecessor, and so on to the END. The Backward Pass, as before, is the mirror image of the Forward Pass, and starts from the END Box, and works backward to the START Box using a negative summation (subtractive) process. Critical activities have the same Early and Late Start Times. Total Float for a non-Critical activity is the difference between the Early and Late Start Times.

Fig. 20.3 is the Precedence Diagram for the "Building a New Home" problem 1.1 of Chapter 1. An Actual Start and an Actual Finish Time section has been added to the Activity Box for Project Monitoring purposes. Note that Random Numbering of the activities may be used: Activity #9, "Frame Second Floor to Roof," precedes Activity #8, "Frame and Shingle Roof." The activity numbers have been taken from the list of Problem 1.1. On actual projects, it is just as easy to create a loop in the logic with a Precedence Diagram as it is with an Arrow Diagram, so sequential numbering of the activities is recommended.

The claims that with the Precedence Diagram it is easier to define the sequence of work, and that the Plan may be more quickly drawn cannot be substantiated by the author. This statement is admittedly from an Arrow Diagram biased view, since the author has prepared over 100 Arrow Diagrams representing over $300 million capital cost projects, and has constructed only 2 Precedence Diagrams, by client request, on a capital cost of $7 million. It is suggested that the reader study the "Building a New Home" Diagram in Chapter 8 or 18 with its Precedence Diagram of Fig. 20.3. There are 24 activities in each one, be they drawn over an arrow or in a box. It does take longer to draw a box than an arrow, but this is trivial. However, there are only 7 dummies in the Home Diagram of Chapter 8, and 33 interconnecting lines in the Precedence Diagram above. Also, compare the arrows of Fig. 2.13, The Arrow Diagram Plan for Manufacture of Rocket Engine Sections, in Chapter 2 with its Precedence Diagram in Fig. 20.1, above.

312

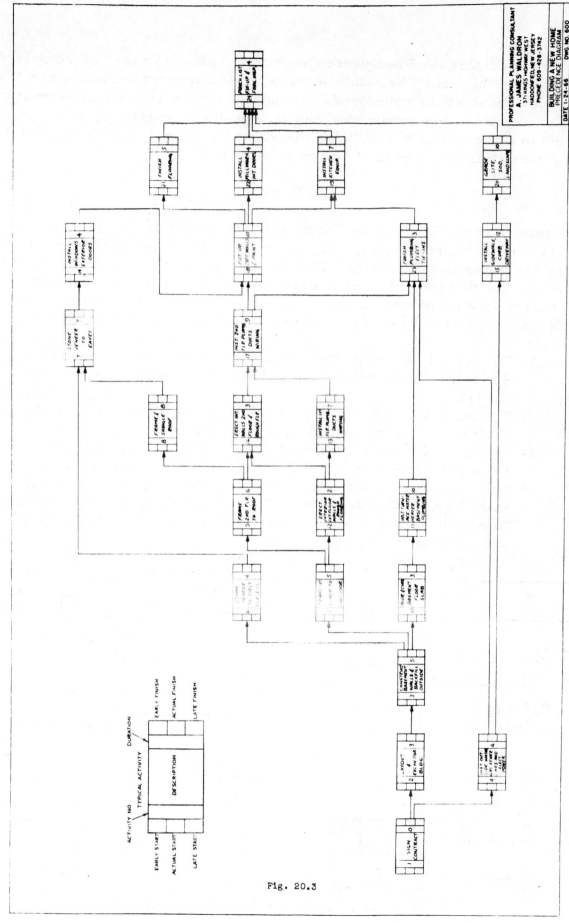

Fig. 20.3

It is claimed that the Precedence Diagram is easier to read, as compared to the Arrow Diagram. The author prefers not to comment on this, since it is a subjective evaluation, not objective. Just let it be noted that in several civil litigation cases, the author has had no trouble in explaining to judges and juries in a short "Chalk Talk" on a blackboard in a court, the fundamentals of the Arrow Diagram, and the function of a dummy.

Lag Factors

One feature of the Precedence Diagram is the ability to insert a Lag or Lead Factor on an interconnecting line, in order to represent finer definitions of work for a particular activity. This feature allows dependencies other than upon the 100% completion of preceding activities and thus minimizes the necessary amount of segmenting of activities. One such system is designed to function with four types of lag factors; shown in Fig. 20.4 below.

1) "S" lag - The activity may not begin before a time equal to the Early Start Time of the preceding or "lagged" activity, plus the lag interval (a).
2) "C" lag - The activity may not begin before a time equal to the Early Finish Time of the preceding activity plus the lag interval (b).
3) "F" lag - The activity may not finish before a time equal to the Early Finish Time of the preceding activity plus the lag interval (c).
4) "Z" lag - The activity may not begin before a time equal to the Early Start Time of the preceding activity plus the lag interval <u>and</u> may not finish before a time equal to the Early Finish Time of the preceding activity plus the lag interval (d).

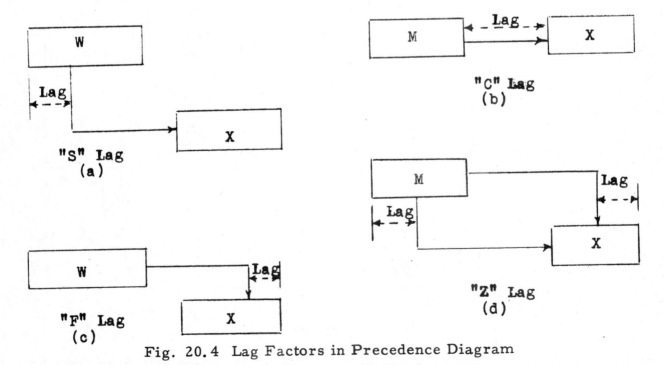

Fig. 20.4 Lag Factors in Precedence Diagram

Critique

Except for the fact that the Precedence Diagram <u>may</u> be easier to read by the uninitiated at first glance (the author doesn't concede this, since it is a subjective aspect of the individual reading the chart), the system does not appear to offer great improvements over the Arrow Diagram. Remember, the concept of this text is to develop a simple, flexible information system for the user. If a system takes longer to draw (but initially in the 'raw' form, and redrawn for presentation purposes), obtain a computer schedule and all its valuable sorts or listings; it then contradicts the concept herein. There may evolve, in the future, a hybrid system of activity arrows and activity nodes that will further simplify the technique of Planning. As of this date, the author is not convinced of the superiority of the Precedence Diagram over the Arrow Diagram, particularly since he considers the logic diagram, be it Arrow Diagram or Precedence Diagram, a work sheet that is necessary to generate the project's schedule. Once that is done, the Project Control Chart, either the Time Scale Chart of Chapter 11, or the Continuous Milestone Chart of the next chapter, is constructed and <u>used on the project.</u> THE ARROW DIAGRAM OR THE PRECEDENCE DIAGRAM SHOULD NEVER BE ISSUED ON A PROJECT, but used to formulate the time oriented, weather and constraint adjusted Project Control Chart.

One of the fallacies here is the automatic assumption of the validity of all activities starting at their Early Start Time (ES). Refer to comments on "Free Float" in Chapter 7. This claim of fewer activities is meant to compare the procedure in Arrow Diagramming where additional activity arrows and dummies, representing a finer detail of work, would be required. Unless extreme care is taken and symmetrical Lag Factors are added to the Precedence Diagram, particularly in repetitive operations, fallacious schedules will be obtained, because the Lag Factor goes right through a chain of activities, and becomes accumulative.

Data Processing

The use of the computer for calculating a schedule has been discussed in Chapter 16. For the Arrow Diagram it was suggested that the computer be considered when there are 100 to 120 arrows because of the speed, accuracy, and low cost of the machine usage. Precedence Diagramming is definitely at a disadvantage in data processing; more cards per activity are required; key punching is slower due to format; computer running time is higher for the same number of activities. Updating is more difficult. The Precedence Diagram does not draw Time Scale schedule too well. Manpower Leveling is much more difficult. The matrix chart approach of Chapter 11, used the I-J number (based on sequential numbering) to allow rapid checking of precedent activities in determining the schedule, is relatively simple. A single activity number will not allow this easy method of determining precedence, even if sequential activity numbers are used in the Precedence Diagram.

CHAPTER 21

THE CONTINUOUS MILESTONE CHART

This book has stressed the fact that the control of a project goes on hour by hour, and control requires a reference against which a comparison of actual performance is made. The philosophy is that the Arrow Diagram, or the Precedence Diagram, is a work sheet, needed to formalize the logic of the project plan, and to generate the schedule. Once the schedule is obtained, a Time Scale Chart as shown in Chapter 11 as an interconnected bar chart, is to be reproduced, and used as the Project Control Chart. Then the Arrow Diagram and the Precedence Diagram are to be folded up and filed away.

Limitations of the Time Scale Chart

The above philosophy is all well and good, but when we discussed updating in Chapter 19, we suddenly discovered the logic sequence changed in the actual execution of the project, as in Fig. 18.5. The immediate question arises; if we accept the premise that most logic sequences in the original plan are feasible; not the only sequence with which the work could proceed, and the real goal of the technique herein is to determine key milestones (a point where meaningful accomplishment of a work phase is reached) by dates, then won't the Time Scale Chart of Chapter 11 have to be redrawn every so often if it is to be a useful tool?

The answer to the above question is yes! This explains why the Time Scale Chart has not received extensive use--there is just too much drafting, or re-drafting of the Time Scale Charts on some projects where the sequences are either changed, or weren't too well thought out at the construction of the initial plan. To keep the Time Scale Chart current and relevant to the project sometimes takes a great deal of effort. Quite a few organizations initially adopted the Time Scale Chart as a Project Control Chart, then when they discovered they were working for the system, instead of the system working for them, by an inordinate amount of redrafting effort, they withdrew from the Time Scaling, and went back to the Arrow Diagram and its digital schedule.

Efforts to ameliorate the resistance to the Arrow Diagram (confusing to the uninitiated) were: the development of the Precedence Chart, which is a form of a Work Flow Chart, and is more readable to the field and lower echelon personnel; the use of X-Y computers to computer draw the Time Scale Chart, with its large costs and non-chain orientation of the various activity bars.

The Continuous Milestone Chart

The Author went through the same phases of acceptance, re-working and, finally, withdrawal from the Time Scale Chart. After re-assessment, since the Time Scale format is most powerful in its readability and utility by any level of supervision, some experimentation was tried, and a form of the Time Scale Chart has been developed that allows sequences of activities to change in any manner during actual execution, with no need for any redrawing of the chart, in order to determine the effect on milestone dates. The advantage of such a chart are manifold, and obvious. We have called this format a CONTINUOUS MILESTONE CHART, since it readily affords a continuous graphic, visual comparison of the project's status with that desired. The only time any redrawing of this chart would be required is if new activities are added to the project. More than likely this would be a change order situation, and the cost of any re-drafting should be part of the Change Order costs.

Construction of the Continuous Milestone Chart

1. A Time Scale Chart is set up in the usual columnar form, as described in Chapter 11, and shown here in Fig. 21.0 below. All aspects of Constraint periods, Limited Weather periods are pertinent to this chart.

2. A Project Control Line is drawn from the upper left hand corner down to the lower right hand corner, at any convenient angle. The most common angle is 45°.

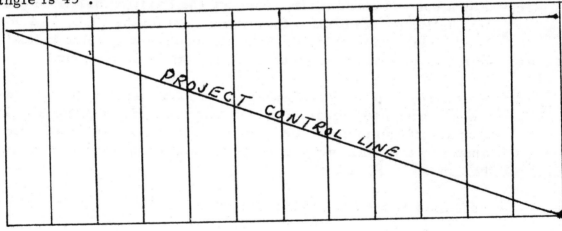

Fig. 21.0

3. As in the basic Time Scale Chart preparation of Chapter 11, the Forward and Backward Pass of the initial Project Arrow Network plan is made. In order to make this technique relevant to updating a project, as described in Chapter 18 of this book, we will use the basic information from the time bounded network of Fig. 18.4.

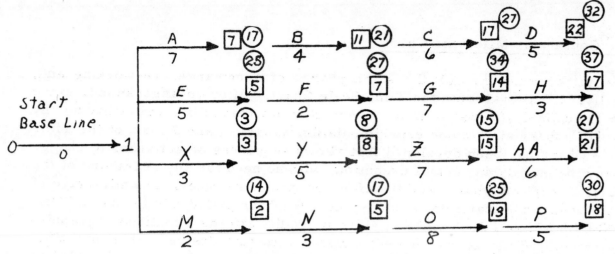

Fig. 18.4 Initial Project Schedule

In the area below the Project Control Line (hereinafter called PCL) first show, and emphasize the CRITICAL PATH, then show all the non-Critical activities. The following rules of construction apply to all activities:

a. Locate the activity's Late Finish Date from the basic schedule, on the PCL. This establishes the vertical location of that activity on the Continuous Milestone Chart (hereinafter called the CMC).

b. From that point on the PCL, go left horizontally, and display that activity at its Early Start and Early Finish Time.

c. For a non-critical activity, connect the Early Finish time end of that activity's bar back to the control point on the PCL, by a horizontal dashed line. This dashed line becomes a graphic indication of the float time associated with that activity's logic chain.

d. Connect each activity to its predecessors by vertical dashed lines. These "equivalent" dummies maintain the logic sequences. Note that actual dummies on the Arrow Diagram will automatically be shown by this rule, since a zero time activity always is a vertical bar on a Time Scale chart.

Fig. 21.1 is the first step in this construction.

As the reader studies Fig. 21.1, he will note some of the activities in Fig. 18.4 are not shown. Missing are activities B, N, O, F. These activities have Late Finish Times identical with some activities already on the CMC (B has a LF date of Day #21, as does AA; N has a LF of Day #17, as does C). These cannot be shown on the same horizontal position on the CMC, since they would be obscured by the already depicted activity.

There are two approaches to the multiple Late Finish Date problem. One is to drop a segment of the PCL down to a clear section of the chart, in parallel to the main PCL, and use that point to vertically locate the obscured activity.

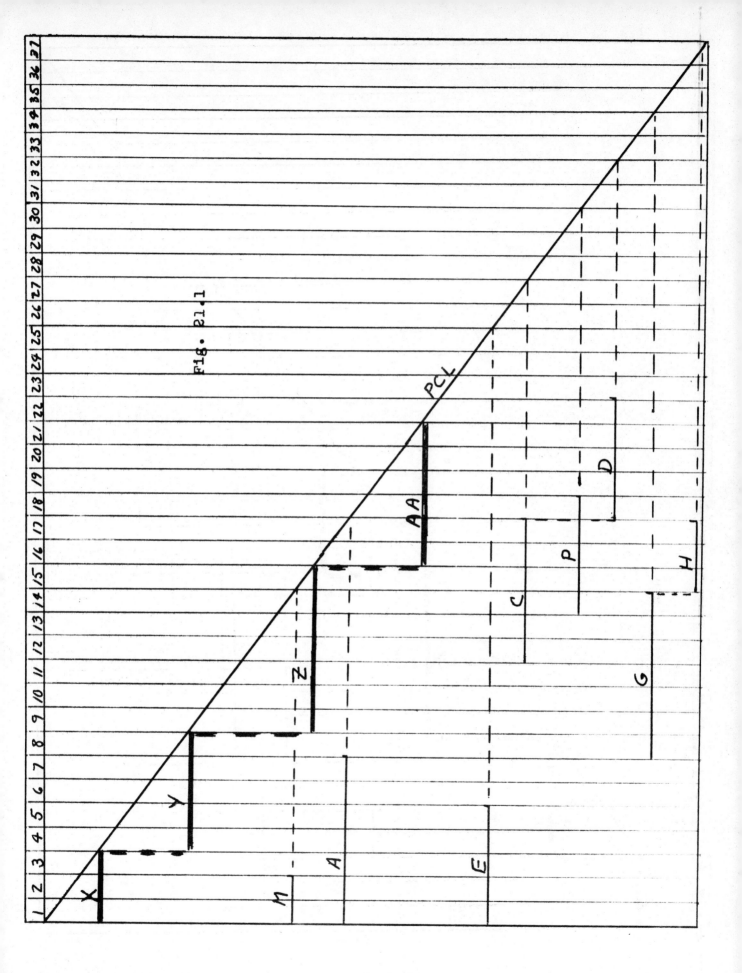

Fig. 21.1

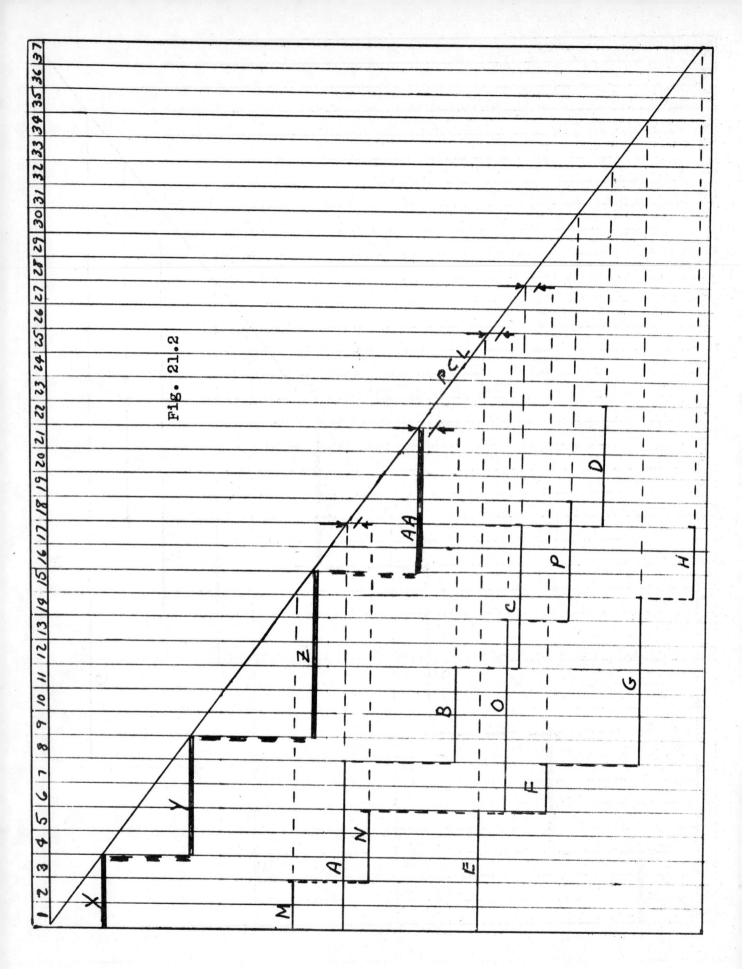

Fig. 21.2

320

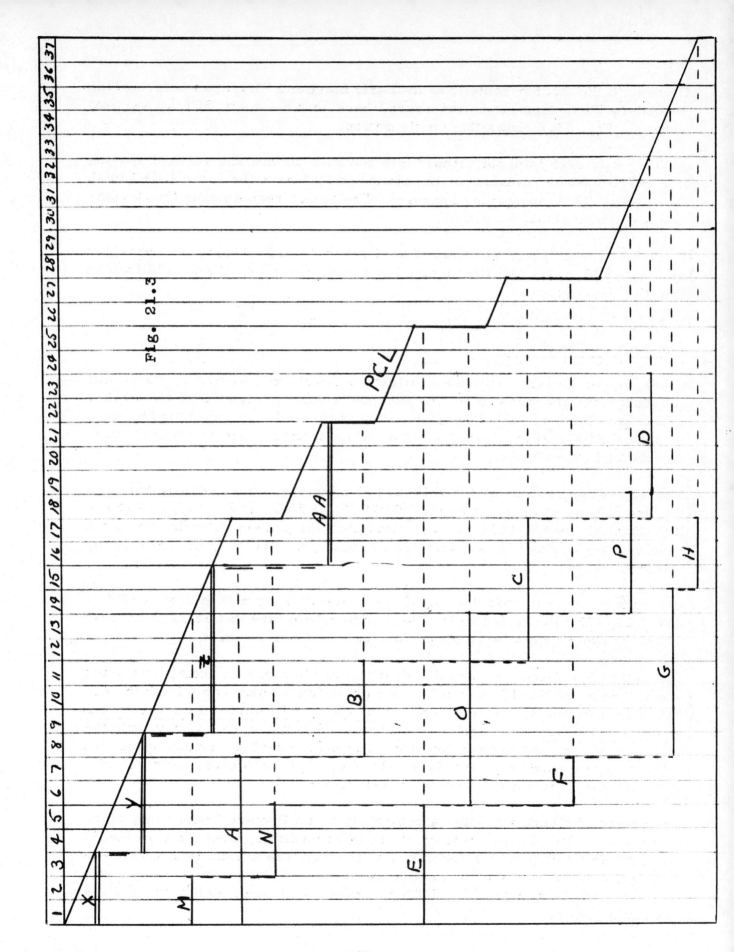

Fig. 21.3

Go over to the left horizontally to its Early Start and Finish, as before, This secondary PCL segment is referenced back to t h e main PCL by vertical arrows. Fig. 21.2 shows this configuration.

Another approach is to put "steps" into the PCL to suit the vertical spacing (the chart must be readable) of the activities that have the same Late Finish Date. Fig. 21.3 depicts this approach. The author recommends this formation for use on actual projects.

Incidentally, the "ideal" schedule of the Distributed Float basis, as described in Chapter 12, is an excellent method of constructing a Continuous Milestone Chart.

The Project Trend Line

Superimposed on the Continuous Milestone Chart are the actual starts and finishes of the activities, including gaps and restarts, regardless of how they differ from the original. This information is marked on a reproducible, or a print of the CMC at the project site, or taken from reports at a management center, and shown there.

The actual completions of CRITICAL activities are connected together, and the line of connection becomes the PROJECT TREND LINE. Fig. 21.4 shows the actual starts and finishes, and simulates the project situation up to the update status of Day #10, as depicted in Figs. 18.5 and 18.9 in the Updating Chapter, 18.

It should be apparent that the goal of the Project Manager is to keep the PROJECT TREND LINE (noted as PTL hereinafter) below the PROJECT CONTROL LINE: in other words, stay out of the upper half of this chart.

The author has found it essential to differentiate between an actual completion and a "projected finish", which is a predicted finish (may come from the update Forward and Backward Pass). Projected or Predicted Finishes have a propensity to slip. A projection of finishes of not more than 30 days into the future is shown at any time (usually at an update status review) by dashed lines past that update status date. In Fig. 21.4, projected finishes after Day #10, the Update date, are shown dashed.

The projected finishes do allow a projection of the Project Trend Line, which is important for project control; it shows immediately whether a project is trending ahead or behind schedule. The Project Trend Line in Fig. 21.4 is healthy; it shows that by Day #14, Critical activity Z should finish 1 day ahead of its original schedule. The PTL is projected to drop below the PCL, a desirable situation.

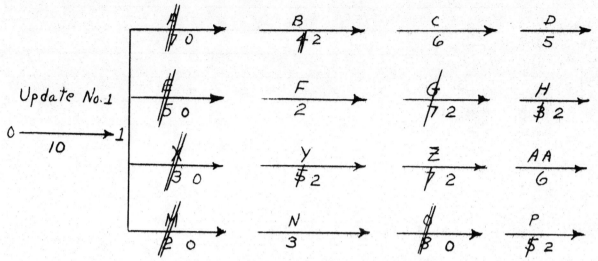

Fig. 18.5 Updated Original Schedule

The Update Forward Pass for Fig. 18.5 is shown in Fig. 18.9 below.

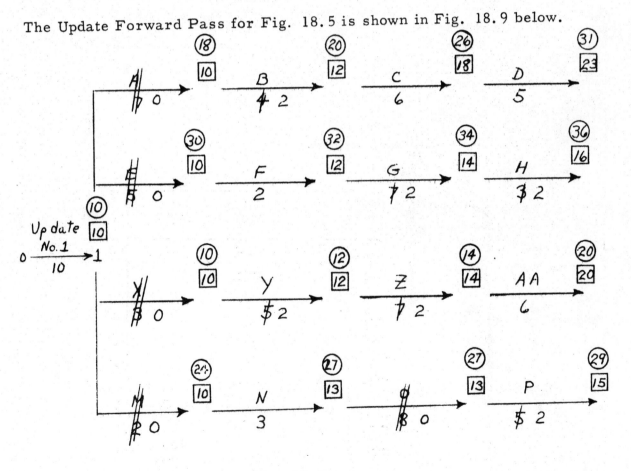

Fig. 18.9

323

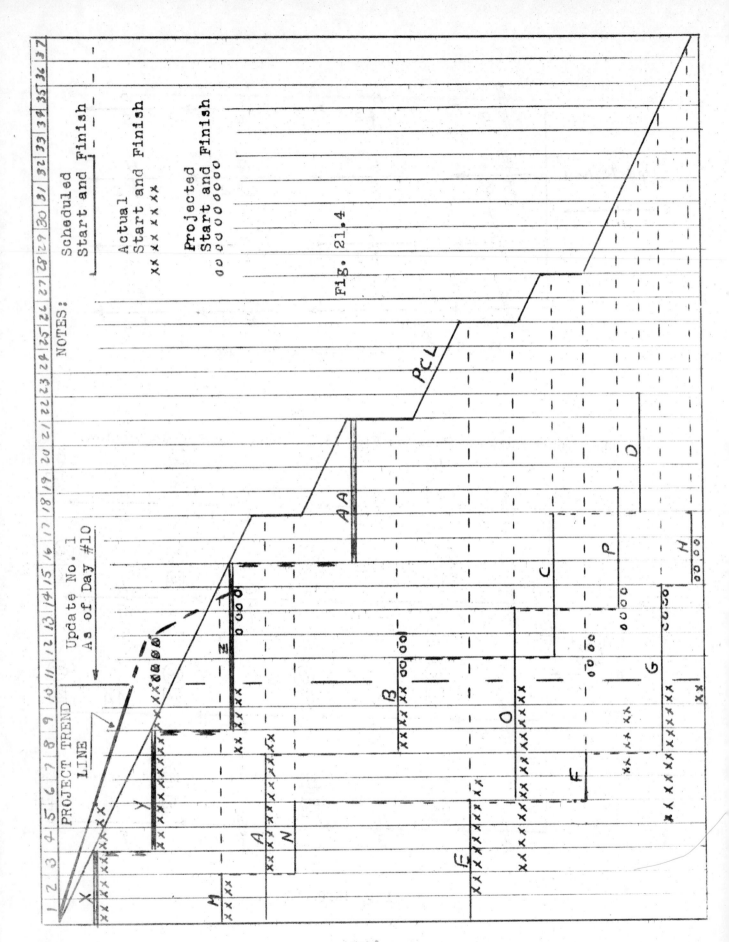

Fig. 21.4

324

Other Aspects of the Continuous Milestone Chart

1. Grossing of Activities by Chains or Paths

It is not necessary to show each activity at its own particular vertical position on the CMC. On a large project this approach might generate large, and/or many CMC's (though at this writing a $4.5 million dollar hospital expansion program is successfully being handled with 1200 activities individually shown on four CMC's of reasonable size).

Activities on a chain in the basic Project Network Diagram that are commonly related, such as all the work done by one sub-contractor, or all the work on a particular chain in one physical area, can be grossed, i.e., that chain of activities is displayed on the CMC in one horizontal row, with the Late Finish Time of the last activity of that chain determining the control point on the PCL, consequently determining the vertical location of that chain on the CMC. In Fig. 21.5, we show the initial CMC for our prevalent textbook problem, "Building A New Home". (Problem 1.1, and examples and problems in Chapters 2, 6, 8, 11, and 18.)

Figure 21.5, shows the chain of activities 5-70, 70-90, and 90-100 on the same horizontal row, on the basis that these activities are to be performed by one sub-contractor. Note also that Milestone Dates are as effectively shown on the CMC as on the Time Scale Chart of Chapter 11.

2. The Cost Domain Chart

The vertical ordinate of the CMC, which is unused in the Time structure, may be used to superimpose the Cost Domain chart described in Chapter 19. This is the display of costs if all activities started Early (The Cash Flow Curve), and also if all activities started at their Late Start and Finish Time. This format is shown in Fig. 21.6, with the data from the Cost Control Chapter, 19, (particularly Fig. 19.1) superimposed on the CMC.

3. Manpower Histogram and Leveling Charts

Figure 21.7 shows both the histogram for Laborer distribution, if all Laborer activities started Early (information is from Problem 11.3, Craft #2), and the effect of manpower leveling (note re-scheduled starts) based on the matrix approach to manpower leveling in Chapter 11.

4. A Project Summary Chart

The Continuous Milestone Chart is an excellent vehicle to summarize the status of a large project, in size, cost and resource assignments. On the theory that information should be more and more summarized as it flows upward in the pyramid of management hierachy, theCMC can be so

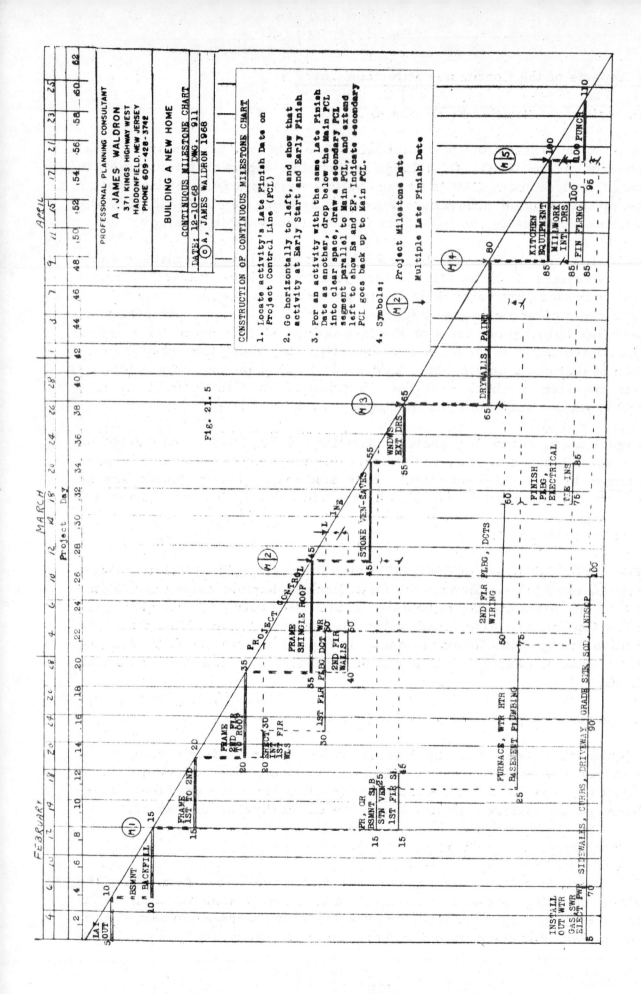

Fig. 21.5

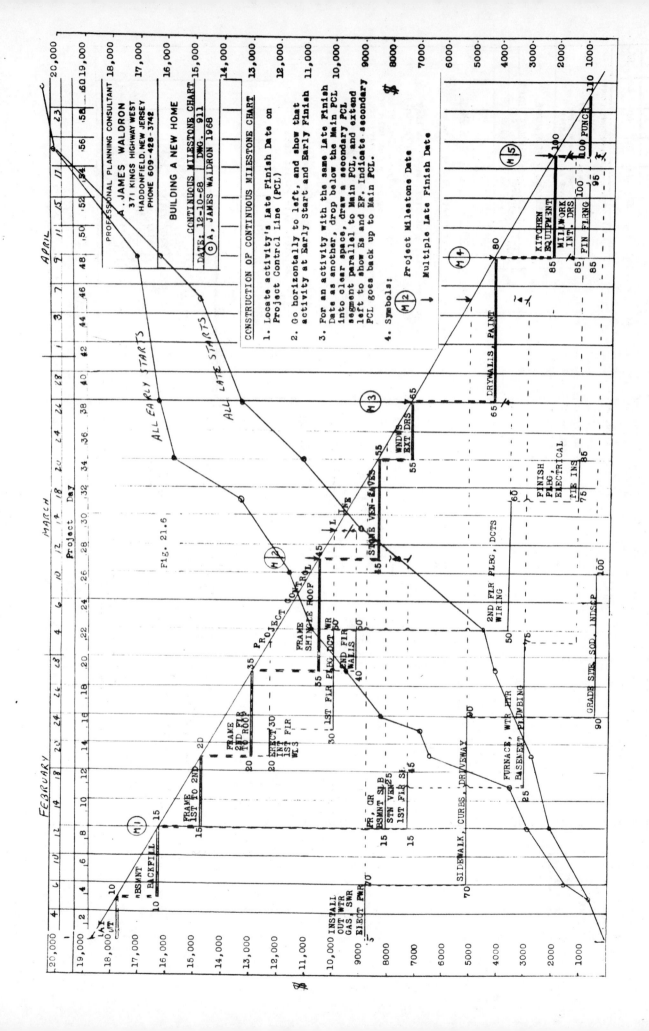

Fig. 21·6

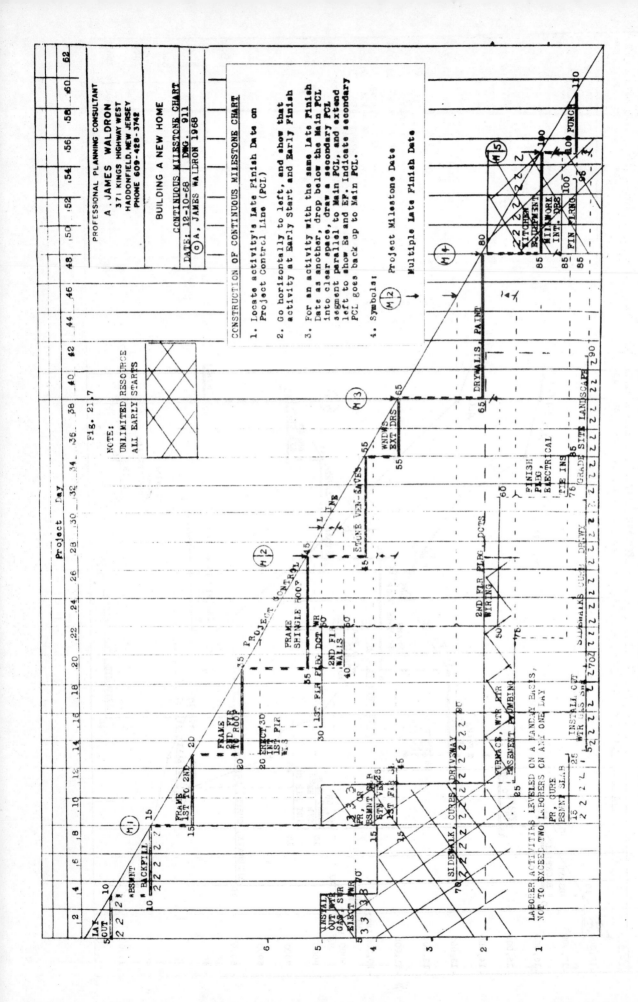

Fig. 21.7

summarized that major milestone dates, with major work phases summarized into five or ten gross activity bars, can be displayed. Thus, a $500 million Nuclear Power Plant construction project can be reduced to an 8" x 11-1/2" chart in a periodic status report to the president of the utility company owner. The Project Summary Chart can have the Cost Domain, and accumulated and/or committed costs depicted, as in Fig. 21.6, and any resource histogram of manpower or manhour curves superimposed. The Project Trend Line of time performance, cost totals, and manhours by trades can be readily displayed; this will show the exact status of the project, at any time, for any resource of time, men, money, and machines. The format for a Project Summary chart should be that of Fig. 21.2.

5.   A Project Trend Formula

For those readers who like to mathematically formalize their reports, a trend number can be generated if the Project Control Line is 45°. Fig. 21.8 shows the Dimensional Analysis approach, based on the isoceles triangle.

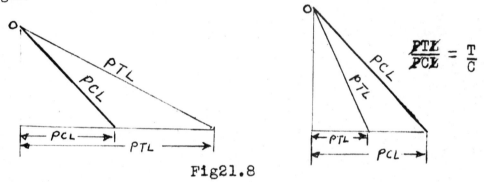

$$\frac{PTL}{PCL} = \frac{T}{C}$$

Fig21.8

The Trend to Control ratio, $\frac{T}{C}$, would be equal to 1 if the project is on schedule; $\frac{T}{C}$ of less than 1 means the project is ahead of schedule, and $\frac{T}{C}$ over 1.0 means a schedule slippage. The decimal values are percents of the original total number of days of the project, from the original CMC.

PROBLEMS

21.1   Draw the Continuous Milestone Chart for Problem 11.4, Chapter 11. Include the affects of Weather Restraints.

21.2   The following is the historical record of performance on The Build A New Home Project, up to the date of the update on Day #20 (refer to Chapter 18 on Updating) From the records below construct the Project Trend Line on the Continuous Milestone Chart (Use Fig. 21.5). An

activity starts at the beginning of the day listed. Activities
in progress at the date of Update are asterisked in the ACTUAL
FINISH column:

ACTIVITY NO.	ACTUAL START (DAY #)	ACTUAL FINISH (DAY #)	PROJECTED FINISH (DAY #)
5 - 10	2	4	
10 - 15	4	7	
15 - 20	6	11	
20 - 35	10	15	
20 - 30	16	17	
35 - 45	16	*	23
30 - 50	15	*	21
40 - 50	19	*	21
5 - 70	19	*	22
45 - 55	17	*	22
15 - 25	12	16	
15 - 45	15	18	
25 - 75	15	*	24

CHAPTER 22

USE OF NETWORK PLANNING IN CLAIMS AND LITIGATION

There comes a point in some projects where it is necessary for the executor of that project to seek an equitable solution to the problems of extended time and concomitant increased costs, due to circumstances beyond his control, or claimed beyond his control. In the name of justice and equity, he petitions the owner, or the agency controlling the project, for recourse. If the contract has the provisions, a system and procedure is established for such a claim, and the executor is bound by such a provision. If no such procedure exists, or if the results of such a petition are unsatisfactory, the claimant turns to more legalistic procedures such as arbitration, or ultimately to litigation. This chapter deals with an effective approach of graphic display of such claims, and the several aspects thereof.

In any claim for additional time, and for additional cost reimbursement, there is an attempt to show a reference schedule, and then a comparison to an actual schedule, i.e., this is the way it was versus the way it should have been. In the author's experience in five litigable cases, the predominant factor was the lack of an adequate and valid reference for the claimant to base his petition upon.

The technique described in this book, that of a method of developing a feasible reference plan and schedule; translating that schedule graphically to the Project Control Chart and/or the Continuous Milestone Chart, is an excellent basis for the establishment of the necessary primary document--the feasible reference project schedule!

The Reference Schedule

As pointed out before, project control is analagous to environmental control. Just as room temperature is controlled through a thermostat wherein the actual temperature is measured by a bi-metallic element and compared to a reference desired temperature through a dial setting that mechanically positions linkages so that deviations of actual from reference cause a physical movement. That differential is measured mechanically, and the deviation calls for the exertion of energy, either hot or cold air, until the environmental is brought back to the desired value.

In a claim, the same situation exists--there must be an acceptable reference which is compared to the actual. The Arrow Diagram, its resultant schedule, and the translation of that schedule to a time oriented graphic display of the

Time Scale Chart or the Continuous Milestone Chart, becomes that necessary reference.

The necessary reference for any claim is obtained in several ways:

a. It was part of the original bid and/or contract documents, and was is-- sued by the owner or the agency controlling the project. The fact that it was issued by the owner, or by an owner-retained consultant, or by the agency controlling the project does not necessarily mean it was a valid, or even feasible schedule. As a matter of general advise to the reader; if he encounters a project where the owner has had pre- pared a Network Diagram and typical digital schedule, as in Table 8.0 on page 69, and it is either part of the bid documents, or is issued at the beginning of the project, the first step is to construct the Time Scale Chart of that schedule, as described in Chapter 11. It is pointed out again, that most users of the technique stop with the Arrow or Precedence Diagram and the digital schedule. They fail to take the next small step of converting the schedule to the time oriented Pro- ject Control Chart. The logic of the schedule is more easily analyzed in the Time Scale chart for feasibility and applicability to the specific project conditions. Weather and other constraints are specifically to be found and studied for reasonableness.

IF, after such a study, the reader cannot accept the owner's ostensible but non-feasible schedule, he is to follow the prime rule of project management, GO ON RECORD of objecting to whatever does not ap- pear to be feasible. The author has been involved in one case where the owner issued a Network Plan and digital schedule by computer printout at the beginning of a $10 million, multi-story building con- struction project. The project ran two years late! When the author, as an expert witness, translated the specification Arrow Diagram and its computer printout schedule into a Time Scale Chart, it was readily apparent that the so-called specification schedule was not feasible, or even practical. There was an obvious assumption of unlimited re- sources--there would have had to be four simultaneous Mechanical and Electrical crews performing simultaneous roughing work. There was obvious incompetency on the part of the owner's planner, in that finish work, such as Ceramic Tile, plastering, etc., started months before the building was closed in, and there was no provision for temporary heat and enclosures. Another glaring error was the instal- lation of the elevator guide rails months before the interior masonry walls were installed half way up the building (see INAUTHENTICITY in Chapter 23). From the bid documents of the Architect's drawings and specifications, a "feasible" plan was reconstructed (see "c" below) that showed the original feasible project schedule would be within four months of the actual completion. The extension of time was readily seen to be caused by specific delays encountered during the life of the project. The sad part of this story was that the General Contractor accepted the specification schedules without analysis or complaint until the project was a year late. Then he began to wonder about the validity of the schedule he accepted.

Rule #1. Convert any Project Schedule received into a Time Scale Chart, if it is a Network Diagram type, and evaluate the feasibility of that Time Scale schedule against the specification requirements, contractual and physical constraints. If not satisfied with analysis, GO ON RECORD.

b. The executor of the project had an original Network Diagram. Again, the feasibility of that schedule and plan must be evaluated, via the Time Scale Chart. In several instances, such documents claimed by the executor to be the Project Schedule, proved to be too gross (too much duration time assigned to vaguely defined activities), so much so that they were valueless in use as a reference. Other questions the outside consultant must have answered are: Was it specifically issued to the owner and/or his representative? Was it accepted? Was it specifically rejected? If so, what were the grounds of rejection? Documentation of transmittal letters or forms, and correspondence are vital for this important point. The fact that the Project Arrow Diagram and its schedule was issued, and never formally accepted by the owner or his representative, does not preclude that it was invalid. Another aspect is an actual case wherein a General Contractor on a construction project issued a Network Plan and Schedule, and had it rejected in writing as unacceptable. Consequently he dropped the whole technique from use on the project. At a much later date, in court, it was established by an outside expert witness that it had indeed been a feasible schedule, and the reason for rejection was picayune. It ultimately served admirably as the reference in the satisfactory settlement of a large claim.

If the case exists where the executor of the project constructed a Network Plan and schedule for his own use, and never issued it formally to the owner of the project, then it must be examined and certified as a "feasible" schedule by an outside expert before it can be used as a reference in a claim.

c. The most common situation is the one where no Network Plan and its schedule was constructed for the project. The project schedule was the typical bar chart, which proves to be totally inadequate as a reference on which to structure a claim.

It will be necessary for an outside, impartial and competent consultant to construct the feasible reference schedule from the bid and/

or original contract documents, a post facto feasible schedule without any influence anticipation from the claimant. This "feasible" reference plan and schedule will take a great deal of effort. It consists first of constructing the logical sequences of operations, which are specifically defined. Naturally, this requires experience in the type of project being structured for a claim. It also requires knowledge of trade practices in the geographical area of the claim. For example, the placing of terrazzo floors follows immediately after the building is closed in, and the interior masonry partitions erected in the Northeast part of the country; in the South this work is considered one of the last finish operations. In a high rise building, roughing work proceeds upward floor by floor, but at some point in final finish work, the work starts at the top floor, and flows downward, so that a completely finished and furnished floor can be locked off. If the "expert" preparing the feasible plan has any doubts about sequences and interrelations of activities, then he is obligated to retain the services of a sub-contractor or specialist in this line of work. If the feasibility of the reference claim can be attacked, then the basis of the claim is lost. The reference Network should be reviewed with the claimant and his technical and legal staff, with someone from the claimant playing the part of the Devil's Advocate in attacking the logic of the submitted plan. This insures as thorough a review of the plan's feasibility as possible.

The second step is the collection of adequate estimates for the activity durations on the feasible plan. There are some excellent estimating manuals published that will serve as this source of information. One such manual is the annual "CONSTRUCTION PRICING AND SCHEDULING MANUAL" published by the F. W. Dodge Corp. It not only gives unit costs that are realistic, but also reasonable crew sizes and durations of activities based on unit work items.

It is obviously important that the "feasible" schedule be well researched, documented and cross-referenced, so that it will stand up in court. No matter how the claim is going to be presented, the "expert" constructing it should do so on the premise that he will have to testify in court as to its reasonableness.

After the construction of the reference "feasible" schedule and the collection of valid duration estimates, the consultant then analyzes the output schedule to see if it meets all contractual requirements and criteria of completion date, specific milestone dates, restraints, etc. There are going to be times when the analysis will reveal that the claimant did not have a reasonable chance of attaining his contractual dates with his approach (remember, there is no guarantee of happy answers with the technique). However, it does inform the claimant of the weakness of his case before he gets involved in expensive and time consuming litigation.

A word about consultants in Network Planning. There are as many incompetents in this field as in any other profession, and there is no way of preventing them from operating. There are no regulatory bodies controlling or registering them; there are no professional examinations or minimum experience requirements; no educational background criteria that is mandatory. Membership in any "technical" society only proves that the person was able to raise the nominal initial fee. The only criterion the claimant can rely upon is reputation.

The procedure that the author uses in the preparation of a claim exhibit via the Network Planning Technique is to first inform the claimant, by Registered Letter, that he will examine the claimant's records and construct a reasonable reference schedule, obtain the schedule, and analyze it. However, it is stressed that the results cannot be predicted nor guaranteed. There is a likelihood that the results may indicate that the claimant never could have made the contract milestone dates and completion date. This clause in the Registered Letter both establishes the impartiality of the consultant, and will be part of the exhibit of his credentials as a dis-interested "expert"; it establishes the fact that the study is reimbursible, regardless of the outcome. Again, a negative report on the validity of the analyzed claim is of some value to the claimant by pointing out weaknesses in the claimant's case.

If the feasible reference schedule is within the contractual requirements, it is then translated into the Time Scale Chart of Chapter 11, and/or the Continuous Milestone Chart of Chapter 21, complete with Weather and other Contractual Constraints.

The Actual Schedule

After the construction of the acceptable Reference Schedule for the contested project, an Actual Schedule is produced. Actual Start times, gaps in performance, Actual Finish times, changes, delays, conditions of Force Majeure (unavoidable Acts of God) and other situations are collected. The source documents for this information are Daily Time Sheets, project Logs, Diaries, correspondence, transmittal letters and forms, minutes of project progress meetings, etc. There is considerable documentation search and cross referencing involved.

The Actual Schedule is graphically displayed by superimposing over the Reference Schedule, by acetate or clear plastic sheets overlayed on the Reference Time Scale Chart and/or the Continuous Milestone Chart, or by colored markings or tapes overlayed on prints of the Time Oriented Reference Schedule. This actual schedule is first prepared as a time duration, time extension, and/or time gaps, for comparison to the original feasible Reference Schedule Time Scale Charts. In addition, separate exhibits may be prepared to display actual man-hours and actual costs compared to the original estimates and/or contract costs. This may be done on separate sheets, or as additional overlays on the basic Reference Schedule. In general, keep the display as simple as possible to avoid the lessening of the impact of a valid claim by too confusing a presentation. The author prefers separate charts for time, manhours, and costs.

Delays

There are two types of delays usually encountered on a project.

a. Proportional Delay: The extension of the contract or project end date by the same amount of time units this delay extends an activity past its original Late Finish Date.

b. The Quantum Delay: The extension of a contract or project end date by a much larger amount of time units than this delay extends the Late Finish Date of an activity beyond its original (reference) Late Finish Date, due to causes and/or situations outside the Network Plan and beyond the control of the claimant. As an example of a "Quantum" Delay, consider the well publicized commercial "shuttle" flights between the major airports of New York City and Washington, D. C. Assume there is a "shuttle" flight every hour on the hour. If you arrive at the gate at one minute before the 10:00 a.m. flight you are on the 10:00 a.m. flight. If you arrive at the gate at one minute past the hour, you have a "quantum" delay of 59 minutes until the next flight. Thus, if an order for the fabrication of structural steel is delayed by one day, that may be sufficient to miss a mill run, and cause an eight week delivery delay in the structural steel. This one day delay could occur in a precedent activity on this chain of operations: a one day delay in the return of the approved drawings by the Engineer may have the same result.

Another typical "quantum" delay is the delay of an activity by a certain amount past its Late Start or Finish Date, and consequentially push a subsequent weather affected activity into the inclement weather period of Winter, with a much larger resultant of lost time, and a greater extension of the contract end date. The largest "quantum" delay the author has seen was the failure to deliver some equipment to a dock for shipment to a Nuclear Power Plant in the Antarctic. A one day delay caused a quantum delay of one year, the annual supply ship schedule for the opening of McMurdo Sound from ice conditions.

c. Concurrent and Serial Delays: To portray the effect of concurrent delays over the life of a project, we have shown in Fig. 22.0 a Network Plan with duration estimates. Fig. 22.1 shows the activities of chains D, E, F and G, H, K, at their Late Start and Finish Times on a

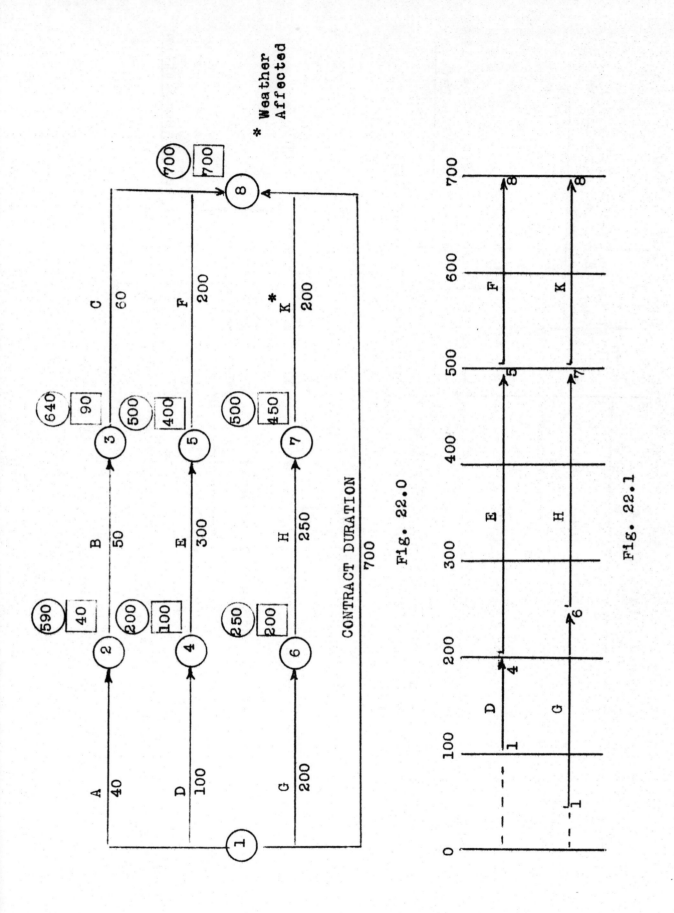

Fig. 22.0

Fig. 22.1

* Weather
  Affected

CONTRACT DURATION
700

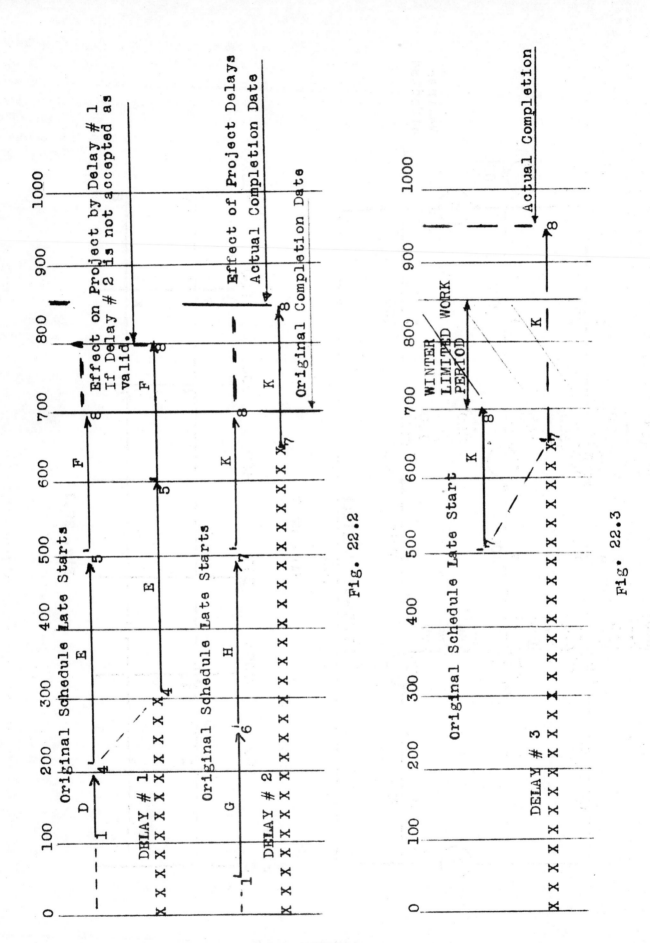

Fig. 22.2

Fig. 22.3

338

Time Scale Chart. Fig. 22.2 has superimposed two concurrent delays; Delay "1" pushes activity E past its Late Start Date by 100 days, while Delay "2" pushes activity K past its Late Start Date by 150 days. Delay "2" causes the greatest consequential effect on the project end date. Note that Delay "1", primarily affecting activity E, caused F to be serially delayed. Thus the delays and their consequential effect on the project may be examined one at a time, and peeled off one at a time like the layers of an onion.

A "quantum" delay is shown in Fig. 22.3. Here, Delay "3" pushes activity K, a Winter affected activity, past its Late Start, and into a period of the year where it becomes inefficient, as discussed in Weather Allowances in Chapter 11. This situation effectively increased its normal duration so that the project is further extended.

Note that some point in time during the life of the project a delay may be going on that the executor is unaware of, since it will affect a much later activity. In Fig. 22.2, at Day #200, the Executor may be aware of the beginning of the effects of Delay #1 on activity E, but unaware of Delay #2, which is going to affect, at a later date, activity K, whose Late Start Date is not for another 300 days. Delay "2" causes the greatest consequential effect on the project.

Note that exception should be taken to standard clauses in contracts which state that unless the effect of a delay is reported within seven (7) days after its occurrence, or its start, no allowance will be made for that delay in the project schedule. The executor should have the right to submit his claim for an extension as soon as he is able to evaluate fully the consequential and serial effects of a delay; in some cases this could be years after that delay.

Rule #2. KEEP RECORDS

Rule #3. GO ON RECORD as soon as a situation is ascertainable about the possible consequential effects of a delay.

Accelerated Schedules

There is also the case where the completion date of a project is mandatory for the owner; consequently it is imposed on the executor of the project through Penalty or Liquidated Damage clauses. Typical projects are the construction of a new school to be completed before the influx of students for the school year; a Utility Company having a new power plant going on line, and meeting a commitment to their power pool for those megawatts at a specific date; the delivery of a space system component or sub-system to a space agency in time for a launch date that is controlled by astronomical factors, etc.

Unfortunately, at some point in the active life of this project, delays of whatever cause and type occur. At some point, an accelerated schedule is enforced. Toward the end of the project, the executor discovers to his horror that his costs are greatly exceeding his estimate, or his contract allowance. Some vehicle is needed to show graphically the cause or causes of delays at this point, and the effects in additional costs. The contract end date is to remain the same original end date, since it is mandatory. The vehicle best suited for this analysis and display is the Continuous Milestone Chart.

Fig. 22.4 is a qualitative graphic display of what has been observed on several projects by the author. At some point in time it is decided that an accelerated schedule is necessary to attain fixed milestone or end dates of a project. For whatever reason, schedule slippage occurs up to time t_2 in Fig. 22.4. At this point, additional resources and premium time are authorized or enforced, in order to make up the lost time. As a theoretical ideal, the manhours should level out at the original 100% manhours tendered or estimated. If the ordinate of Fig. 22.4 were considered manhours, the area A_4 under the dotted extension of the "Actual Manhours" curve would equal the sum of the areas of A_1 plus A_2. However, when the ordinate of Fig. 22.4 is considered labor costs, the actual labor cost curve does not follow the manhour area balancing curve, but is considerably higher, represented by Area A_3 on Fig. 22.4. First of all, there is the increase in labor costs by the amount of premium time, time and a half or double time, as the case may be. Secondly, there are inefficiencies that are inherent in an accelerated schedule that add to the costs.

Please note in our discussion to this point, there are no additional delays encountered in the accelerated period. If they should occur, they will be treated as above and below, by the further compounding of inefficiencies, and the new total cost would be higher.

Before proceeding to the discussion of inherent inefficiencies, note should be taken of Area A_1 in Fig. 22.4. This is the difference between the normal scheduled production and the actual production, before acceleration. This is the result of delays, changes and additions to the scope of work that occur during the period from t_1 to t_2. This area could be the subject of a claim for excess overhead costs, if determinable and excusable delays in this period can be identified. This represents the indirect costs of staffing by the executor of the project in anticipation of a larger work force or productivity, based on the original Reference Schedule. This staffing includes salaried supervisory personnel, technical personnel, support personnel, tools, equipment, facilities, etc. This staff and facility commitment could not be terminated out of hand by the executor, or transferred to other projects, since this commitment was part of the contract. A Joint Venture is particularly limited in the re-assignment of these resources. These resources were available to handle the larger work force, or production, which did not materialize. Ergo, a percentage of the associated indirect costs can be considered an in-

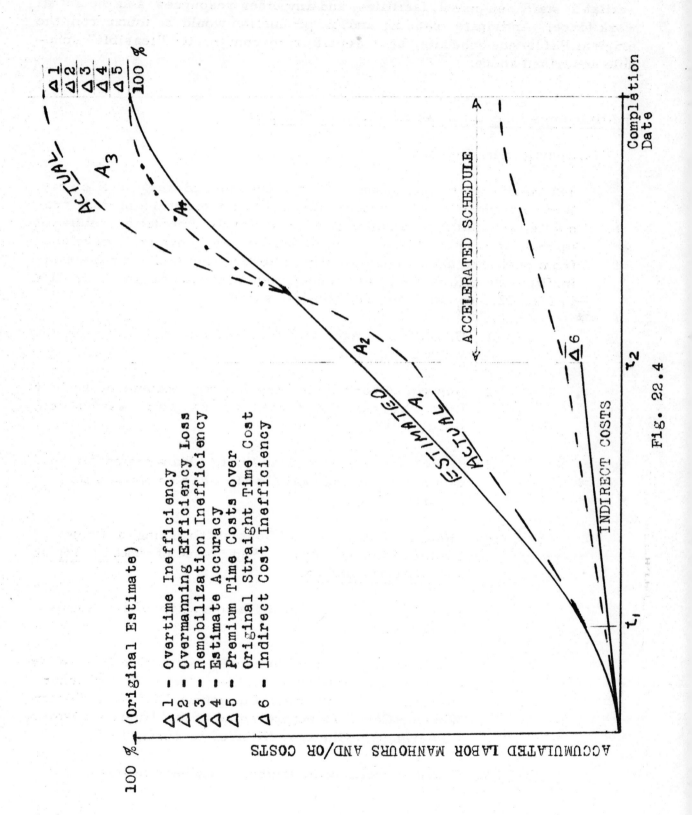

$\Delta 1$ – Overtime Inefficiency
$\Delta 2$ – Overmanning Efficiency Loss
$\Delta 3$ – Remobilization Inefficiency
$\Delta 4$ – Estimate Accuracy
$\Delta 5$ – Premium Time Costs over
Original Straight Time Cost
$\Delta 6$ – Indirect Cost Inefficiency

Fig. 22.4

efficiency, subject to reimbursement. This percentage can be found in the project records, since the daily or weekly time sheets will indicate the available staff, equipment, facilities, and any other resources, and the actual work force. Anticipate manning and/or production would be found from the original Reference Schedule, be it actual or re-constructed "feasible" schedule described above.

Inefficiencies Inherent In An Accelerated Schedule

A. Overtime Inefficiency

This is the most well-known and best documented work inefficiency. When a normal work period is extended to an overtime period, productivity diminishes, and there is sufficient documentation published to make this quantifiable. Figure 22.5 is an "envelope" developed from published data, to help an arbitrator or judge decide a reasonable inefficiency allowance in an Accelerated Schedule claim. The data in Fig. 22.5 comes from the following sources:

1. U. S. Department of Labor Bulletin #917, "Hours of Work and Output."

2. Overtime Work Efficiency Survey, Manual of Labor Units, National Electrical Contractors Association, August, 1962.

3. "How Much Does Overtime Cost," Management Methods Commitee, Mechanical Contractors Association of America.

4. "How To Estimate The Cost of 'Crashing' a Project," Thomas Dinning. April, 1968, issue of Heating, Piping and Air Conditioning.

5. Boiler Construction Specifications, City of Los Angeles, California.

6. "Overcoming The Problems of Construction Scheduling On Large Central Station Boilers." by L. V. O'Connor, Director of Construction Equipment Division, Foster Wheeler Corp. To be presented at the American Power Conference in Chicago, Illinois, in April, 1969.

7. Client sources of the author, which are proprietary.

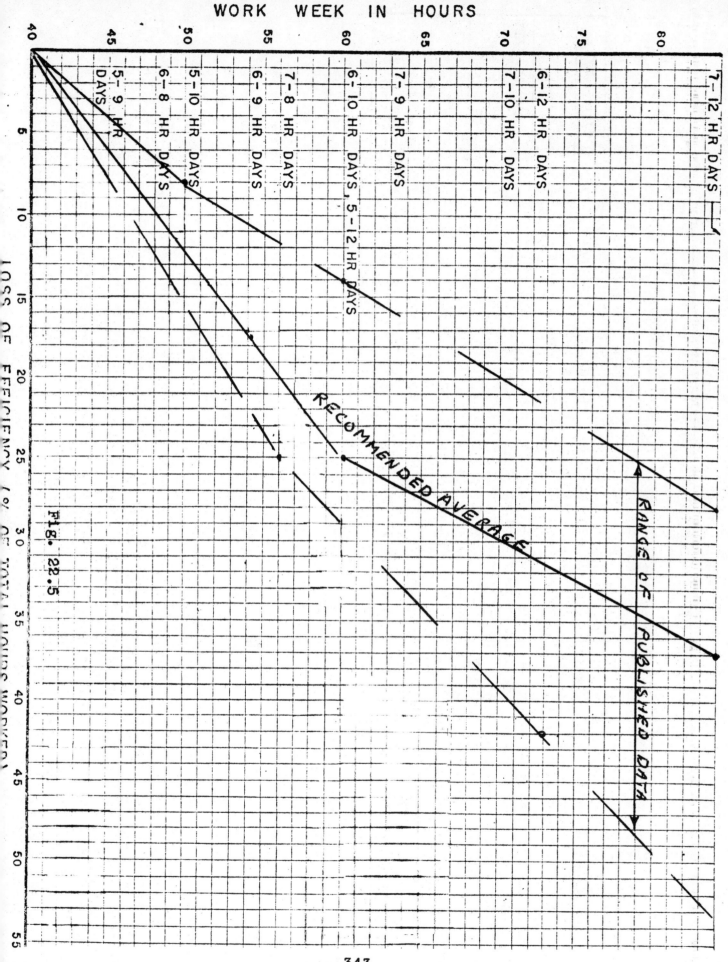

WORK WEEK IN HOURS

7 - 12 HR DAYS
6 - 12 HR DAYS
7 - 10 HR DAYS
7 - 9 HR DAYS
6 - 10 HR DAYS, 5 - 12 HR DAYS
7 - 8 HR DAYS
6 - 9 HR DAYS
5 - 10 HR DAYS
6 - 8 HR DAYS
5 - 9 HR DAYS

RECOMMENDED AVERAGE

RANGE OF PUBLISHED DATA

LOSS OF EFFICIENCY (% OF TOTAL HOURS WORKED)

Fig. 22.5

343

LOSS OF EFFICIENCY TABULATION SHEET

NUMBER OF DAYS/NUMBER OF HOURS PER DAY

SOURCE	5/9	5/10	5/12	6/8	6/9	6/10	6/12	7/8	7/9	7/10	7/12
NECA	--	12-1/2%	25%	12-1/2%	--	15-1/2%	29%	23%	--	22.2%	37-1/2%
CITY OF LA	7%	13%	--	12.3%	18%	27%	--	--	--	--	--
FOSTER WHEELER	6-1/2%	13%	--	12.3%	18%	27%	--	--	--	--	--
U. S. DEPT. OF LABOR #917	5%	8%	14%	3-1/2%	12%	17%	25%	7%	12-1/2%	21%	28%

TABLE 22.0

344

The envelope of Fig. 22.5 covers the range of data published. The average was also developed from the published data mentioned above. The proprietary client data falls within the envelope and fairly closely matches the average shown.

Table 22.0 is a tabulation of published data of the various inefficiencies versus the overtime work period.

In the area of legal precedence, the fact that overtime work results in decreased productivity has been judicially recognized. Since federal government construction contracts usually contain a clause requiring the contractor to accelerate the work, e. g. , work overtime, work additional shifts, increase manpower, etc. , to overcome nonexcusable delays, a considerable amount of litigation has resulted involving situations where the contractors claimed that justification for such acceleration did not exist. The cost of overtime work including the associated loss of efficiency, has long been recognized as a part of the "impact" costs of accelerated construction work. A comprehensive discussion of impact costs is contained in Impact Costs Of Acceleration, Volume 25, No. 2, Federal Bar Journal (Spring 1965).

The United States Court of Claims, in Maryland Sanitary Manufacturing Corp. vs. U. S. , 119 C. Cls. 100 (1951), took judicial notice of the fact that efficiency was impaired by working extensive overtime. This case involved the manufacture of 105-mm shells during World War II. The Government required the Contractor to go into "all-out production" which involved two shifts of 12 hours each, 7 days per week. The Contractor had planned to work two shifts of 10 hours each, 6 days per week. In deciding to award the Contractor the cost of labor inefficiency resulting from the increased work hours, the Court considered a report by the Bureau of Labor Statistics (Bulletin No. 917, dated May 21, 1947). This report showed that, for a similar manufacturing operation, efficiency of labor dropped 22. 3% after the work week was changed from five 8-hour days to six 10-hour days.

The cost of overtime inefficiency in connection with construction has been recognized by the Armed Services Board of Contract Appeals (ASBCA). In Lew F. Stilwell, Inc. , ASBCA No. 9423, 1964 BCA 4128, the Government accelerated modifications to guided missile facilities which resulted in a work week of 72 hours or more for various specialty sub-contractors. The board awarded the Contractor the cost of overtime inefficiency in amounts up to 65. 9% of straight time labor costs. Other percentages awarded included 30. 3%, 38. 7%, 39. 0%, 40. 0%, 44. 4%, and 52. 0%. The highest percentages involved mechanical construction.

J. W. Bateson Co. , Inc. , ASBCA No. 6069, 1962 BCA 3529, involved construction of the Air Force Academy. The Government required the Contractor to work a 7-day week ranging from 8 to 10 hours per day. The Contractor made a claim for acceleration costs including the cost of overtime inefficiency. The Board summarized the testimony of the Contractor's President as follows:

> Its President testified that, in the experience of his company, overtime hours seldom accomplish any additional productive work. They work overtime voluntarily only when the added pay therefor is necessary to attract sufficient workmen. He stated that the working of overtime hours produced the loss of efficiency through fatigue and also a volitional tendency to slow down on the part of the men. Appelant claims for this loss of efficiency 25% of the premium wages paid and 10% of the gross payroll. To this is added 11% for taxes and insurance, 10% overhead, and 10% for profit.

(N. B. Application of the above 25% of premium pay and 10% of gross payroll results in an overall loss of efficiency of 33% of all hours worked during the 7-10 hour day work week.) Although the Board did not find the claimed percentages supported by the evidence, it stated, "We are persuaded that there is a loss of efficiency attendant upon extended work. The extent thereof would vary dependent upon the daily hours and number of days per week added to the regular schedule. The Board did not actually award the cost of overtime inefficiency because it concluded that the Contractor would have had to work some overtime at its own expense to maintain schedule if the acceleration order had not been issued.

B. Inefficiency Due to Overmanning

In a normal schedule, the initial crew size and manpower loading is, if anything, kept slightly below "normal", with supervision required to produce optimum performance. As the project progresses a percentage is added to maintain at least the normal requirements. However, as a history of manpower attendance develops of labor turnover and "no-shows" an increase to the crew is made. There are going to be days when the normal turnover does not occur, impressing surplus manpower on the project. In an accelerated schedule, additional manpower is added over the norms. This additional manpower becomes inefficient due to physical conflicts of more men of different trades working in the same area; there is more competition for the same facilities, equipment and services; fewer productive workers being drawn from the available labor pool; and more dilution of supervisory control (the formwork foreman may now have eight crews under his supervision, as contrasted with a reference schedule of six crews), etc.

The only quantifiable data comes from Mr. L. V. O'Connor's paper, "Overcoming The Problems Of Construction Scheduling, Etc." mentioned above. Fig. 22.6 is taken from that paper, and is offered as a basis for negotiation or arbitration.

C. Inefficiency Due To Remobilization

After a work stoppage, due to whatever causes, restarting the interrupted work activity will produce an inefficiency. This will be due to several factors such as new members on the work crew, time delay in re-obtaining tools and material normally kept at a specific project site, or returning construction equipment to the activity site from another project location. Fig. 22.7 again comes from the only published source, Mr. O'Connor's paper, "Overcoming The Problem of Construction Scheduling, Etc.," mentioned above, but is again confirmed from proprietary sources of the author's client.

Fig. 22.7 is based on unit costs ($/sq. ft. of formwork, $/cu. yd. of concrete in place, $/lb. of reinforcing steel in place, $/lb. of pipe in place, etc. These are all direct labor costs that are obtained from the project cost records.

Thus, considering an arbitrary cost-incurring activity of 10 work day duration after re-start from a work stoppage (labor disturbance, weather, etc.) that leveled out at a 100% unit cost per unit installed over the last 65% of its duration, we could expect a range of efficiency from 60% to 92%, with an average of 85% efficiency on the first day after re-start (10% of that activity's remaining duration). In other words, an average of 15% of that activity's manhour costs the first day after re-start is a remobilization inefficiency.

It is to be noted that the "Remobilization" data offered is identical to well publicized "Learning Curves" of "Manufacturing Progress Functions" that show improved productivity with time (such as concrete pours, welds, etc.) The type of curve shown is an "Ascending Learning Curve." The difference between the Learning Curve and the Manufacturing Progress Function is that the Learning Curve is measurable and directly related to the direct worker, whereas the Manufacturing Progress Function must be attributed to the "learning" or efficiency of the support organization around the worker, such as direct supervision, supply, management, etc.

D. Inefficiency Due To Remanning

In the preparation of a report on a claim concerning the construction of a hydro-electric plant, the author noted a phenomonon that was related to the Learning Curve configuration that leads him to suspect there is another inefficiency occurring on projects of any type of

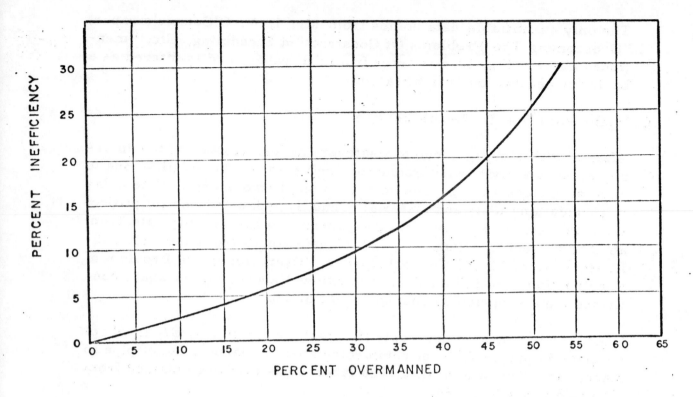

Fig. 22.6

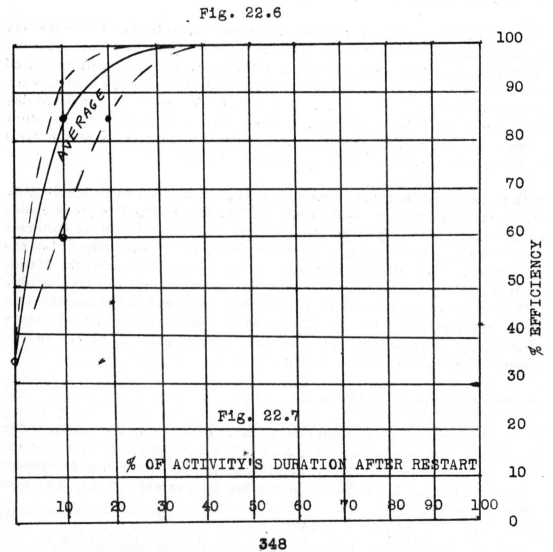

Fig. 22.7

348

schedule, normal or accelerated, but not presently quantifiable. In his study of the concrete formwork activities (straight forms, repetitive activities), and from the excellent cost records the contractor maintained on the project he noted that with time, the unit costs diminished along the learning curve, as in Fig. 22.8, where 100% is the asymptote of the unit cost curve. However, a study of the payroll records revealed that there was in the last twelve months of the period covered, an average of one completely new crew each month. This was due to an unstable condition at the location of the project which ultimately became the subject of a government agency report. It was an unusually high turnover rate.

The thought arises, therefore, if the original crew and its supervision (remember, the foreman is also from the labor pool) had stayed on the project, should not the unit costs for this type of work gone even further lower on its asymptote, for productivity should have been better than that obtained. So, there is a re-manning inefficiency, which would be as shown in Fig. 22.9. However, this is a hypothetical assumption, since there is nothing available as quantifiable data published, to the best of the author's knowledge. It seems as though this would be an excellent topic for study, research, or the basis of a doctoral thesis.

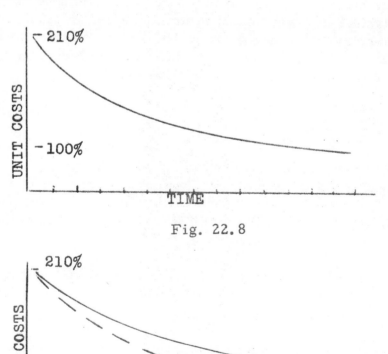

Fig. 22.8

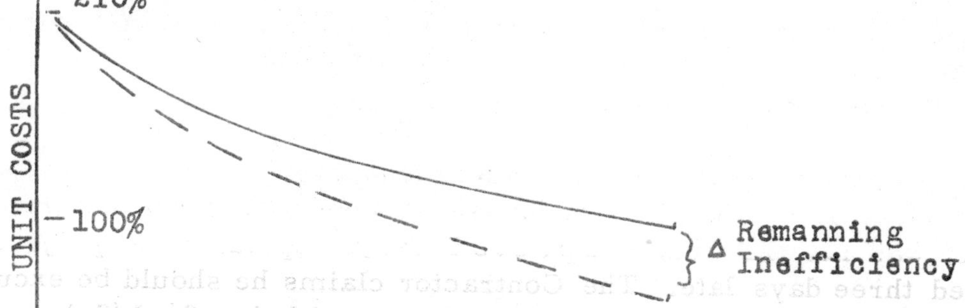

Fig. 22.9

Another interesting aspect of Fig. 22.8, particularly in the construction field, is that few people in construction bid a project that has repetitive tasks on a learning curve basis, by accepting some average between the maximum unit cost and the asymptotic unit cost; most bid on the maximum. In manufacturing, the learning curve is a widely used basis for estimating and bidding repetitive work packages.

E. Estimate Accuracy

The accuracy of the estimates in the Reference Schedule has to be considered in any study of inefficiencies of an accelerated schedule; its effect is an efficiency factor for the estimator. Most organizations will consider a plus or minus 2% as a reasonable range for their estimate accuracies.

There is a growing awareness that inefficiencies are inherent in any accelerated schedule. More and more organizations are including inefficiency data in their proposals and incorporating them in the terms and conditions of contracts.

The data herein is offered as a guideline for either mutual negotiation on a claim by the involved parties, or for the use of an Arbitrator in determining a reasonable and equitable adjustment to the claim. It should be noted that a unit cost contract, typical in heavy construction projects, is particularly insensitive to inefficiencies in an accelerated contract schedule, and unless prior provision is made for the categorized inefficiencies, a contractor is vulnerable for additional costs on this type of project.

A Simulated Analysis Of A Claim

We will use the problem of "Building A New Home" of Problem 1.1 in Chapter 1; the schedule developed for it in Chapter 2; Chapter 6 (Figure 6.0); Chapter 8 (Problem 8.3), the Time Scale charts thereof in Chapter 11, manpower as in problem 11.3; and the Milestone Chart of Chapter 21; as the basis of an exercise in the preparation of a claim analysis.

You are asked, as a consultant, to prepare for a plaintiff contractor, an analytical report and graphic display that shows the situations encountered during the course of "Building A New Home."

A study of the project records and correspondence reveals that there was a $200 per day Liquidated Damage clause for every day beyond Day #59 that the project was not finished, inspected and accepted. The project actually finished three days late. The Contractor claims he should be excused from the Liquidated Damage clause, in fact he is entitled to 26-1/2 days extension due to Change Orders and other situations over which he had no control, and that he also claims he should be reimbursed an additional $2477 for the costs of delays and changes for which he was not culpable. The Architect, the Owner's agent, states that due to the poor performance of the Contractor, he should not be excused from the Liquidated Damage Clause. The Contractor is entitled to the agreed upon amount of $1,000 for Change Order #1, but this should be offset by $600 penalty for the three (3) day lateness, for a net reimbursement of $400. The Architect notes there had been a problem in selection and approval of a specific design of the windows (probably an Owner caused situation), but once it was resolved he (the Contractor) failed to accelerate the schedule at a propitious time, by going on overtime, as specified in the contract when called for by the Architect. In fact, the Architect is on record that the Contractor was undermanned on a Critical Activity, and ignored the Architect's request for more craftsmen for the particular activity.

Project Narrative History

The contract records and correspondence file gives the following history:

1. Change Order No. 1, issued on Day #10, authorizes the installation of a new multi-colored asphaltic compound for the driveway, which creates a distinctive psychedelic appearance. Since some of the original specified driveway material was installed, Change Order No. 1 also authorizes demolition, or removal of the partially installed work. A supplement to the project contract is negotiated for Change Order No. 1. Both parties sign, and the additional amount of $1,000 is added to the contract price. In addition, the Contractor is on written record for requesting an eight day extension to the end date of the contract due to this additional work. This request was apparently ignored by the Architect, because there is no record of acceptance or rejection by the Architect on this request.

2. The furnace was delivered to the project on Day #16. The Contractor had selected Company's B model, on the assurance of the manufacturer that it met the "or equal" intent of the specifications. The Architect's inspector rejects it, since its capacity does not meet specifications. The Contractor claims a mis-interpretation of the specifications, and objects to the Architect's interpretation of his own specifications. These objections are in writing. A new furnace, the model specified, is ordered and delivered by Day #28. It takes 16 days to install with two men, compared to the original estimate of 10 days with two men. The Contractor claims in writing an additional 12 mandays compensation of $360, and six days extension to the contract end date.

3. The Contractor submitted the shop drawings for the window units on Day #10. There is no approval received until Day #30, and then it is "Approved As Noted," which effectively changes the design of the

window from the unit specified to another model (from the same manufacturer). This causes a delay of the delivery of the window units until Day #42, with the new window frames arriving at the site on Day #38. There is no change in the cost of the new model window.

4. On Day #31, the Stone Masons have gone as far as they can on the Stone Veneer without the window frames. They stop, and do not start again until Day #39.

5. On Day #39, the Architect complains, on record, that there are only two Stone Masons on the project, whereas previously there had been four masons. It is recorded that the new window frames cause some additional mason work. The Architect requests the Contractor to increase this craft to four masons. The Contractor ignores this written request, and finishes the Stone Masonry midway in Day #45.

6. The windows are started to be installed on the morning of Day #43. The Contractor realizes that he is behind schedule and authorizes overtime for the Drywall and Painting activity (#65-80). He also increases the crew size from four men to five men; and goes on two hours overtime a day. This premium time activity finishes on Day #52.

7. The Kitchen Equipment sub-contractor agrees he can start as soon as some of the Drywall is erected in the Kitchen area. He starts activity #85-100 on Day #51, and finishes it by Day #57.

8. The Punch List, Inspection and Acceptance activity, #100-110, takes as long as originally estimated, and finishes three (3) days late, on Day #62.

9. The Contractor submits a final statement for the amount of $2,477.00 and requests the equivalent of 26-1/2 days extension. The statement breakdown is as follows:

   a. $1,000.00 for Change Order #1
   b. $360.00 for the additional work on the furnace installation.
   c. $157.00 for the additional work on the Stone Veneer due to the new windows.
   d. $720.00 for the overtime for the Carpenters and painters on the Drywall activity #65-80.
   e. $240.00 for the additional overhead costs (indirect) for the extra three (3) days worked at the end of the project.
   f. The claim also requests excuse from the 3 days Liquidated Damage clause, because of "entitlement" to the following contract extensions:

1. 8 days due to Change Order #1.
2. 10-1/2 days for Work Stoppage on
    Stone Veneer due to the Window
    Design change.
3. 8 days due to the Window Delay.

This is the situation you are requested to clarify, and justify.

A. The Feasible Schedule

The first step is the establishment of a feasible schedule for this project, either by analysis of what was originally issued on the project (convert to a Time Scale Chart for rapid analysis of the logic of sequences, and reasonable estimated durations for the activities); or construct a feasible project plan, by the Arrow Diagramming technique, from the original contract or bid documents. If you are not qualified in the field of the project (construction, manufacturing, research, etc.), obtain qualified outside help, such as a contractor or manufacturer who would have no vested interest in the project, the Owner and the Executor of the project. Prepare this diagram with the thought that you will have to verify and testify to its reasonableness. Be prepared to REFERENCE activity arrow and logic transfer agent (dummy) to a specific document and/or reference. Use accepted references as standards wherever possible, e.g., in construction, use the annual F. W. Dodge Co.'s "Construction Pricing and Scheduling Manual" for duration estimate support.

For our case history we will use the Arrow Diagram for "Building A New Home," Fig. 22.10; its digital schedule, Fig. 22.11, and its Continuous Milestone Chart, Fig. 22.12. Of course, in your actual presentation, the Continuous Milestone Chart will be weather adjusted, and calendar dated (the blank row on top of Fig. 22.12) as described in Chapter 11. The Cost Domain (obtained from the data on this problem in Chapter 19) is constructed on the Milestone Chart. This would be obtained from the project estimates. In addition, the indirect costs (Project Overhead) are assumed to be $80 a day, and also added to Fig. 22.12.

B. The Actual Schedule

Actual performance of starts and finishes of all the activities are obtained from project diaries, progress reports and any other pertinent records. Again, be prepared to document all references for each actual activity. The Actual Schedule is made as an overlay, and color coded. This is done either on a print of the Feasible Schedule, or by a transparent overlay. Madison Avenue graphic display techniques are valuable here. Delays, Changes, etc., are coded differently from actual starts and finishes. Any discontinuities or gaps in the actual project schedule should stand

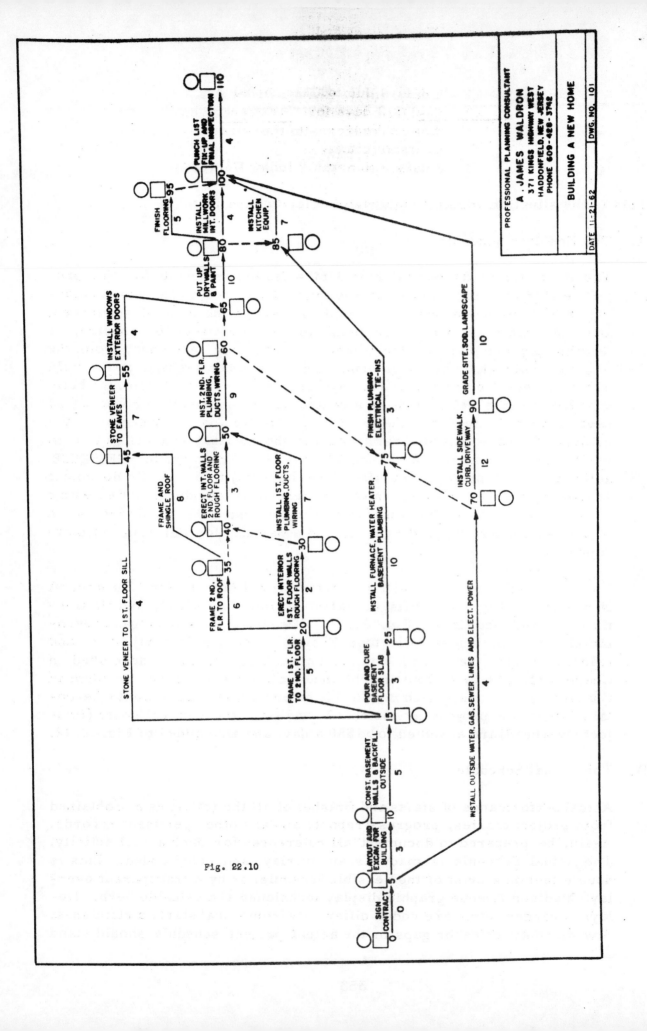

Fig. 22.10

PROFESSIONAL PLANNING CONSULTANT

A. JAMES WALDRON
371 KINGS HIGHWAY WEST
HADDONFIELD, NEW JERSEY
PHONE 609-428-3742

BUILDING A NEW HOME

DATE 11-21-62   DWG. NO. 101

BUILDING A NEW HOME
CRITICAL PATH METHOD NORMAL TIME
TOTAL TIME - 59 DAYS   TOTAL COST - 20640

I	J	DURATION	COST	DESCRIPTION	ES	EF	LS	LF	TF
0	5	0	0	SIGN CONTRACT	0	0	0	0	*
5	10	3	700	LAYOUT AND EXCAVATE	0	3	0	3	*
5	70	4	950	INSTALL OUTSIDE WATER GAS SEWER POWER	0	4	29	33	29
10	15	5	1320	CONSTRUCT BASEMENT WALLS BACKFILL	3	8	3	8	*
15	20	5	850	FRAME FIRST TO SECOND FLOOR	8	13	8	13	*
15	25	3	470	POUR AND CURE BASEMENT	8	11	12	15	24
20	45	5	2200	STONE VENEER TO FIRST FLOOR SILL	8	13	23	27	15
20	30	2	430	INSTALL FIRST FLOOR WALLS ROUGH FLOOR	13	15	20	22	24
20	35	8	1200	FRAME SECOND FLOOR TO ROOF	13	21	35	43	7
25	40	8	350	INSTALL FURNACE WATER HEATER BASEMENT PLUMBING	11	19	13	19	24
25	75	0	0	DUMMY	15	15	26	26	11
30	50	7	650	INSTALL FIRST FLOOR PLUMBING DUCTS WIRE	15	22	22	29	7
30	40	0	0	DUMMY	19	19	19	19	11
35	45	8	1130	FRAME AND SHINGLE ROOF	19	27	27	34	7
35	50	0	0	DUMMY	19	19	26	26	7
40	50	4	490	INSTALL SECOND FLOOR WALLS ROUGH FLOOR	22	26	29	34	7
45	55	7	2060	STONE VENEER TO EAVES	27	34	34	41	7
50	60	9	480	INSTALL SECOND FLOOR PLUMBING DUCTS WIRE	22	31	29	38	7
50	65	7	370	INSTALL WINDOWS EXTERIOR DOORS	31	38	38	45	7
55	65	0	0	DUMMY	31	31	38	38	7
55	75	0	0	DUMMY	31	31	45	45	14
60	65	0	0	DUMMY	34	38	38	38	7
60	75	10	840	PUT UP DRYWALLS PAINT	38	48	38	48	14
65	80	0	0	DUMMY	4	16	33	45	41
65	75	0	0	DUMMY	4	4	45	45	29
70	90	12	1320	INSTALL SIDEWALKS CURB DRIVEWAY	31	34	45	48	14
70	85	4	490	FINISH PLUMBING ELECTRICAL TIE-INS	48	55	48	55	29
75	85	0	0	DUMMY	48	52	51	55	3
80	95	5	730	FINISH FLOORING	48	53	50	55	2
80	100	4	490	INSTALL MILLWORK INTERIOR DOORS	48	48	48	48	3
80	100	7	1870	INSTALL KITCHEN EQUIPMENT	31	48	45	55	*
85	100	10	750	GRADE SITE SOD LANDSCAPE	16	26	45	55	29
95	100	0	0	DUMMY	53	53	55	55	2
100	110	4	500	PUNCH LIST INSPECTION	55	59	55	59	*

Fig. 22.11

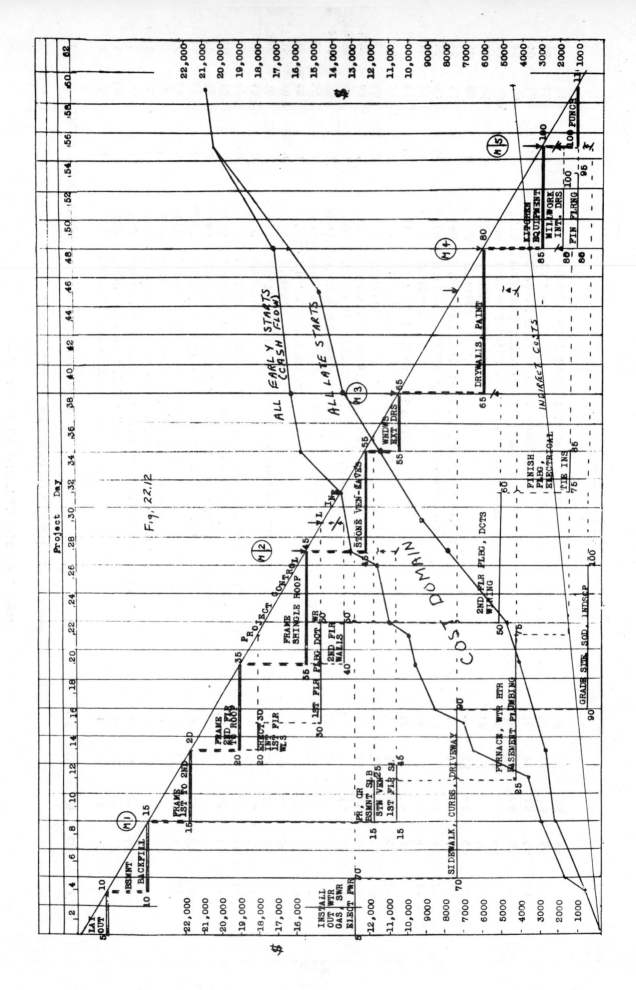

Fig. 22.12

out, and they should be labeled or coded for separate discussion.

Fig. 22.13 represents the Actual Schedule with all conditions shown. For clarity, actual starts and finishes are identified by I-J numbers. Incidentally, Fig. 22.13 would be typical of a field copy of an original Continuous Milestone Chart, with marked up starts and finishes as shown on a reproducible of the original.

There is now all the evidence assembled on one document necessary for analysis of the claims and counter claims in the Project Narrative History. In preparing this type of exhibit for an actual claim, it is recommended that the exhibits be as simple and clear visually as possible. This may require separate overlays or prints showing the conditions chronologically.

Your analysis of the claims are as follows:

1. The effect of Change Order No. 1 is seen visually. The agreed upon extra cost of $1,000 can be seen in the jump in the Cost Domain curves. Even though activity #70-90 increased in duration, it did not pass its original Late Finish date. The contractor is entitled to the $1,000 additional cost, but no extension of the project end date.

2. From the records, there is nothing to refute the inspector's claim that the wrong and undersized furnace was originally delivered. Unless the manufacturer wishes to participate in this claim and offer conclusive proof that his originally purchased then rejected model does meet all the specification requirements, this claim will have to be disallowed. There is no justification in any case for an extension of the contract end date since the furnace was installed before its original Latest Finish date.

3. The evidence here is against the Architect. A transmittal form indicates the window drawings were submitted on Day #10, and the manufacturer asserts that the "Approved As Noted" comments on the returned shop drawings did effect a change in model, with a consequential delay in delivery. The consequential effects of this delay after Day #30 must be evaluated.

4. From an analysis of the Architect's drawings, it is apparent that the Stone Masons had gone as far as they could and the shutdown of that work on Day #31 is reasonable.

5. By examining the Actual Schedule of Fig. 22.13, it will be noted that reduction of time in the final work of activity #45-55 (as requested in the Architect's demand for more Stone Masons on that activity) would not have improved the project status, since it still took four (4) days

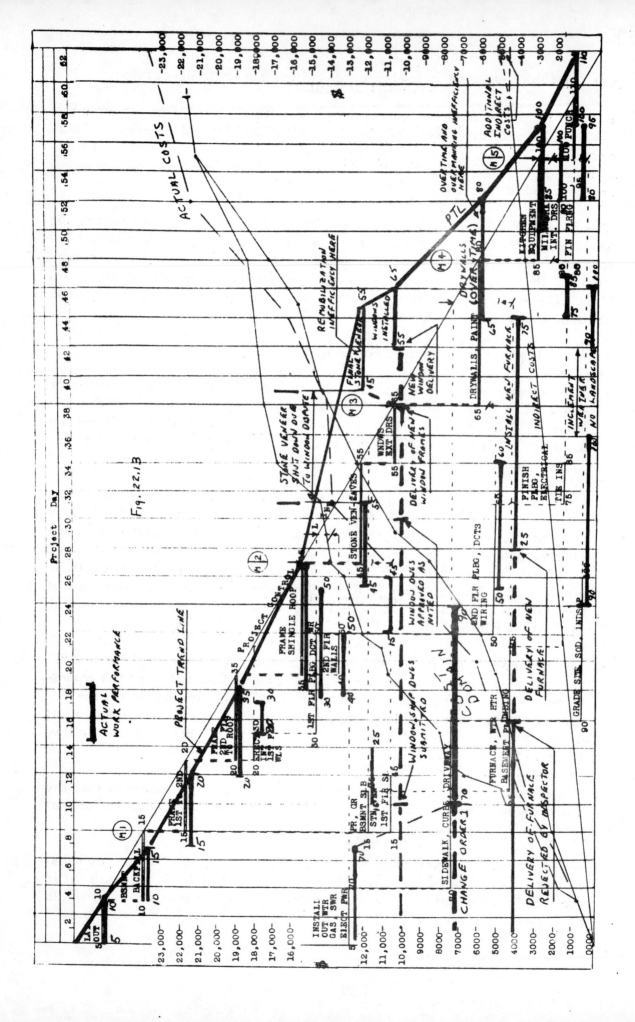

Fig. 22.13

for the window installation after their receipt on Day #42. To test this, shorten the second half of the actual activity #45-55, and note that the Project Trend line still goes out to the end of Day #46, the actual completion of the window installation activity #55- **65**

By using the remobilization Inefficiency Curve, Fig. 22.7, and noting that there were 6-1/2 days consumed on the actual work on the second half of the interrupted activity #45-55, the inefficiency effects can be collected. Assuming that the first 1/2 day of activity #45-55 is 9% $\frac{(0.5)}{6.5}$,

the average inefficiency is 20% from Fig. 22.7. This means 0.2 x 2 mandays, or 0.4 mandays was wasted on the first day's effort of activity #45-55. The second half of that first day $\frac{(1.0)}{6.5}$ was 17% of the

remaining time, and that caused 10% inefficiency, or 0.1 x 2 mandays, or 0.2 mandays lost. Thus, the contractors actual time remaining on activity #45-55 in the second half should have been 10 mandays. The original estimate was 4 men for 7 days (from problem 11.3). The actual work in the first half of activity #45-55 was 3 men for six days, for 18 mandays. 28 mandays minus 18 mandays leaves 10 mandays for the second half of interrupted activity #45-55. The extra work, due to the new window design, caused two additional mandays (the two men one day more), over the 10 mandays left, plus the 0.6 mandays of inefficiency. There is also a remanning inefficiency here, since new masonry crews were used. Unfortunately, there is nothing published to use to obtain a quantification of that inefficiency.

The $157 for the additional work, which includes 0.6 mandays inefficiency for remobilization, appears justifiable.

6. Since the Overtime situation for activity #65-80 is a consequential effect of the delays due to the precedent window delivery and final installation, the additional costs are justifiable. In the claimed additional cost of $720 for this premium time activity, 6% of it is due to an overtime inefficiency (Fig. 22.5, using five 10-hour work days), and 13% for overmanning (Fig. 22.6). The $720 appears reasonable.

7. A field decision was made that tends to substantiate the Contractor's claim that he did exert diligent prosecution of the work. The original logic of the Arrow Diagram had the Kitchen Equipment being installed after all the Drywall was installed and painted (activity #85-100 follows activity #65-80). The project records show that some lost time was made up by starting the Kitchen Equipment sub-contractor earlier than his original logic sequence; he was started as soon as that part of the Drywall that controlled his work, the kitchen drywall, was finished. In other words, there was an overlap in work, made by a field supervisor, that bettered the logic of the Project Arrow Diagram.

8. Since the Punch List, Inspection and Acceptance activity #100-110 met the original estimate (and perhaps a specification clause for a four day period for this activity) there is no evidence to fault its performance. The actual start was delayed consequentially due to the window change and delivery delay.

9. From the data studied, it does appear that the Contractor did exert reasonable diligence in pursuing his work. His lateness in finishing the contract can be shown to be a consequential effect of delay over which he had no control, and for which he was not culpable. He appears to be entitled to excuse from the Liquidated Damage clause. It does appear that his claims for (a) $1,000 for Change Order No. 1 on the Driveway, (c) $157 for extra Stone Veneer work plus remobilization inefficiency, (d) $720 for Overtime work on activity $65-80, which contains 6% inefficiency for Overtime, and 13% Overmanning inefficiency, and (e) the $240 for additional Indirect Costs due to the three (3) day extension of the project, are all allowable.

His claim (b) for $360 for the furnace "change is not supportable, and is disallowed. His claim for the equivalent of 26-1/2 days extension is disallowed.

Thus, your report would allow $2,117 in extra costs to the Contractor, and excuse from the Liquidated Damage clause.

GENERAL RECOMMENDATIONS

1. KEEP RECORDS.

2. ESTABLISH AS SOON AS POSSIBLE AN ACCEPTABLE PROJECT SCHEDULE THROUGH THE NETWORK PLANNING TECHNIQUE AND ITS RESULTANT CONTINUOUS MILESTONE CHART.

3. GO ON RECORD AS SOON AS POSSIBLE WHEN TRENDS INDICATE A LEANING TOWARDS SLIPPAGE. SET UP A REASONABLE LEAD TIME OF NOTICE (AT LEAST ONE WEEK ON REAL PROJECTS) BEFORE YOU REACH A LATE START OR LATE FINISH OF AN ACTIVITY.

4. KEEP RECORDS.

CHAPTER 23

CURRENT ABUSES OF THE NETWORK PLANNING TECHNIQUE

With over eight years of direct and continuous experience in the field of Network Planning as a consultant, the author has observed many requirements specified, and systems installed for the use of this technique. Unfortunately, it appears that in many instances, the abuses exceed the uses. This is all the more unfortunate since the technique is essentially a simple, explicit and flexible one; these attributes should make it a powerful technique.

In an attempt to eliminate these prevalent abuses, it was felt that an effort to categorize them would be of assistance to the reader. For the neophyte, this categorization should list the pitfalls to be avoided. To the experienced user, this categorization may strengthen his suspicions, and arguments, to correct conditions that frustrate the effective use of the technique.

Prevalent Abuses

1. Inauthenticity. The wrong people are originating, estimating and issuing the initial project schedule. This is basically a line manager's technique, yet too often the staff personnel are usurping the line manager's prerogatives. Whoever is to manage the execution of the work, or whoever normally estimates the work, is the person or persons who are responsible for the preparation of the Network Plan, and for the time duration estimates, resource assignments and cost estimates for the activities. In a Construction group, this means the Project Manager, the Construction Superintendent, the Estimator, and the Buyer should participate. In an Engineering or Architectural project, the Engineering Project Manager, the Project or Co-ordinating Architect, the Chief Draftsman, and the various specialty engineers (Mechanical, Electrical, Structural, etc.) should all participate.

    The above suggestion of the origination of the Project Network Plan and its schedule does not imply that this effort cannot be delegated to others at a lower managerial level, or to outside consultants, or to a staff group. However, the fundamental prerogative of the line manager is full review and approval of the logic, resource allocations, and duration estimates, with particular note of weather affected activities, if pertinent.

    It is noteworthy that the above comments imply a single, or "key" man concept--there is one Project Manager. Some organizations do not function this way; they have project responsibility distributed over two or more individuals. A typical case is the project management concept in the field of military electronics, where there is an Engineering Project Manager, responsible for technical performance and production, and a Contract Administrator, responsible for project costs, schedules, changes, etc. Obviously, there has to be concurrence in a multi-distributed management responsibility group.

Another aspect of inauthenticity occurs when the Owner of a project has had prepared a Network Plan and Schedule, either by his own staff, or an outside consultant; and this schedule is to be enforced by the selected Contractor. The author has observed that either the durations are conservatively estimated (a "safe" schedule has been given to the Owner), or unrealistically low in durations, implying high resource assignments or assumptions, that may not be realistically available in the project geographical area. Very few Owner prepared Network Plans and Schedules appear to have given realistic consideration to manpower availability and/or productivity. Another shortcoming noticed is disregard for weather effects on weather affected activities--the same production rate is shown for concrete pours, welding, laying of buried pipe, masonry, etc., for both Summer and Winter work.

The reader will note in Chapter 25, that under the section, "Owner Prepared Network Plan and Schedule" it is stated that the Contractor should be allowed to reject all or part of the specification Network Plan. If such a rejection should occur, then it is encumbent on the originator of the Project Network Plan and Schedule to analyze the submitted Contractor schedule for feasibility.

2. Inadequacy

a. Time and Information. There is a trend to the allowance of a very short time to prepare the Project Network Plan and Schedule. The Contractor is allowed an unreasonably short time to submit his initial Plan and Schedule, and this is complicated by inadequate or incomplete drawings and/or other information being available at the time of Notice to Proceed, or Contract Award. Typically, drawings are revised narratively by addenda to the specifications during the bid preparation period.

A reasonable formula for the time of preparation os an adequate Initial Network Plan and Schedule is as follows:

$$T = 12 \text{ plus } 0.4 \text{ (Capital Cost of Project in Millions) } (23.0)$$

where T is time in working days, and Capital Cost of Project is the budget estimate for the project, and does not include real estate costs.

Thus, for someone knowledgeable in Network Planning, and with a background in the type of project being planned, it would be reasonable to expect a $5 million construction project to take 12 plus 0.4 (5), or 14 working days.

b. Display. The majority of people using Network Planning techniques stop with the issuance of a Network Plan, and its digital output schedule. It is such a simple next step to construct the Time Scale Chart. and the continuous Milestone Chart. These are control tools readable by any level of supervision on the project.

3. Massivity - an excessive amount of information required in both the input and output data.

a. Input Abuses. In the input, the "Level Of Indenture," or fineness of detail sometimes is far beyond the practical needs for the schedule. The activities are of excessively fine or small definition. A recent concrete and steel two story building, of relatively simple design, requiring the contractor to use Critical Path Scheduling, had a capital cost of $900,000. The specifications required a minimum of 500 activities to be shown, exclusive of dummies. This magnitude of activities should be that for a $3 million construction project. The contractor was forced to break down many activities into one hour duration work packages.

b. Output Abuses. The number of copies of the output schedule grows excessively and successively larger, and the number of sorts of the output increases. Besides the basic schedule output format as shown in Chapter 8, there are burgeoning requirements for sorts by Early Start Time, Late Finish Times, by Float, and combinations thereof, by sub-contractor and/or internal Department Code numbers.

The remedy for these abuses is to have the specifications call for a reasonable level of indenture, as mutually agreed upon by the Sponsor and the Contractor. Distribution of the schedule outputs should be limited, with only Summary Reports (in the form of a Summary Continuous Milestone Chart, no larger than an 11" x 17" sheet) being circulated, and the completely detailed schedule available in Central Files.

4. Network Diagrams With Bids. There are an increasing number of organizations, commercial as well as governmental agencies, that are requiring a Network Plan and Schedule to be submitted along with a Cost Proposal. This is based on the premise that it allows the bid evaluator to see if a reasonable approach to project execution is also being proposed by the bidder. There does not appear to be any increase in time

for the preparation of the Network Plan supplemented bid, which puts an increasing onus of effort and costs on the bidder. This situation impresses the author as an excessive requirement. It must be assumed that if an organization is qualified to make a Cost Bid, that it must have also been qualified to manage the project. It is recommended that the bid specifications inform the bidders that the successful bidder will be required to submit a Project Plan and Schedule within a reasonable period after Notice to Proceed.

In the near future, there will occur an interesting legal problem. This will occur when Bidder A is not low bidder, but the Owner wants the successful bidder, B, to use A's Network Plan and Schedule. New legal precedents will be established, at some cost to Bidder A. It is recommended that if the reader must issue a Network Plan and Schedule with a bid, that he protect it with the International Copyright Convention ©.

5. Superfluity. Another growing abuse is the requirement that a detailed schedule, via the Network Technique, for the first 60 or 90 days of effort, be issued by the Contractor in a relatively short period of time after Contract Award, Notice to Proceed, or Approved Work Order. The concept is to be able to observe the initial phases of work, and maintain project control from the beginning.

In a construction or remodeling project, the initial phase of work is mobilizing, site preparation, and foundation work, and perhaps demolition. When the total project Critical Path is finally obtained, it may be discovered that the Vendor Drawings, delivery and installation of equipment is Critical, and controls Project completion. Time spent in preparing the initial 60 or 90 day plan and schedule has been lost for expediting the real Critical Path.

The above example is from an actual case history. It is the author's contention that the project is best served if a complete, reasonably detailed initial Network Plan and Schedule is constructed and analyzed. After acceptance, this Master Plan can be summarized, or sectioned, to show any initial phase of the project.

6. Rigidity. Once the project Network Plan is constructed, it is assumed that the project will never deviate from the sequences shown. This is a common abuse. Once a Plan has been constructed, and the schedule generated, analyzed, and approved, it is too frequently assumed to be the only way the project can be executed. The shibboleth, "A good plan should never be changed, just followed," attributes an omniscience to the planner that is extremely rare.

The turmoil and effort that go into manipulating and revising the original plan and schedules appear quite impressive, but the concept of this technique is flexibility to provide explicit, accurate and timely information for decision making. It is <u>not</u> necessary or even advisable to redraw the Arrow Diagram every time a change occurs. It must be remembered that when the plan was originally constructed, many of the sequences were arbitrary, based on available knowledge, or lack of it, and experience. There are usually several methods of executing any project, many of them relatively equal in feasibility. The Continuous Milestone Chart of Chapter 21 was designed to handle sequence changes without redrafting. It is in Updating the initial, and approved Project Schedule, that the Rigidity Syndrome appears most frequently. If a trench is to be dug in sections, each section represents a reasonable amount of trenching activity. Section 1, then 2 may be finished on the CPM Schedule, but upon starting Section 3, process pipes in the ground may be found that are not shown on the Engineering Drawings. The project does not stop because the chart states Section 4 does not start until Section 3 is finished. The field superintendent would jump to Section 4 and continue digging that section until the problem at Section 3 is resolved. It should not be necessary to redraw the Network Plan.

Note the stress on attainment of "Milestone" Dates in Chapter 11 - the goal of the technique. Individual sequences are of minor importance (since there may be several methods of executing the work) as compared to the milestone - the accomplishment of discrete packages or phases of work. This philosophy is repeated in the chapter on Updating (Chapter 18).

7. <u>Periodicity</u>. A schedule must be produced, or updated, at even calendar periods. This abuse, also in the Updating phase of Scheduling, is a carry over from the previous generation of Management Control Systems. When the bar chart was the tool, frequent periodic reporting and updating (either monthly, quarterly, or -weekly) was normal and was based on the approximation technique of breaking a system into small enough elements, so that somewhat reliable trends could be detected. That tool and method is now antiquated.

The technique of Network Planning and scheduling, a formal logic method, identifies key or milestone dates, where significant activities must be started or completed. These never occur at regular intervals. In a 10 month duration project, 10% of the work is not necessarily finished the first month, nor 20% by the end of the second month, etc. The Sponsor should realistically consider the frequency of periodic updating required.

The author has found, after serving as a consultant and project monitor on over 300 projects, that from two to twelve milestones are all that are necessary to control a project. This range is from a six month to a five year project duration. Network Scheduling has been extolled for its virtue of helping people become more efficient managers. Why doesn't this apply to the paper work on a project? Only radical deviations or trends from an approved plan and its schedule need be reported.

The salient feature of the Time Scale Project Control Chart is the rapid and graphic manner in which the Key or Milestone Dates on any project may be identified. In between Updates, Line Management has the Time Scale Chart to observe the Start and Finish Dates of all activities, and compare actual performance with the original plan and schedule. With sound managers, the project does not go unobserved, or uncontrolled between Updates. Reporting the statement of Chapter 11, Updating a Schedule is essentially a re-estimate, and is not necessarily related to Project Control. Control goes on continuously.

Conclusion

The most common abuse today is a combination of "Rigidity" and "Periodicity." To a Sponsor whose history is replete with missed and slipped schedules, Network Planning and Scheduling may appear the be-all and end-all. Yet it must be understood that this technique of scheduling can't solve problems-- people solve problems. It is only an information model, and is only as good as the information and experience that goes into it. Its essence is logic, and flexibility, and must be used in that context.

Network Planning and Scheduling is a logically simple and flexible method of rapidly building an accurate model of any type of project. Keep it that way!

CHAPTER 24

VALUE ENGINEERING AND NETWORK PLANNING

Recently, a "new" technique has been exploited as an answer to Cost Effectiveness, and is called "Value Engineering." Touted as new science, the benefits received from it are claimed to be truly fabulous; an optimum cost for a facility, device, or system for a specified period of time is to be realized. Before a critique of the technique and its claims is made, a simple exercise in the principles of Value Engineering is in order.

The technique works this way:

Here is a picture of an object. What is it?

Fig. 24.0

What do you do with the object in Fig. 24.0? If you answered "write with it," you gave an incorrect answer. What you do with it is shown in Fig. 24.1.

Fig. 24.1

The correct answer is: you make marks on paper with it! Writing is a specific form of making marks on paper. This is the difference between "function" and "use." Recall in Chapter 2, you learned the "function" of the dummy; it transfers arrow heads from its tail to its head. Then, in Chapter 3 you learned several "uses" of that function. In Fig. 3.11, there are four uses of the dummy arrow: Dummy #70-100 is a Safety Requirement; Dummy #110-210 is a resource (Mechanical Crew) transfer; Dummy #110-120 is a logic isolator; and Dummy #115-120 is a Numbering Dummy.

Value Engineering works in this manner: Concerning a product, device, system, etc., the following questions are asked:

1.   What is it? (Identification)
2.   What Does It Do? (Function)
3.   How Is It Used? (Use)
4.   How Much Does It Cost? (Cost)

and the key question:

5.   What else can also be used (it may have a different function, but a similar use) and how much does this alternate cost?

VALUE ENGINEERING IS NOTHING MORE THAN GOOD ENGINEERING, AND IT SHOULD HAVE BEEN DONE IN THE FIRST PLACE. As it is used now, it is really RE-ENGINEERING. It is an effort to try to get people to think.

The relationship of Value Engineering to Network Planning should be obvious; they are both systems of getting people to think. The biggest asset of Network Planning is that it gets the right people doing the right thing--thinking about their project at the beginning, and then graphically sorting out the potential project problems. By "building a model" on paper via the Arrow Diagram, problem areas will quickly pop up and be subject to analysis, review and most importantly, decision making before the fact, not after.

CHAPTER 25

MISCELLANEOUS

The following specification section has been, in some form or other, used by government agencies, in their contractual requirements for construction projects. The U.S. Army Corps of Engineers is increasing the requirements of network planning and scheduling on projects which they are involved for information. With just a few modifications, it will serve any organization seeking to enforce network planning on their projects. It should be included in the original Proposal or Invitation to Bid Specifications.

Construction Progress Schedule Analysis

In addition to the requirements set forth in Section GC-5 "Progress Charts and Requirements for Overtime Work", the contractor shall prepare and submit for approval a network analysis of the construction progress schedule within __ days after receipt of the Notice to Proceed. This method of analysis is generally referred to as the Critical Path Method (CPM), Critical Path Scheduling (CPS), Critical Path Analysis (CPA), Least Cost Estimating and Scheduling (LESS), or Program Evaluation and Review Technique (PERT).

This analysis shall include as a minimum a graphic representation of all significant activities and events involved in the construction of the project, and a written statement explanatory thereof, if necessary, for a complete understanding of the diagram.

The network graphic representation (Arrow Diagram) and statement must clearly depict and describe the sequence of activities planned by the contractor, their interdependence, and the time estimated to perform each activity. The statement shall describe the construction methods proposed, and basis of performance time selected for each activity, in terms of size of crews, shifts and equipment to be utilized.

In developing the project network the contractor may use either the arrow notation clearly depicting activities and events, or a circle and connecting arrow notation of an acceptable manual method.

The initial schedule generated from the network shall contain the following minimum information for each activity: The early activity start time, the latest activity start time, the earliest activity finish time, the latest activity finish time, total float or slack.

If the contractor elects to use time units of "working days", he shall include with the initial schedule, a calendar schedule to show calendar dates for each activity.

Periodic status reports from the approved Network Plan will be required as determined by the contracting officer. These reports will indicate the expected completion dates of all current activities.

In the progress of the project, any changes in the original network plan desired by the contractor must be approved by the contracting officer before being put into effect.

Costs incurred by the contractor for training his personnel, for utilizing professional planning consultants, and for any computer rental time are considered part of the contractor's general and administrative expenses and are not directly reimbursible on this contract.

If, in the opinion of the contracting officer, the contractor falls behind the progress schedule, the contractor shall take such steps as may be necessary to improve his progress, and the contracting officer may require him to increase the number of shifts, and/or overtime operations, days of work, and/or the amount of construction planned, and to submit for approval such supplementary schedule or schedules in this form as may be deemed necessary to demonstrate the matter on which the agreed rate of progress will be regained, all without additional cost to the government.

The failure of the contractor to comply to the requirements of the contracting officer under this provision shall be grounds for the determination by the contracting officer that the contractor is not prosecuting the work with such diligence as will insure completion within the time specified. Upon such determination, the contractor may terminate the contractor's right to proceed with the work, or any separable part thereof, in accordance with the delay - damages part of the contract.

Position of the Contractor or Executor of the Project

Dependent on the fiscal nature of the particular contract (Fixed Price, Cost Plus Fixed Fee, Cost Plus Incentive Fee), the contractor or executor of the contract may wish to protect his contractual status and obligations with respect to submittal of the network plan for approval. On a fixed price contract, the contractor should retain some perogatives in changing his plan to take advantages of favorable business conditions, that do not affect the technical aspects of the project (a manufacturer sub-contracting a power supply unit, rather than fabricating it, a contractor arranging to pour a first floor slab before the basement slab by use of shoring and mud sill, etc.). The following clause is quite frequently added to the network drawing for the protection of the executor or contractor of the project:

Network Note

"This plan is for information only. The XYZ Company reserves the right to modify or change this plan to suit project conditions, changes in scope of work, conditions of emergency or force majeure, or advantageous business conditions that do not affect the technical aspects of the project."

An Owner Prepared Network Plan and Schedule

In the case where the Owner has had prepared a Network Plan and Schedule for his project, before Contract Award, the contract specifications should have the following conditions contained therein:

1. The specifications should state that the referenced drawings and schedules represent a feasible schedule, and that the Contractor will be required to meet the contractual end date and intermediate Milestone Dates shown.

2. The specifications should ask for concurrence by the Contractor, in writing, if he accepts the feasible schedule.

3. The specifications should also allow the Contractor a period of time to review the feasible schedule, and that he has the right to reject all or part of the enclosed feasible schedule.

4. If the Contractor rejects all or part of the feasible schedule, then he must, within a period of ___ days after Notice to Proceed, submit his revisions, or completely new Network Plan and Schedule for review and acceptance by the Owner.

5. The accepted schedule, be it the original feasible schedule or the Contractor's revised schedule, must be in the form of a Time Scale Project Control Chart, and/or a Continuous Milestone Chart.

It is encumbent on the originator of the original contract specification Network Plan and Schedule to review the submitted logic and durations of the Network Plan if the Contractor decides to reject any or all of the specification feasible plan, and make recommendations to the Owner.

The following is a recommended format for inclusion in Contract Specifications for an Owner prepared Network Plan and Schedule.

PROGRESS SCHEDULE

A. THE INITIAL SCHEDULE

1. The General Contractor will be required to employ the techniques of the CRITICAL PATH METHOD (CPM) for Planning, Scheduling, Reporting and Evaluating the work under this contract, and to utilize this technique in the control of Construction Progress. All Contractors are required to adhere to the Owner Approved Project CPM Schedule.

2. The Owner has had prepared CRITICAL PATH Time Scale Schedule Charts, Drawings ___ ___ ___ and ___, of approximately ___ activities and ___ dummies. The activities represent the salient construction phases and include, but are not limited to shop drawings, procurement activities and construction processes. The activities are arranged to represent a logical sequence of construction. Each activity has been assigned a time duration estimated in working days with reasonable crew sizes. A computer schedule has been developed showing the Project Duration, the Critical activities, the Earliest and Latest Time each activity can start and finish, and the amount of Float Time for each chain of activities. The Time Scale Schedule Charts, Drawings ___ ___ ___ and ___ showing Critical Activities, and the Early Start and Finish Time for each activity, has been constructed from the computed schedules. Copies of the Computed Schedules are available upon request.

3. The Time Scale Schedule Charts represent one feasible method of completing phases of this contract by the required dates. Within 15 calendar days of Contract Award, or Notice to Proceed, whichever occurs first, the General Contractor, after review of the Arrow Diagram Plan and Time Scale Schedule Charts with any other Contractors, will either:

a. Accept, by a statement in writing, the aforementioned Owner Approved Time Scale Schedule Charts (with their sequences and durations) and agree to adhere to this schedule, OR:

b. The General Contractor shall submit for review and approval by the Owner and the Architect,

1. A Project Arrow Diagram in graphic form, which shows the order in which the various Prime Contractors propose to execute their work. This submitted Arrow Diagram shall not have less than ___ activities, excluding dummies. It shall include, but not be limited to procurement, shop drawings, fabrication, and deliveries of materials and equipment, and on-site operations (including acceptance testing), whether performed by the Contractor's own forces, or by his sub-contractors. This Arrow Diagram shall show the duration of each activity in working days, with a minimum duration of one (1) working day, and a maximum of 20 working days for any on-site operation. Material and Equipment Delivery Activities are excepted from

the above maximum duration. The Plan and Schedule shall be in sufficient detail to meet the approval of the Owner and Architect.

2. A tabulated schedule, either manually or machine generated, will be supplied. It will list each activity by Description, I and J Numbers (Begin and End Event Numbers), and with the following data:

   a. The Earliest Time the individual activity could Start.

   b. The Earliest Time the individual activity could Finish.

   c. The Latest Time the individual activity could Start.

   d. The Latest Time the individual activity could Finish.

   e. The Total Float Time of each activity.

   If the tabulation is not in Calendar Dates, then a Calendar Conversion Table will also be furnished.

3. Time Scale Schedule Charts, similar to Dwgs. ___ ___ ___ and ___ will also be furnished, indicating on a Calendar Dated scale the Critical Path (emphasized) and all other activities at their Early Start and Finish Times. Minimum graduations on the Time Scale Chart will be two work days. Weather affected activities will be noted.

The Tabulated Schedule and the Time Scale Chart submitted must indicate Beneficial Occupancies at the dates indicated on Dwgs. ___ ___ ___ and ___ or earlier. The Timely completion of the various phases of the contract and the construction methods used is the responsibility of the General Contractor. No claims will be allowed for extra compensation on the part of the General Contractor, or his subcontractors, for overtime or additional personnel or equipment in obtaining these completion dates.

B. UPDATING THE SCHEDULE

1. The General Contractor shall maintain the approved progress schedule at the job site, and shall post on the Time Scale Schedule Charts the actual percentage complete for each activity on the last Friday of each month. Activities which are behind schedule on a particular report date shall be so indicated by special notation or by the use of color codes. An activity shall be deemed to be behind schedule whenever, in view of the currently approved duration for that activity, it appears that the activity cannot be completed by its latest completion date.

In addition, the General Contractor upon the direction of the Owner and/or the Architect, shall periodically submit an Updated Schedule, in tabulated form with the data required as listed in para, A3. b. 2 above. It shall indicate the elapsed time from the start of the Project to the date of The Updated Schedule Report; those activities completed by the date of the Updated Schedule Report; the remaining time durations for those activities in progress at the time of the Updated Schedule Report; any additions and/or deletions of activities from the original Approved Schedule.

If the submitted Updated Schedule reflects a slippage in the required completion dates, a narrative report shall accompany the Updated Schedule Report indicating the cause of the delays; and a description of what corrective action will be taken to improve the schedule back to the required completion dates. The Time Scale Schedule chart will be revised accordingly and submitted for approval.

C. COPIES

The number of reproducible drawings, and prints thereof, and copies of the Tabulated Schedule shall be the same as the Shop Drawing Requirements for this Project. as described in Section _____ of the Specifications.

Sub-Contractor Clause

Below is a typical clause inserted into Sub-contractor contracts, when the General Contractor is using Network Planning on the Project. (Courtesy of Mr Michael Salter, Pres. Stewart Williams Inc. Augusta Maine)
ARTICLE _____

TIME OF COMMENCEMENT AND COMPLETION

TIME OF COMMENCEMENT:

Is _____ - Date of letter of "Notice To Proceed" by the _____ _____, or else as is ordered by_____ in accordance with the established "Critical Path" Schedule. The Contractor agrees to give the Sub-contractor five (5) days notice when to begin work at the site.

TIME OF COMPLETION:

After the Contractor instructs the Sub-contractor to start an operation, the Sub-contractor agrees to substantially complete the work hereunder and any operations thereof, within the Time of Completion which is the number of working days assigned to a specific operation, as shown or contained in this project's originally established or updated "Critical Path" Schedule. An operation is defined as any Submission of Shop Drawings and/or Sample; Fabrication; Delivery; Storage; and Installation, complete or partial, of any task, phase or portion of the Sub-contractor's work. The assignment of the number of working days to an operation shall be, as provided by the Sub-contractor during the initial establishment or preparation of the "Critical Path" Schedule and any subsequent updating requested by the Contractor, or, in the absence of the Sub-contractor's assignment of working days, the assignment of the working days shall be by the Contractor.

The Sub-contractor hereby agrees that the Contractor shall be the sole interpreter of the Critical Path Method Schedule in instructing the Sub-contractor to start an operation, and thereby the Contractor shall determine the use and/or assignment of the float time, if any, available to complete the sub-contractor's operation. In the consideration of Liquidated Damages and in order to insure that the project completion date of _____, be realized, the Sub-contractor agrees that he shall do all things necessary, as requested by the Contractor, to start and finish a critical operation on the assigned respective start and finish dates and to finish a non-critical operation on or before the assigned latest finish date.

EXTENSION OF TIME:

Should the Sub-contractor be delayed in the erection or completion of the work by the act, neglect, or default of the Contractor, by fire, strikes, lockouts, and embargoes or other causes for which the Sub-contractor is not responsible, then the time fixed for the completion of the work shall be extended for a period equivalent to the time lost by reason of any or all of the causes mentioned, which extended period shall be determined by the Contractor. But no such allowance shall be made unless the claim for extension is presented to the Contractor in writing.

LIQUIDATED DAMAGES:

Time of completion is of the essence of this contract. An important consideration in the award of this contract is the assurance of the Sub-contractor that he has or can obtain both men and material to complete the work within the time stated.

Inasmuch as it is understood that the Contractor will suffer damages amountin to at least $_____ per day for each working day that the work herein contemplated remains uncompleted beyond the date mentioned, it is agreed that the Contractor will deduct from any amount due the Sub-contractor the sum of $_____ for each working day's delay in completion, such amount to be considered as actual liquidated damage suffered by the Contractor and in no event to be considered as a penalty, and in the event the amount due the Sub-contractor shall not be sufficient to pay such liquidated damages, the Sub-contractor shall and will pay upon demand of the Contractor any balance of such liquidated damages remaining due to the Contractor.

APPENDIX

Glossary of Terms
Symbols & Formulae
Bibliography
Standard Normal Distribution Function
Network Schedule Forms

GLOSSARY OF TERMS

Project	Any company, organization or department undertaking consisting of many activities and events required to attain a unique goal.
Activity	Any definable and time-consuming task, operation or function to be executed in a project.
Event	A point in time indicating beginning or ending of one or more activities. It is a milestone, representing that point in time when some meaningful specified accomplishment is either to be started or completed.
Critical Path	That particular sequence of activities in a flow chart which establishes the length of time for the accomplishment of the end event of a project.
Float Path	Those sequences of activities and events that do not lie on the critical path or paths.
Planning	The establishing of the project activities and events, their logical relations and inter-relations to each other and the sequence in which they are to be accomplished.
Scheduling	The assigning of project dates to the project plan.
Control	The ability to determine project status as it relates to the selected time plan and schedule; and the ability to compensate rapidly for design changes, delays, additions deletions, and so forth.
Scheduled Date or Scheduled Event Time	A specific target or milestone date, determined outside of the project network plan, and inserted into the plan at that particular Event, to give information as to which activities are affected, and by how much.
Distributed Float	The re-assignment of path Float (total Float) to non-critical activities, in proportion to a weighing factor assigned to each activity.

SYMBOLS & FORMULAE

a   = Most optimistic activity elapsed time estimate

m   = Most realistic activity elapsed time estimate

b   = Most pessimistic activity elapsed time estimate

t_e or D   = Expected time duration of activity

b-a   = Range of activity estimates

σ = Deviation of estimate $\dfrac{b-a}{6}$

σ^2 = Variance of estimate $\left[\dfrac{b-a}{6}\right]^2$

Z   = $\dfrac{T_{SOE} - T_{EOE}}{\sigma_{C.P.}}$

Pr   = Probability

T_{SOE} = Scheduled Event Time of Objective (final) Event

T_{EOE} = Earliest Event Time of Objective (final) Event

$\boxed{T_E}$ = Early Event Time

$\left(T_L\right)$ = Latest Event Time

$\boxed{T_E}\,i$ = Early Event Time at tail (start) of activity arrow

$\boxed{T_E}\,j$ = Early Event Time at head (finish) of activity arrow

$\left(T_L\right)i$ = Latest Event Time at tail (start) of activity arrow

$\left(T_L\right)j$ = Latest Event Time at head (finish) of activity arrow

$\left\langle T_S\right\rangle$ = Scheduled Event Time

$$\left(T_L\right)_j - \boxed{T_E}_i - D \quad = \quad \text{Activity Slack or Total Float}$$

$$\left\langle T_S \right\rangle_j - \boxed{T_E}_i - D \quad = \quad \text{Scheduled Activity Slack}$$

$$\boxed{T_E}_i = ES \quad = \quad \text{Early Activity Start Time}$$

$$\boxed{T_E}_i + D \quad EF \quad = \quad \text{Early Activity Finish Time}$$

$$\left(T_L\right)_j = LF \quad = \quad \text{Latest Activity Finish Time}$$

$$\left(T_L\right)_j - D \quad LS \quad = \quad \text{Latest Activity Start Time}$$

$$\text{Activity Cost Slope} \quad = \quad \frac{\text{Crash Cost } - \text{ Normal Cost}}{\text{Normal Time} - \text{ Crash Time}} \quad \frac{\$_C - \$_N}{D_C - D_N}$$

$$\triangle T_D \quad = \quad \text{Distributed Event Time}$$

$$X \quad = \quad \text{Event Weight Factor}$$

$$Y \quad = \quad \text{Position Value Factor}$$

COMMON DIAGRAMMING ABBREVIATIONS

A	Area		CONVTR	Convector
ABV	Above		CP	Complete
AC	Air Condition		CT	Ceramic Tile
ACCESS	Accessories		CVR	Cover
ACOUST	Acoustic			
ACT	Activate		D	Dummy
AD	Approve, deliver		D or d	Duration
AFD	Approve, fabricate, deliver		DEL	Deliver
AL	All		DEMOL	Demolish
ALT	Alteration		DIFF	Diffuser
ALUM	Aluminum		DK	Deck
APPL	Approval		DMPRF	Damp proof
ASMBLY	Assembly		DR	Door
ASP	Asphalt		DRINKG	Drinking
			DRN	Drain
BAL	Balance		DUCTWK	Ductwork
BALC	Balcony		DWG	Drawing
BD	Board			
BKFL	Backfill		E	East
BKFLG	Backfilling		EF	Early finish
BLDG	Building		EFRP	Excavate, form reinforce, pour
BLKG	Blocking			
BLT	Bolt		EIB	Excavate, install, backfill
BM	Beam			
BRG	Bearing		ELEC	Electric
BRK	Brick		ELEV	Elevator
BSE	Base		ENERG	Energize
BSMT	Basement		EQUIP	Equipment
			ERCT	Erect
C/B	Columns and beams		ES	Early start
CER	Ceramic		EXC	Excavation
CP	Cap		EXP	Exposed
CL	Column line		EXT	Exterior
CLG	Ceiling		EXTG	Existing
CLKG	Caulking			
CNTL	Control		F	For
CO	Cutoff		FAB	Fabricate
COSTG	Costing		FD	Fabricate, deliver
COL	Column		FDN	Foundation
COMP	Complete		FFG	Fill, fine grade
CONC	Concrete		FINL	Final
COND	Conduit		FL	Floor
CONN	Connection		FLL	Fill
CONST	Construct		FLSHG	Flashing
CONT	Continue		FM	Form

FRPS	Form, reinforce, pour, strip		MH	Manhole
FTG	Footing		MLLWK	Millwork
FX	Fixture		MISC	Miscellaneous
			MK	Make
GLAZG	Glazing		MSNRY	Masonry
GRD	Grade		MTL	Metal
GRDR	Girder		MTR	Motor
GRDG	Grading			
GRLL	Grill		N	North
GRATG	Grating		NLR	Nailer
GUT	Gutter		NT	Not
HD	Head		OH	Overhead
HDWRE	Hardware		OPNG	Opening
HI	High			
HM	Hollow metal		PARTN	Partition
HTR	Heater		PC	Precast
HU	Hookup		PERIM	Perimeter
			PH	Penthouse
I	Iron		PHS	Phase
I/C	In ceiling		PILG	Piling
IFW	In floor work		PIPG	Piping
INCLDG	Including		PKG	Parking
INSTL	Install		PL	Plate
INSTLG	Installing		PLCP	Pile cap
INSUL	Insulation or insulate		PLG	Plug
INT	Interior		PLSTC	Plastic
ITMS	Items		PLSTR	Plaster
			PLTFM	Platform
JC	Janitor closet		PLUMBG	Plumbing
			PNL	Panel
LAYG	Laying		PNT	Paint
LF	Late finish		PNTG	Painting
LN	Line		POURG	Pouring
LO	Low		PRES	Pressure
LS	Late start		PRF	Proof
LT	Light		PRM	Primary
LTH	Lath		PROT	Protection
LVL	Level		PRS	Piers
			PVG	Paving
MACH	Machinery			
MECH	Mechanical		RAD	Radiant
MEMBRN	Membrane		RAILG	Railing
MEZZ	Mezzanine		RD	Road
			REBAR	Reinforcing

REING	Reinforcing	TO/R	
REL	Relocate	TPG	
REQD	Required	T/R	
RESIL	Resilient	TR	
RESTL	Reinforcing steel	TRANSF	
REMV	Remove	TRD	
RFG	Roofing	TST	
RISR	Riser	TWR	
RM	Room		
RR	Railroad	UG	
RSC	Rolling steel curtain	ULG	
RUBB	Rubber	UTIL	
RUFF	Rough	US	
S	South	VB	
SBSTNTLY	Substantially	VENTILTR	
SDWK	Sidewalk	VEST	
SETTG	Setting		
SEWR	Sewer	W	
SHT	Sheet	WASHG	
SIDG	Siding	WK	
SLB	Slab	WLKWY	Walkway
SOG	Slab on grade	WLL	Wall
SPDRL	Spandrel	WNDW	Window
SPRNKLR	Sprinkler	WP	Waterproofing
SS	Structural steel	WTR	Water
SS	Substation		
ST	Start		
ST	Street		
STD	Stud		
STL	Steel		
STM	Steam		
STR	Stair		
STRP	Strip		
STRUCT	Structural		
SUB	Submit		
SUPT	Support		
SURF	Surface		
SUSP	Suspension		
SWTCHGR	Switchgear		
SYS	System		
TEMP	Temporary		
TFT	Total float time		
TK	Tank		

Hamlin, F., How PERT Predicts for the Navy, Armed Forces Management, July 1959.

Proceedings of the PERT Coordination Task Group Meeting (s), SPO, Bu Weps, USN, 1) 17-18 March 1960, 2) 16-17 August 1960.

"Breakthrough", Special Projects Office; Department of the Navy-Film on principles of PERT and its use on the Polaris program. Merit Productions of California, 10044 Burnet Avenue, San Fernando, California (28 minutes, $150).

Breitenberger, Ernst, "Development Projects as Stochastic Processes," Technical Memo K-33/59, December 1959, U.S. Naval Weapons Laboratory, Dahlgren, Virginia.

Malcolm, D. G., Roseboom, J.H. et al, "Application of a Technique for Research and Development Program Evaluation," Operations Research, Vol. 7, pp. 646-669, 1959.

Fulkerson, D.R., An Out-of-Kilter Method for Minimal Cost Flow Problems, Santa Monica, California, Rand Corp., 1959 (Paper P1825).

Fulkerson, D. R., A Network Flow Computation for Project Cost Curves, Man. Sci., 1961, 7, 167-178.

Eisner, H., A Generalized Network Approach to the Plannings and Scheduling of a Research Program, Ops. Res., 1962, 10, 115-125.

Ford, L. R., and Fulkerson, D.R., Maximal Flow Through a Network, Can. J. Math., 1956, 8, 399-404.

"Industry Borrows POLARIS Planning," Product Engineering, June, 1958.

Instruction Manual and Systems and Procedures for the Program Evaluation System (PERT). Special Projects Office, Bureau of Naval Weapons, Department of the Navy, Washington, D.C.

Jarrigan, M.P., "Automatic Machine Methods of Testing PERT Networks for Consistency," Technical Memo K-24/60, August, 1960; U.S. Naval Weapons Laboratory, Dahlgren, Virginia.

Martino, Dr. R.L., How Critical Path Scheduling Works, Canadian Chemical Processing, February, 1960.

Martino, Dr. R.L., "What's the Shortest Path in Project Planning?" Executive Magazine, August, 1960.

Clark, C.E., "The Optimum Allocation of Resources Among the Activities of a Network." Journal of Industrial Engineering. January - February, 1961; pp. 11-17.

Neiman, Ralph A. and Learn, Robert N., "Mechanization of PERT System Provides Timely Information," Navy Management Review, August, 1960.

Neiman, R.A. and Learn, R.N., "Mechanization of the PERT System on NORC," Technical Memo K-19/59, August, 1959 (Rev. April 1960), U.S. Naval Weapons Laboratory, Dahlgren, Virginia.

PERT Data Processing Handbook for Technicians, Special Projects Office, Department of the Navy, Washington, D.C., June, 1960.

PERT Summary Report Phase I, July, 1958, Special Projects Office, Bureau of Ordnance, Department of the Navy, Washington, D.C.

Program Evaluation Procedure Instructions, Headquarters, Air Research and Development Command, USAF, November, 1960.

Program Planning and Control System, Special Projects Office, Bureau of Naval Weapons, Department of the Navy, Washington, D.C.

Kelley, J.E., Jr., "Parametric Programming and the Primal-Dual Algorithm," Operations Research, Vol. 7, No. 3, 1959, pp. 327-334.

Kelley, J. E., Jr., "Critical-Path Planning and Scheduling: Mathematical Basis, "Operations Research, Vol. 9, 1961, pp. 296-320.

Kelley, J.E., Jr., and M.R. Walker, and J.S.Sayer, "Critical Path Scheduling," Factory, July, 1960.

Klass, P.J., "PERT/PEP Management Tool Use Grows," Aviation Week, November 28, 1960.

Malcolm, D.G., et al, "Application of a Technique for R & D Program Evaluation," SP-62, Systems Development Corporation, Santa Monica, California; March, 1959.

Martino, R.L., "New Way to Analyze and Plan Operations and Projects Will Save You Time and Cash," Oil/Gas World, September, 1959.

Engineering News-Record, "New Tool for Job Management," January 26, 1961.

Fazar, W., "Program Evaluation and Review Technique,"The American Statistician, April, 1959.

Fazar, W., "Program Evaluation and Review Technique," Statistical Reporter, Bureau of the Budget, January, 1959.

Fazar, W., "Progress Reporting in the Special Projects Office," Navy Management Review, April, 1959.

Ford, L.R., Jr., and Fulkerson, D.R.,"A Simple Algorithm for Finding Maximal Network Flows and an Application to the Hitchcock Problem," Canadian Journal of Math., Vol. 9, 1957, pp. 210-218.

Freeman, R.J., "A Generalized Network Approach to Project Activity Sequencing," IRE Transactions on Engineering Management, September, 1960.

Voress, H. E., Houser, E. A., Jr. and Marsh, F. E., Jr., Critical Path Scheduling, a Preliminary Literature Search. U. S. Atomic Energy Commission, Division of Technical Information, October 1961, (TID-3568).

Progress in Resources Planning Through PERT, Technical Information Series R-60EML46, General Electric Light Military Electronics Department, Utica, New York, 1960.

Project PERT, Phase II, Special Projects Office, Bureau of Naval Weapons, Department of the Navy, Washington, D. C.

Reeves, Eric, "Critical Path Speeds Refinery Revamp," Canadian Chemical Processing, October, 1960.

Sayer, J. S., et al, "Critical Path Scheduling," Factory, July, 1960.

Anonymous, "Now Industry Schedules by Computer," Control Engineering, pp. 16-17, January, 1962.

General Electric Company Bulletin #CPB 184, "GE-225 and CPM for Precise Project Planning."

Remington Rand Univac Division Bulletin #PX 1842, "Fundamentals of Network Planning and Analysis."

Boehm, Carl, "Helping the Executive Make Up His Mind", Fortune, April, 1962.

DOD and NASA Guide, PERT COST Systems Design, Office of the Secretary of Defense, Washington 25, D.C.

Heller, J., and Logemann, G. "An Algorithm for the Construction and Evaluation of Feasible Schedules", Man. Sci., 1962, 8, 168-183.

Miller, Robert W. "How to Plan and Control with PERT" Harvard Business Review, March-April, 1962.

Waldron, A. James, "The USE of PERT in Scheduling Manufacturing Operations", SAE Paper 658 A-National Production Meeting, Society of Automotive Engineers, Cleveland, Ohio, March 12, 1963.

Ferdinand K. Levy, Gerald L. Thompson, and Jerome D. Wiest, "The ABCs of the Critical Path Method," HBR September-October 1963, p. 98.

Secretary of Defense and National Aeronautics and Space Administration, DOD and NASA Guide: PERT COST Systems Design (Washington, Government Printing Office, June 1962).

National Aeronautics and Space Administration, NASA PERT and Companion Cost System Handbook (Washington, Government Printing Office, October 1962).

David M. Stires and Maurice M. Murphy, Modern Management Methods: PERT and CPM (Boston, Materials Management Institute, 1962).

Daniel D. Roman, "The PERT System: An appraisal of Program Evaluation Review Technique, "The Journal of the Academy of Management, April 1962, p. 57.

J. W. Pocock, "PERT as an Analytical Aid for Program Planning--Its Payoff and Problems", Operations Research, November-December 1962, p. 893.

Ivars Avots, "The Management Side of PERT," California Management Review, Winter 1962, p. 16.

John G. Barmby, "The Applicability of PERT as a Management Tool," IRE Transactions on Engineering Management, September 1962, p. 130.

Hilliard W. Paige, "How PERT - Cost Helps the General Manager," HBR November-December 1963, p. 87.

White, Glenn L. "Complete Project Management." The NAHB Journal of Home-building. February 1964, pp. 32-35.

"How To Pick A Computer Consultant" Special Report, Construction Methods. and Equipment (McGraw-Hill). September, 1965.

"Do It Yourself CPM," Construction Methods and Equipment (McGraw-Hill), January, 1966.

L. R. Shaffer, J. B. Ritter, W. I. Meyer. "The Critical Path Method," McGraw-Hill Book Co.

"CPM In Construction" - A Manual For General Contractors. Published by The Associated General Contractors of America. 157 E. Street, N. W., Washington, D. C., 20006.

Waldron, A. James, "The Use And Abuse Of Critical Path Scheduling." ASME Paper No. 67-PEM-26, presented at the Plant Engineering and Maintenance Conference, Detroit, Mich., April 11, 1967.

Jennett, Eric. "Experience With And Evaluation Of Critical Path Methods." Chemical Engineering, February 10, 1969.

TABLE OF VALUES OF THE STANDARD NORMAL DISTRIBUTION FUNCTION (Continued)

z	0	1	2	3	4	5	6	7	8	9
.0	.5000	.5040	.5080	.5120	.5160	.5199	.5239	.5279	.5319	.5359
.1	.5398	.5438	.5478	.5517	.5557	.5596	.5636	.5675	.5714	.5753
.2	.5793	.5832	.5871	.5910	.5948	.5987	.6026	.6064	.6103	.6141
.3	.6179	.6217	.6255	.6293	.6331	.6368	.6406	.6443	.6480	.6517
.4	.6554	.6591	.6628	.6664	.6700	.6736	.6772	.6808	.6844	.6879
.5	.6915	.6950	.6985	.7019	.7054	.7088	.7123	.7157	.7190	.7224
.6	.7257	.7291	.7324	.7357	.7389	.7422	.7454	.7486	.7517	.7549
.7	.7580	.7611	.7642	.7673	.7703	.7734	.7764	.7794	.7823	.7852
.8	.7881	.7910	.7939	.7967	.7995	.8023	.8051	.8078	.8106	.8133
.9	.8159	.8186	.8212	.8238	.8264	.8289	.8315	.8340	.8365	.8389
1.0	.8413	.8438	.8461	.8485	.8508	.8531	.8554	.8577	.8599	.8621
1.1	.8643	.8665	.8686	.8708	.8729	.8749	.8770	.8790	.8810	.8830
1.2	.8849	.8869	.8888	.8907	.8925	.8944	.8962	.8980	.8997	.9015
1.3	.9032	.9049	.9066	.9082	.9099	.9115	.9131	.9147	.9162	.9177
1.4	.9192	.9207	.9222	.9236	.9251	.9265	.9278	.9292	.9306	.9319
1.5	.9332	.9345	.9357	.9370	.9382	.9394	.9406	.9418	.9430	.9441
1.6	.9452	.9463	.9474	.9484	.9495	.9505	.9515	.9525	.9535	.9545
1.7	.9554	.9564	.9573	.9582	.9591	.9599	.9608	.9616	.9625	.9633
1.8	.9641	.9648	.9656	.9664	.9671	.9678	.9686	.9693	.9700	.9706
1.9	.9713	.9719	.9726	.9732	.9738	.9744	.9750	.9756	.9762	.9767
2.0	.9772	.9778	.9783	.9788	.9793	.9798	.9803	.9808	.9812	.9817
2.1	.9821	.9826	.9830	.9834	.9838	.9842	.9846	.9850	.9854	.9857
2.2	.9861	.9864	.9868	.9871	.9874	.9878	.9881	.9884	.9887	.9890
2.3	.9893	.9896	.9898	.9901	.9904	.9906	.9909	.9911	.9913	.9916
2.4	.9918	.9920	.9922	.9925	.9927	.9929	.9931	.9932	.9934	.9936
2.5	.9938	.9940	.9941	.9943	.9945	.9946	.9948	.9949	.9951	.9952
2.6	.9953	.9955	.9956	.9957	.9959	.9960	.9961	.9962	.9963	.9964
2.7	.9965	.9966	.9967	.9968	.9969	.9970	.9971	.9972	.9973	.9974
2.8	.9974	.9975	.9976	.9977	.9977	.9978	.9979	.9979	.9980	.9981
2.9	.9981	.9982	.9982	.9983	.9984	.9984	.9985	.9985	.9986	.9986
3.	.9987	.9990	.9993	.9995	.9997	.9998	.9998	.9999	.9999	1.0000

TABLE OF VALUES OF THE STANDARD NORMAL DISTRIBUTION FUNCTION

z	0	1	2	3	4	5	6	7	8	9
-3	.0013	.0010	.0007	.0005	.0003	.0002	.0002	.0001	.0001	-.0000
-2.9	.0019	.0018	.0017	.0017	.0016	.0016	.0015	.0015	.0014	.0014
-2.8	.0026	.0025	.0024	.0023	.0023	.0022	.0021	.0021	.0020	.0019
-2.7	.0035	.0034	.0033	.0032	.0031	.0030	.0029	.0028	.0027	.0026
-2.6	.0047	.0045	.0044	.0043	.0041	.0040	.0039	.0038	.0037	.0036
-2.5	.0062	.0060	.0059	.0057	.0055	.0054	.0052	.0051	.0049	.0048
-2.4	.0082	.0080	.0078	.0075	.0073	.0071	.0069	.0068	.0066	.0064
-2.3	.0107	.0104	.0102	.0099	.0096	.0094	.0091	.0089	.0087	.0084
-2.2	.0139	.0136	.0132	.0129	.0126	.0122	.0119	.0116	.0113	.0110
-2.1	.0179	.0174	.0170	.0166	.0162	.0158	.0154	.0150	.0146	.0143
-2.0	.0228	.0222	.0217	.0212	.0207	.0202	.0197	.0192	.0188	.0183
-1.9	.0287	.0281	.0274	.0268	.0262	.0256	.0250	.0244	.0238	.0233
-1.8	.0359	.0352	.0344	.0336	.0329	.0322	.0314	.0307	.0300	.0294
-1.7	.0446	.0436	.0427	.0418	.0409	.0401	.0392	.0384	.0375	.0367
-1.6	.0548	.0537	.0526	.0516	.0505	.0495	.0485	.0475	.0465	.0455
-1.5	.0668	.0655	.0643	.0630	.0618	.0606	.0594	.0582	.0570	.0559
-1.4	.0808	.0793	.0778	.0764	.0749	.0735	.0722	.0708	.0694	.0681
-1.3	.0968	.0951	.0934	.0918	.0901	.0885	.0869	.0853	.0838	.0823
-1.2	.1151	.1131	.1112	.1093	.1075	.1056	.1038	.1020	.1003	.0985
-1.1	.1357	.1335	.1314	.1292	.1271	.1251	.1230	.1210	.1190	.1170
-1.0	.1587	.1562	.1539	.1515	.1492	.1469	.1446	.1423	.1401	.1379
-.9	.1841	.1814	.1788	.1762	.1736	.1711	.1685	.1660	.1635	.1611
-.8	.2119	.2090	.2061	.2033	.2005	.1977	.1949	.1922	.1894	.1867
-.7	.2420	.2389	.2358	.2327	.2297	.2266	.2236	.2206	.2177	.2148
-.6	.2743	.2709	.2676	.2643	.2611	.2578	.2546	.2514	.2483	.2451
-.5	.3085	.3050	.3015	.2981	.2946	.2912	.2877	.2843	.2810	.2776
-.4	.3446	.3409	.3372	.3336	.3300	.3264	.3228	.3192	.3156	.3121
-.3	.3821	.3783	.3745	.3707	.3669	.3632	.3594	.3557	.3520	.3483
-.2	.4207	.4168	.4129	.4090	.4052	.4013	.3974	.3936	.3897	.3859
-.1	.4602	.4562	.4522	.4483	.4443	.4404	.4364	.4325	.4286	.4247
-.0	.5000	.4960	.4920	.4880	.4840	.4801	.4761	.4721	.4681	.4641

| ACTIVITY | | DURATION | DESCRIPTION | EARLIEST | | LATEST | | TOTAL FLOAT |
I	J	D		START T_{E_i}	FINISH $[T_E]_i + D$	START $[T_L]_j - D$	FINISH $[T_L]_j$	$[T_L]_j - [T_E]_i - D$
0	5	0		0	0	0	0	0
5	10	3		0	3	0	3	0
5	70	4		0	4	29	33	29
10	15	5		3	8	3	8	0
15	20	5		8	13	8	13	0
15	25	3		8	11	32	35	24
15	45	4		8	12	23	27	15
20	30	2		13	15	20	22	7
20	35	6		13	19	13	19	0
25	75	10		11	21	35	45	24
30	40	0		15	15	26	26	11
30	50	7		15	22	22	29	7
35	40	0		19	19	26	26	7
35	45	8		19	27	19	27	0
40	50	3		19	22	26	29	7
45	55	7		27	34	27	34	0
50	60	9		22	31	29	38	7
55	65	4		34	38	34	38	0
60	65	0		31	31	38	38	7
60	75	0		31	31	45	45	14
65	80	10		38	48	38	48	0

ACTIVITY		DURATION	DESCRIPTION	EARLIEST		LATEST		TOTAL FLOAT
I	J	D		START	FINISH	START	FINISH	

ACTIVITY		DURATION	DESCRIPTION	EARLIEST		LATEST		TOTAL FLOAT
I	J	D		START	FINISH	START	FINISH	

Use the following forms to convert from a Project Day to a Calendar Date. Day #1, the START BASELINE, is entered as a 1 in the proper calendar date box, and each successive Project Day is entered accordingly to a 5 Day, 6 Day or 7 Day Work Week. Holidays are X'd out of the Project Calendar.

Moveable Holiday -- Good Friday

Year	Date
1967	March 24
1968	April 12
1969	April 4
1970	March 27
1971	April 9
1972	March 31
1973	April 20
1974	April 12
1975	March 28
1976	April 16
1977	April 8
1978	March 24
1979	April 13
1980	April 4

A. JAMES WALDRON
PLANNING CONSULTANT

371 KINGS HIGHWAY WEST
HADDONFIELD, N. J.
609-428-3742

PROJECT _____

_____ CLIENT _____

LOCATION _____ _____

_____ _____

_____ _____

_____ _____

NETWORK SCHEDULE PROJECT CALENDAR

1968		JULY				1968
SUN	MON	TUE	WED	THUR	FRI	SAT
	1	2	3	4	5	6
7	8	9	10	11	12	13
14	15	16	17	18	19	20
21	22	23	24	25	26	27
28	29	30	31			

1968		OCTOBER				1968
SUN	MON	TUE	WED	THUR	FRI	SAT
		1	2	3	4	5
6	7	8	9	10	11	12
13	14	15	16	17	18	19
20	21	22	23	24	25	26
27	28	29	30	31		

1968		AUGUST				1968
SUN	MON	TUE	WED	THUR	FRI	SAT
				1	2	3
4	5	6	7	8	9	10
11	12	13	14	15	16	17
18	19	20	21	22	23	24
25	26	27	28	29	30	31

1968		NOVEMBER				1968
SUN	MON	TUE	WED	THUR	FRI	SAT
					1	2
3	4	5	6	7	8	9
10	11	12	13	14	15	16
17	18	19	20	21	22	23
24	25	26	27	28	29	30

1968		SEPTEMBER				1968
SUN	MON	TUE	WED	THUR	FRI	SAT
1	2	3	4	5	6	7
8	9	10	11	12	13	14
15	16	17	18	19	20	21
22	23	24	25	26	27	28
29	30					

1968		DECEMBER				1968
SUN	MON	TUE	WED	THUR	FRI	SAT
1	2	3	4	5	6	7
8	9	10	11	12	13	14
15	16	17	18	19	20	21
22	23	24	25	26	27	28
29	30	31				

CALENDARS — 1800 TO 2050

INDEX

1800 — 4	1828 — 10	1856 — 10	1884 — 10	1912 — 9	1940 — 9	1968 — 9	1996 — 9	2024 — 9
1801 — 5	1829 — 5	1857 — 5	1885 — 5	1913 — 4	1941 — 4	1969 — 4	1997 — 4	2025 — 4
1802 — 6	1830 — 6	1858 — 6	1886 — 6	1914 — 5	1942 — 5	1970 — 5	1998 — 5	2026 — 5
1803 — 7	1831 — 7	1859 — 7	1887 — 7	1915 — 6	1943 — 6	1971 — 6	1999 — 6	2027 — 6
1804 — 8	1832 — 8	1860 — 8	1888 — 8	1916 — 14	1944 — 14	1972 — 14	2000 — 14	2028 — 14
1805 — 3	1833 — 3	1861 — 3	1889 — 3	1917 — 2	1945 — 2	1973 — 2	2001 — 2	2029 — 3
1806 — 4	1834 — 4	1862 — 4	1890 — 4	1918 — 3	1946 — 3	1974 — 3	2002 — 3	2030 — 2
1807 — 5	1835 — 5	1863 — 5	1891 — 5	1919 — 4	1947 — 4	1975 — 4	2003 — 4	2031 — 4
1808 — 13	1836 — 13	1864 — 13	1892 — 13	1920 — 12	1948 — 12	1976 — 12	2004 — 12	2032 — 12
1809 — 1	1837 — 1	1865 — 1	1893 — 1	1921 — 7	1949 — 7	1977 — 7	2005 — 7	2033 — 7
1810 — 2	1838 — 2	1866 — 2	1894 — 2	1922 — 1	1950 — 1	1978 — 1	2006 — 1	2034 — 1
1811 — 3	1839 — 3	1867 — 3	1895 — 3	1923 — 2	1951 — 2	1979 — 2	2007 — 2	2035 — 2
1812 — 11	1840 — 11	1868 — 11	1896 — 11	1924 — 10	1952 — 10	1980 — 10	2008 — 10	2036 — 10
1813 — 6	1841 — 6	1869 — 6	1897 — 6	1925 — 5	1953 — 5	1981 — 5	2009 — 5	2037 — 5
1814 — 7	1842 — 7	1870 — 7	1898 — 7	1926 — 6	1954 — 6	1982 — 6	2010 — 6	2038 — 6
1815 — 1	1843 — 1	1871 — 1	1899 — 1	1927 — 7	1955 — 7	1983 — 7	2011 — 7	2039 — 7
1816 — 9	1844 — 9	1872 — 9	1900 — 2	1928 — 8	1956 — 8	1984 — 8	2012 — 8	2040 — 8
1817 — 4	1845 — 4	1873 — 4	1901 — 3	1929 — 3	1957 — 3	1985 — 3	2013 — 3	2041 — 3
1818 — 5	1846 — 5	1874 — 5	1902 — 4	1930 — 4	1958 — 4	1986 — 4	2014 — 4	2042 — 4
1819 — 6	1847 — 6	1875 — 6	1903 — 5	1931 — 5	1959 — 5	1987 — 5	2015 — 5	2043 — 5
1820 — 14	1848 — 14	1876 — 14	1904 — 13	1932 — 13	1960 — 13	1988 — 13	2016 — 13	2044 — 13
1821 — 2	1849 — 2	1877 — 2	1905 — 1	1933 — 1	1961 — 1	1989 — 1	2017 — 1	2045 — 1
1822 — 3	1850 — 3	1878 — 3	1906 — 2	1934 — 2	1962 — 2	1990 — 2	2018 — 2	2046 — 2
1823 — 4	1851 — 4	1879 — 4	1907 — 3	1935 — 3	1963 — 3	1991 — 3	2019 — 3	2047 — 3
1824 — 12	1852 — 12	1880 — 12	1908 — 11	1936 — 11	1964 — 11	1992 — 11	2020 — 11	2048 — 11
1825 — 7	1853 — 7	1881 — 7	1909 — 6	1937 — 6	1965 — 6	1993 — 6	2021 — 6	2049 — 6
1826 — 1	1854 — 1	1882 — 1	1910 — 7	1938 — 7	1966 — 7	1994 — 7	2022 — 7	2050 — 7
1827 — 2	1855 — 2	1883 — 2	1911 — 1	1939 — 1	1967 — 1	1995 — 1	2023 — 1	

DIRECTIONS FOR USE

Look for the year you want in the index at left. The number opposite each year is the number of the calendar to use for that year.

1

JANUARY	MAY	SEPTEMBER
FEBRUARY	JUNE	OCTOBER
MARCH	JULY	NOVEMBER
APRIL	AUGUST	DECEMBER

2

JANUARY	MAY	SEPTEMBER
FEBRUARY	JUNE	OCTOBER
MARCH	JULY	NOVEMBER
APRIL	AUGUST	DECEMBER

3

JANUARY	MAY	SEPTEMBER
FEBRUARY	JUNE	OCTOBER
MARCH	JULY	NOVEMBER
APRIL	AUGUST	DECEMBER

4

JANUARY	MAY	SEPTEMBER
FEBRUARY	JUNE	OCTOBER
MARCH	JULY	NOVEMBER
APRIL	AUGUST	DECEMBER

5

JANUARY	MAY	SEPTEMBER
FEBRUARY	JUNE	OCTOBER
MARCH	JULY	NOVEMBER
APRIL	AUGUST	DECEMBER

6

JANUARY	MAY	SEPTEMBER
FEBRUARY	JUNE	OCTOBER
MARCH	JULY	NOVEMBER
APRIL	AUGUST	DECEMBER

7

JANUARY	MAY	SEPTEMBER
FEBRUARY	JUNE	OCTOBER
MARCH	JULY	NOVEMBER
APRIL	AUGUST	DECEMBER

8

JANUARY	MAY	SEPTEMBER
FEBRUARY	JUNE	OCTOBER
MARCH	JULY	NOVEMBER
APRIL	AUGUST	DECEMBER

9

JANUARY	MAY	SEPTEMBER
FEBRUARY	JUNE	OCTOBER
MARCH	JULY	NOVEMBER
APRIL	AUGUST	DECEMBER

10

JANUARY	MAY	SEPTEMBER
FEBRUARY	JUNE	OCTOBER
MARCH	JULY	NOVEMBER
APRIL	AUGUST	DECEMBER

11

JANUARY	MAY	SEPTEMBER
FEBRUARY	JUNE	OCTOBER
MARCH	JULY	NOVEMBER
APRIL	AUGUST	DECEMBER

12

JANUARY	MAY	SEPTEMBER
FEBRUARY	JUNE	OCTOBER
MARCH	JULY	NOVEMBER
APRIL	AUGUST	DECEMBER

13

JANUARY	MAY	SEPTEMBER
FEBRUARY	JUNE	OCTOBER
MARCH	JULY	NOVEMBER
APRIL	AUGUST	DECEMBER

14

JANUARY	MAY	SEPTEMBER
FEBRUARY	JUNE	OCTOBER
MARCH	JULY	NOVEMBER
APRIL	AUGUST	DECEMBER

SOLUTIONS

TO

PROBLEMS

"APPLIED PRINCIPLES OF PROJECT PLANNING AND CONTROL" (2nd Edition)

A. James Waldron

Haddonfield, N.J.

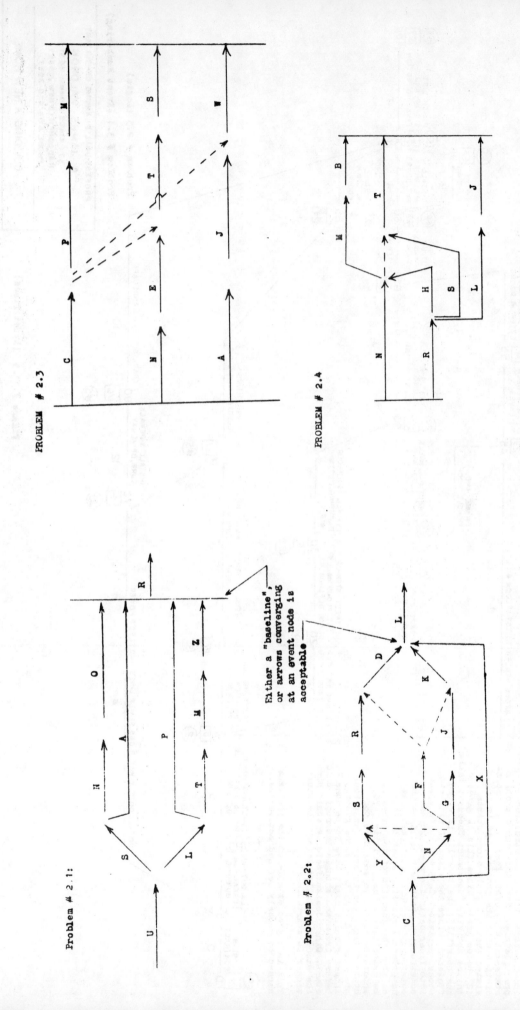

PROBLEM # 2.3

PROBLEM # 2.4

Problem # 2.1:

Either a "baseline",
or arrows converging
at an event node is
acceptable

Problem # 2.2:

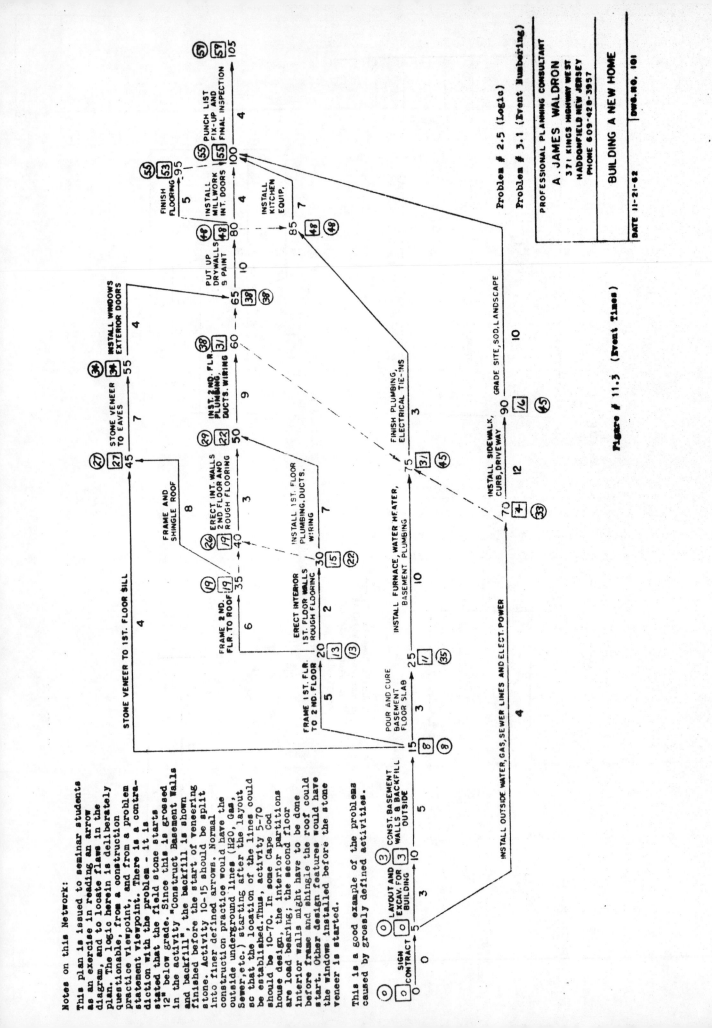

Notes on this Network:

This plan is issued to seminar students as an exercise in reading an arrow diagram, and to locate flaws in the plan. The logic herein is deliberately questionable, from a construction practice viewpoint, and from a problem statement viewpoint. There is a contradiction with the problem - it is stated that the field stone starts 12" below grade. Since this is grossed in the activity "Construct Basement Walls and backfill", the backfill is shown finished before the start of veneering stone. Activity 10-15 should be split into finer defined arrows. Normal construction practice would have the outside underground lines (H2O, Gas, Sewer,etc.) starting after the layout so that the location of the lines could be established.Thus, activity 5-70 should be 10-70. In some Cape Cod house design, the interior partitions are load bearing; the second floor interior walls might have to be done before frame and shingle the roof could start. Other design features would have the windows installed before the stone veneer is started.

This is a good example of the problems caused by grossly defined activities.

PROFESSIONAL PLANNING CONSULTANT
A. JAMES WALDRON
371 KINGS HIGHWAY WEST
HADDONFIELD NEW JERSEY
PHONE 609-428-3957

Problem # 2.5 (Logic)

Problem # 3.1 (Event Numbering)

Figure # 11.3 (Event Times)

DATE 11-21-62   DWG. NO. 101

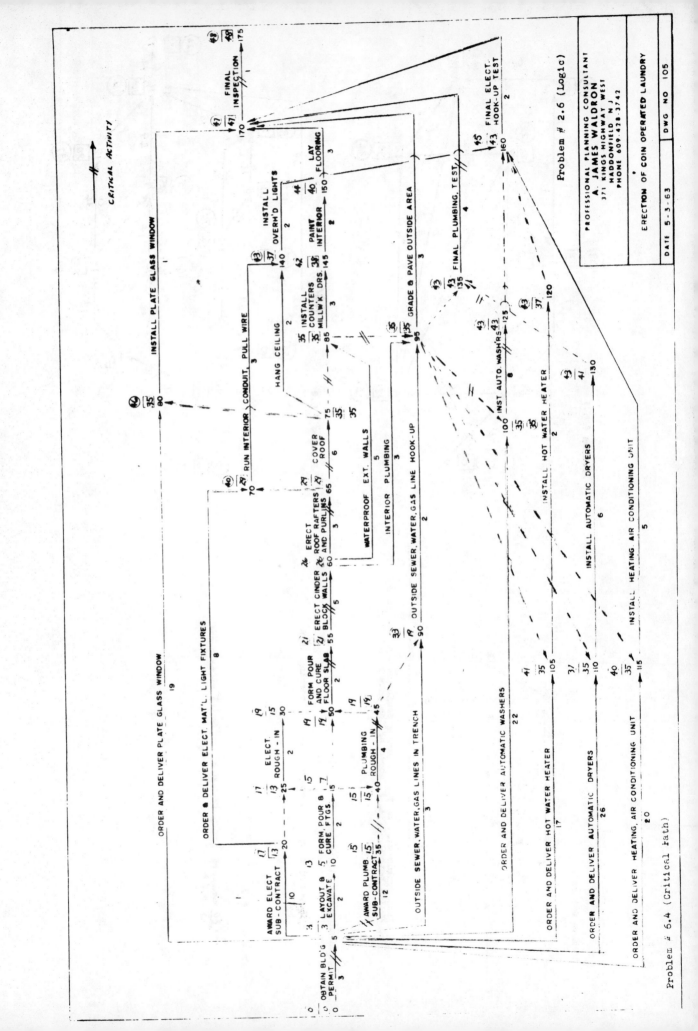

Problem # 6.4 (Critical Path)

Problem # 2.6 (Logic)

PROFESSIONAL PLANNING CONSULTANT
A. JAMES WALDRON
371 KINGS HIGHWAY WEST
HADDONFIELD N J
PHONE 609. 428. 3742

ERECTION OF COIN OPERATED LAUNDRY

DATE 5-3-63 | DWG NO 105

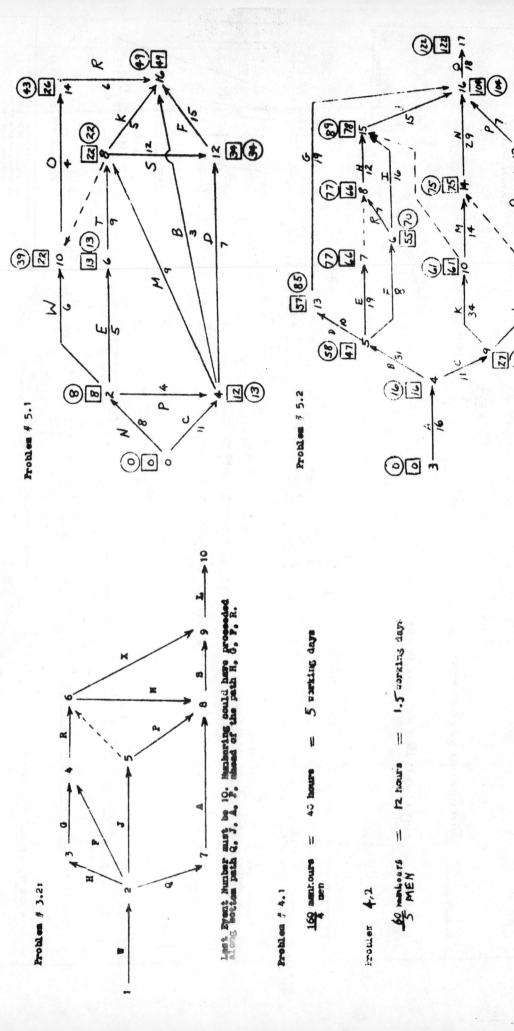

Problem # 5.1

Problem # 5.2

Problem # 3.2:

Last Event Number must be 10. Numbering could have proceeded
along bottom path Q, J, A, P, ahead of the path H, G, P, R.

Problem # 4.1

$\dfrac{160 \text{ manhours}}{4 \text{ men}} = 40 \text{ hours} = 5 \text{ working days}$

Problem 4.2

$\dfrac{60 \text{ manhours}}{5 \text{ MEN}} = 12 \text{ hours} = 1.5 \text{ working days}$

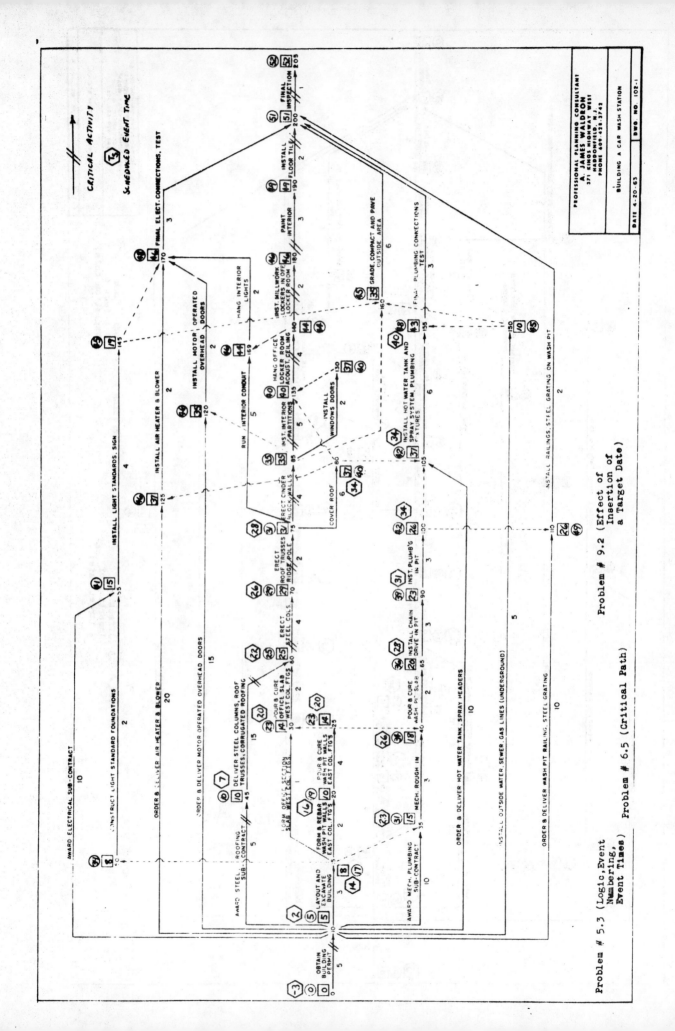

CRITICAL ACTIVITY

SCHEDULED EVENT TIME

PROFESSIONAL PLANNING CONSULTANT
A. JAMES WALDRON
371 KINGS HIGHWAY WEST
HADDONFIELD, N.J.
PHONE 609-428-3742

BUILDING A CAR WASH STATION

DATE 4-20-63   DWG. NO. 102-1

Problem # 5.3 (Logic, Event
Numbering,
Event Times)

Problem # 6.5 (Critical Path)

Problem # 9.2 (Effect of
Insertion of
a Target Date)

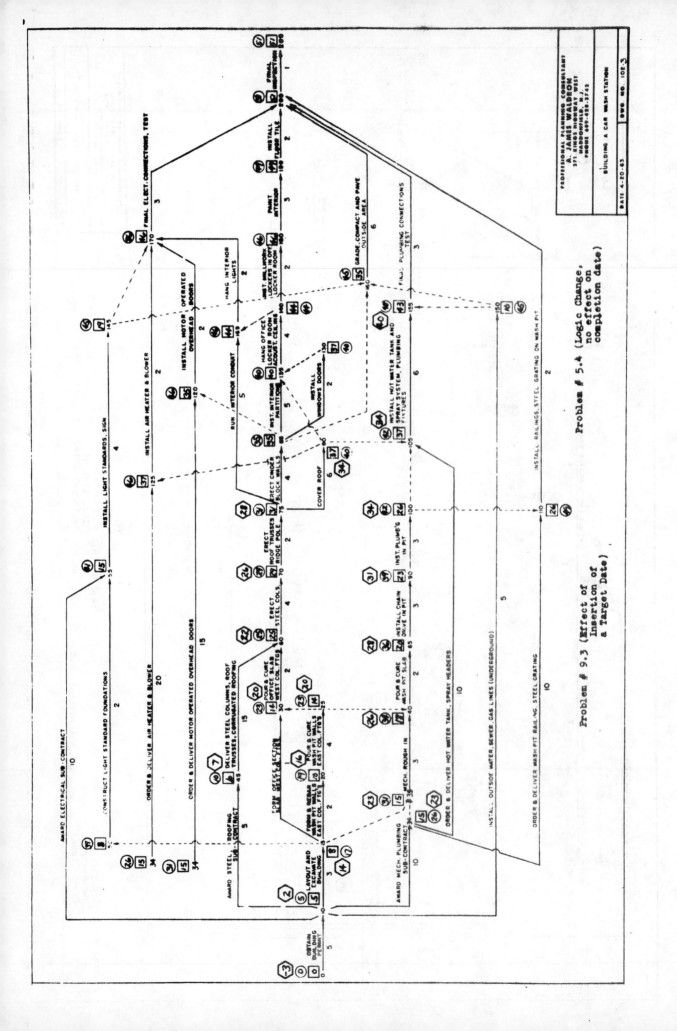

PROFESSIONAL PLANNING CONSULTANT
A. JAMES WALDRON
371 KINGS HIGHWAY WEST
HADDONFIELD, N.J.
PHONE 609-429-3743

BUILDING A CAR WASH STATION

DATE 4-20-63   DWG. NO. 102-3

Problem # 5.4 (Logic Change, no effect on completion date)

Problem # 9.3 (Effect of Insertion of a Target Date)

Problem " 6.1

Operation "D" is not critical, since it does not satisfy all three criteria of a critical activity. The difference between the event times at its head (time unit 25) and the event times at its tail (time unit 4) is not equal to its duration of 14 time units.

Problem " 6.2

The critical path consists of 0-2, 2-6, 6-8, 8-12, 12-16.

Problem " 6.3

The critical activities are A, C, K, L, N, and Q.

Problem # 6.4

See network plan for Problem 2.6

Problem # 6.5

See network plan for Problem 5.3

Problem " 7.1

None.

Problem " 7.2

Activity	Total Float
0 - 2	0
0 - 4	2
2 - 6	1
4 - 6	0
2 -10	25
4 -12	1
4 -12	15
4 -15	34

Activity	Total Float
6 - 8	0
8 - 10	17
8 - 12	0
8 - 16	22
10- 14	17
12- 16	0
14-15	17

Problem " 7.3

Activity	Total Float
A	0
B	11
C	0
D	28
E	11
F	15
G	28
H	11
I	18

Activity	Total Float
J	11
K	0
L	17
M	0
N	0
O	17
P	17
Q	0
R	15

Problem # 7.4

See schedule for Problem 8.5

Problem # 8.1

Schedule for Problem # 5.2

Activity	Duration	ES	EF	LS	LF	FLOAT
A	16	0	16	0	16	0
B	31	16	47	27	58	9
C	11	16	27	16	27	0
D	10	47	57	75	85	28
E	19	47	66	58	77	11
F	8	57	55	62	70	15
G	19	66	76	85	104	28
H	12	55	78	77	89	11
I	16	78	71	73	104	18
J	15	27	93	89	61	11
K	34	27	61	27	61	0
L	17	61	44	44	75	17
M	14	75	75	61	104	0
N	29	111	104	75	97	0
O	36	80	80	61	104	17
P	7	104	87	97	122	17
Q	18	104	122	104	104	0
R	7	55	62	70	77	15

Problem : 8.2

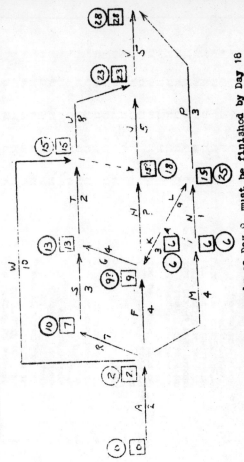

"H" can start as early as Day 9, must be finished by Day 18

BUILDING A NEW HOME
CRITICAL PATH METHOD NORMAL TIME
AJW 11/21/62
TOTAL TIME - 59 DAYS   TOTAL COST - 20640

I	J	DURATION	COST	DESCRIPTION	ES	EF	LS	LF	TF
0	5	0	0	SIGN CONTRACT	0	0	0	0	*
5	10	3	700	LAYOUT AND EXCAVATE	0	3	0	3	*
5	70	4	950	INSTALL OUTSIDE WATER GAS SEWER POWER	0	4	29	33	29
10	15	5	1320	CONSTRUCT BASEMENT WALLS BACKFILL	3	8	3	8	*
15	20	5	850	FRAME FIRST TO SECOND FLOOR	8	13	8	13	*
15	25	3	470	POUR AND CURE BASEMENT	8	11	32	35	24
15	45	4	2200	STONE VENEER TO FIRST FLOOR SILL	8	12	23	27	15
20	30	2	430	INSTALL FIRST FLOOR WALLS ROUGH FLOOR	13	15	20	22	7
20	35	6	1200	FRAME SECOND FLOOR TO ROOF	13	19	13	19	*
25	75	10	350	INSTALL FURNACE WATER HEATER BASEMENT PLUMBING	11	21	35	45	24
30	40	0	0	DUMMY	15	15	26	26	11
30	50	7	650	INSTALL FIRST FLOOR PLUMBING DUCTS WIRE	15	22	22	29	7
35	40	0	0	DUMMY	19	19	26	26	7
35	45	8	1130	FRAME AND SHINGLE ROOF	19	27	19	27	*
40	50	3	490	INSTALL SECOND FLOOR WALLS ROUGH FLOOR	19	22	26	29	7
45	55	7	2060	STONE VENEER TO EAVES	27	34	27	34	*
50	60	9	480	INSTALL SECOND FLOOR PLUMBING DUCTS WIRE	22	31	29	38	7
55	65	4	370	INSTALL WINDOWS EXTERIOR DOORS	34	38	34	38	*
60	65	0	0	DUMMY	31	31	38	38	7
60	75	0	0	DUMMY	31	31	45	45	14
65	80	10	840	PUT UP DRYWALLS PAINT	38	48	38	48	*
70	75	0	0	DUMMY	4	4	45	45	41
70	90	12	1320	INSTALL SIDEWALKS CURB DRIVEWAY	4	16	33	45	29
75	85	3	490	FINISH PLUMBING ELECTRICAL TIE-INS	31	34	45	48	14
80	85	0	0	DUMMY	48	48	48	48	*
80	95	5	730	FINISH FLOORING	48	53	50	55	2
80	100	4	490	INSTALL MILLWORK INTERIOR DOORS	48	52	51	55	3
85	100	7	1870	INSTALL KITCHEN EQUIPMENT	48	55	48	55	*
90	100	10	750	GRADE SITE SOD LANDSCAPE	16	26	45	55	29
95	100	0	0	DUMMY	53	53	55	55	2
100	105	4	500	PUNCH LIST INSPECTION	55	59	55	59	*

Problem # 8.3

Problem # 8.4

ERECTION OF A COIN OPERATED LAUNDRY
DRAWING NO. 105
CRITICAL PATH SCHEDULE

I	J	DUR	DESCRIPTION	ES	EF	LS	LF	TF
0	5	3	OBTAIN BUILDING PERMIT	0	3	0	3	*
5	10	2	LAYOUT & EXCAVATE	3	5	3	5	*
10	15	10	AWARD ELECTRICAL SUBCONTRACT	3	13	27	37	24
20	35	12	AWARD PLUMBING SUBCONTRACT	3	15	7	19	4
35	80	19	ORD & DEL PLATE GLASS WINDOW					
90	95	3	OUTSIDE SEWER H2O GAS LINES	3	6	44	46	27
100	105	22	ORD & DEL AUTO WASHERS					21
105	110	17	ORD & DEL HOT H2O HEATER					10
110	115	26	ORD & DEL AUTO DRYERS					9
115	120	20	ORD & DEL HTG AIR COND UNIT					
5	25		FORM POUR CURE FTGS					
25	30	2	DUMMY					
30	40	8	ORD & DEL ELECT MTL LIGHT FIXT					
40	50		ELECTRICAL ROUGH IN					
50	55	2	DUMMY					
			DUMMY					
		4	PLUMBING ROUGH IN					
			DUMMY					
45	90	2	FORM POUR CURE FLOOR SLAB					
			DUMMY					
		5	ERECT CINDER BLOCK WALLS					
		3	ERECT ROOF RAFTERS PURLINS					
		5	H2O PROOF EXT WALLS					
		6	INTERIOR PLUMBING					
			DUMMY					
		4	COVER ROOF					
		3	RUN INT CONDUIT PULL WIRE					
			DUMMY					
		3	HANG CEILING					
		1	INSTALL PLATE GLASS WINDOW					
			DUMMY					
			DUMMY					
		2	INSTALL COUNTERS MILLWK DRS					
		3	OUTSIDE SEWER H2O GAS LINE HOOKUP					
			DUMMY					
			DUMMY					
			DUMMY					
			DUMMY					
		8	INSTALL AUTO WASHERS					
		2	INSTALL HOT H2O HEATER					
		6	INSTALL AUTO DRYERS					
		5	INSTALL HTG AIR COND UNIT					
			DUMMY					
			DUMMY					
			DUMMY					
			DUMMY					
			DUMMY					
		0	DUMMY					
		4	FINAL PLUMBING TEST					
		2	INSTALL OVERHEAD LIGHTS					
		2	PAINT INTERIOR					
		3	LAY FLOORING					
		2	FINAL ELECTRICAL HOOKUP TEST					
		1	FINAL INSPECTION					

BUILDING A CAR WASH STATION
DRAWING 102-2
A JAMES WALDRON
371 KINGS HIGHWAY WEST   HADDONFIELD N J
CRITICAL PATH SCHEDULE

I	J	DUR	DESCRIPTION	ES	EF	LS	LF	TF
0	10	5	OBTAIN BUILDING PERMIT	0	5	0	5	*
10	15	3	LAYOUT AND EXCAVATE BUILDING	5	8	14	17	9
10	34	10	AWARD MECH PLUMBING SUBCONTRACT	5	15	16	26	11
10	35	5	AWARD STEEL ROOFING SUBCONTRACT	5	10	5	10	*
10	45	10	AWARD ELECTRICAL SUBCONTRACT	5	15	31	41	26
10	55	10	INSTALL OUT H2O SEWER GAS LINES	5	15	40	45	35
10	105	20	FORM REBAR WASH PIT WLS EAST FTGS	8	10	17	19	9
15	110	2	FORM OFF SECT SLAB WEST COL FTGS	8	22	31	31	23
15	150	30	ORD DEL H2O TK SPRAY HDRS	8	31	26	39	31
15	20	2	ORD DEL WASH PIT RAIL STL GRATING	8	9	40	23	9
20	30	0	ORD DEL MTR OPERATED OH DOORS	8	8	17	19	9
25	50	4	ORD DEL AIR HTR BLOWER	10	14	22	34	20
30	35	0	POUR CURE WASH PIT WLS EAST FTGS	14	14	34	25	9
34	105	2	DUMMY	14	16	23	31	14
34	110	0	POUR CURE OFF SLAB WEST FTGS	14	16	31	39	23
34	120	0	DUMMY	15	25	32	37	16
34	125	10	ORD DEL HT H2O TK SPRAY HDRS	15	25	39	42	17
35	40	15	ORD DEL WASH PIT RAIL STL GRATING	15	30	31	49	24
40	65	20	ORD DEL MTR OPERATED OH DOORS	15	35	26	46	16
45	60	2	ORD DEL AIR HTR BLOWER	15	18	31	34	16
50	55	15	MECH ROUGH IN	15	20	34	36	16
55	145	3	POUR CURE WASH PIT SLAB	18	25	10	25	*
60	70	2	DEL STL COLS RF TRUSSES CORR RFNG	10	25	39	41	31
65	90	4	CONST LIGHT STD FDNS	25	29	25	29	*
70	75	3	INSTL LIGHT STANDARDS	20	23	36	39	16
75	80	6	ERECT STL COLS	29	31	29	31	*
75	85	4	INSTL CHAIN DRIVE IN PIT	31	37	34	40	3
80	169	0	ERECT RF TRUSSES RIDGE POLE	31	35	31	35	*
80	125	0	COVER ROOF	31	36	41	46	10
80	135	2	ERECT CINDER BLOCK WALLS	37	37	42	42	5
85	130	0	RUN INTERIOR CONDUIT	37	37	40	40	3
85	160	5	DUMMY	37	37	38	40	3
90	135	2	DUMMY	35	40	35	45	10
100	110	5	INSTL WINDOWS DOORS	35	35	45	42	16
100	105	0	INSTL INTERIOR PARTITIONS	23	26	39	42	16
105	155	0	INSTL PLUMBING IN WASH PIT	26	26	42	49	23
110	200	3	DUMMY	37	43	49	48	5
120	170	2	INSTL HOT H2O TK SPRAY SYS PLUMB	28	28	49	51	23
125	170	2	INSTL RAILINGS GRATING ON WSH PIT	20	22	46	48	26
130	135	0	INSTL MTR OPR OH DOORS	37	39	46	48	9
135	140	4	INSTL AIR HTR BLOWER	37	37	40	40	3
140	169	2	DUMMY	40	44	46	46	2
140	180	0	HANG OFF LKR RM ACOUST CEILING	44	46	44	46	*
145	160	0	DUMMY	19	19	48	46	26
145	170	0	INSTL MLLWK LKRS IN LOCKER ROOM	19	19	48	45	29
150	155	0	DUMMY	10	10	48	48	38
150	160	0	DUMMY	10	10	45	45	35
155	160	3	FINAL PLUMB CONN TEST	43	46	48	51	5
160	200	6	GRADE COMPACT PAVE OUTSIDE AREA	35	41	45	51	10
169	170	3	HANG INTERIOR LIGHTS	44	46	48	48	2
170	200	3	FINAL ELECT CONNS TEST	46	49	48	51	2
180	190	3	PAINT INTERIOR	46	49	46	49	*
190	200	3	INSTALL FLOOR TILE	49	51	49	51	*
200	205	1	FINAL INSPECTION	51	52	51	52	*

BUILDING A CAR WASH STATION
DRAWING 102-1
A JAMES WALDRON
371 KINGS HIGHWAY WEST   HADDONFIELD N J
CRITICAL PATH SCHEDULE

I	J	DUR	DESCRIPTION	ES	EF	LS	LF	TF
0	10	5	OBTAIN BUILDING PERMIT	0	5	0	5	*
10	15	3	LAYOUT AND EXCAVATE BUILDING	5	8	14	17	9
10	35	10	AWARD MECH PLUMBING SUBCONTRACT	5	15	21	31	16
10	35	5	AWARD STEEL ROOFING SUBCONTRACT	5	10	5	10	*
10	45	10	AWARD ELECTRICAL SUBCONTRACT	5	15	31	41	26
10	55	10	INSTALL OUT H2O SEWER GAS LINES	5	15	32	42	27
15	105	20	FORM REBAR WASH PIT WLS EAST FTGS	5	10	39	49	34
15	110	2	FORM OFF SECT SLAB WEST COL FTGS	5	22	31	46	26
15	150	30	ORD DEL H2O TK SPRAY HDRS	5	31	26	45	21
20	30	2	ORD DEL WASH PIT RAIL STL GRATING	5	8	40	23	35
25	50	1	ORD DEL MTR OPERATED OH DOORS	5	8	17	19	9
30	35	4	ORD DEL AIR HTR BLOWER	8	25	22	37	14
34	105	0	POUR CURE WASH PIT WLS EAST FTGS	8	14	31	25	23
34	110	0	DUMMY	8	16	34	36	31
35	40	15	POUR CURE OFF SLAB WEST FTGS	10	25	19	23	9
40	65	20	DUMMY	14	30	23	34	9
45	60	2	ORD DEL HT H2O TK SPRAY HDRS	14	35	34	25	20
50	55	4	ORD DEL WASH PIT RAIL STL GRATING	14	18	31	34	16
55	145	3	ORD DEL MTR OPERATED OH DOORS	15	20	34	36	16
60	70	2	ORD DEL AIR HTR BLOWER	18	25	10	25	16
65	90	4	MECH ROUGH IN	10	20	39	41	*
70	75	3	POUR CURE WASH PIT SLAB	25	29	41	45	16
75	80	6	DEL STL COLS RF TRUSSES CORR RFNG	20	23	36	39	16
75	85	4	CONST LIGHT STD FDNS	29	31	29	31	*
80	169	0	INSTL LIGHT STANDARDS	31	37	34	40	3
80	125	0	ERECT STL COLS	31	35	31	35	*
80	135	2	INSTL CHAIN DRIVE IN PIT	36	37	41	46	10
85	130	0	ERECT RF TRUSSES RIDGE POLE	37	37	42	42	5
85	160	5	COVER ROOF	37	37	40	40	3
90	135	2	ERECT CINDER BLOCK WALLS	35	40	38	40	3
100	110	5	RUN INTERIOR CONDUIT	35	40	35	45	*
100	105	0	DUMMY	26	26	42	42	16
105	155	2	DUMMY	26	43	49	48	23
110	200	2	INSTL WINDOWS DOORS	37	28	49	51	23
120	170	2	INSTL INTERIOR PARTITIONS	20	22	46	48	26
125	170	2	INSTL PLUMBING IN WASH PIT	37	39	46	48	9
130	135	4	DUMMY	37	37	40	40	3
135	140	2	DUMMY	40	44	44	44	2
140	169	0	INSTL HOT H2O TK SPRAY SYS PLUMB	44	46	46	46	*
140	180	0	HANG OFF LKR RM ACOUST CEILING	19	19	48	45	26
145	160	0	DUMMY	19	19	48	45	29
145	170	0	INSTL MLLWK LKRS IN LOCKER ROOM	10	10	48	45	38
150	155	3	DUMMY	10	10	48	45	35
150	160	6	FINAL PLUMB CONN TEST	43	46	48	51	5
155	160	2	GRADE COMPACT PAVE OUTSIDE AREA	35	41	45	51	10
160	200	3	HANG INTERIOR LIGHTS	44	46	49	48	2
169	170	3	FINAL ELECT CONNS TEST	46	49	49	51	2
170	200	3	PAINT INTERIOR	46	49	46	49	*
180	190	2	INSTALL FLOOR TILE	49	51	49	51	*
190	200	1	FINAL INSPECTION	51	52	51	52	*

Problem # 9.1

Problem # 9.2

See network for problem # 5.3.

Activity 65-90, "Install Chain Drive in Pit", has a Scheduled Float of 8 days, as contrasted with the original Total Float of 16 days. The significance of "loss" of Total Float as the result of the insertion of a T_s Target Date means that that the path on which the activity lies is shortened due to the T_s insertion at a common event.

Problem # 9.3

See network plan for problem # 5.4.

The effect of the T_s or Day # 40 at Event # 155 is the same as in Problem # 9.2

Problem # 10.1

A	$(9-1)^2$	64
E	$(16-4)^2$	144
G	$(17-5)^2$	144
M	$(13-5)^2$	64
		416

(The dummy has no effect on probability calculations)

$$\sigma^2 \text{ (Variance)} = \frac{416}{36} = 11.6$$

$$\sigma \text{ (Deviation)} = \sqrt{11.6} = 3.4$$

$$z = \frac{23-27}{3.4} = \frac{-4}{3.4} = -1.18$$

From Standard Deviation Function Table:

P_r (Probability) = 0.119

or 12 "chances out of 100 in reaching Event # 18 by week 23

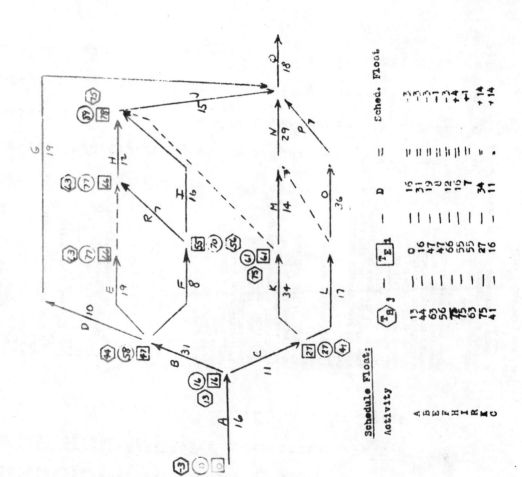

Schedule Float:

Activity	T_E^{1}-T_L^{1}	T_E^{1}		T_E^{h}	D		=	Sched. Float
A	13	0		16	16		=	-3
B	44	16		31	31		=	-3
E	63	47		19	19		=	-3
H	56	66		12	12		=	-1
I	75	55		16	16		=	+4
R	63	55		7	7		=	+1
K	75	27		34	34		=	+14
G	41	16		11	11		=	+14

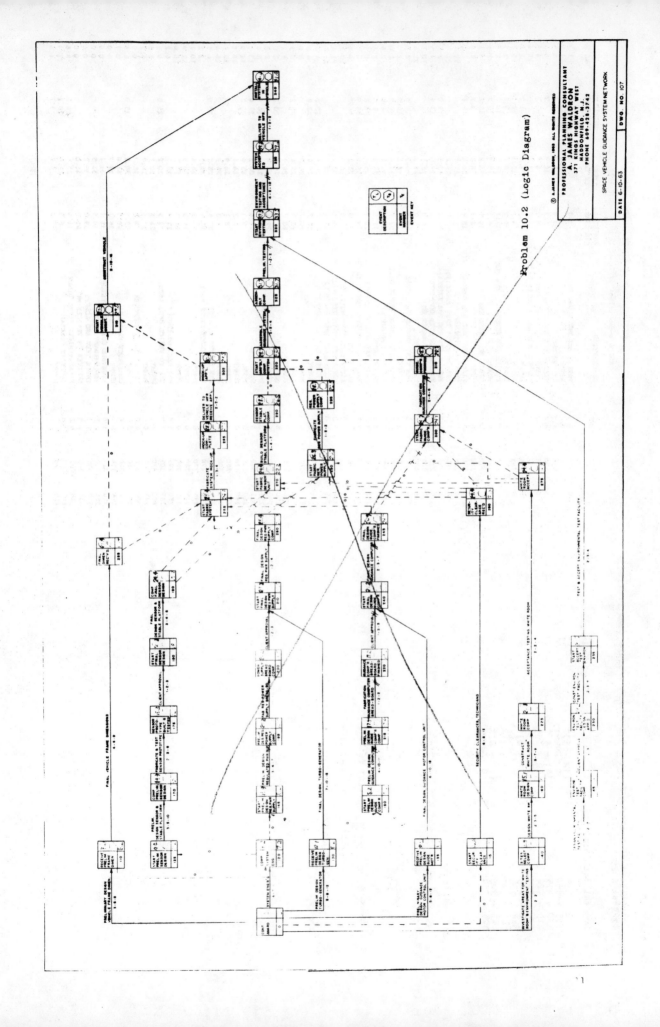

Problem 10.2 (Logic Diagram)

PROFESSIONAL PLANNING CONSULTANT
A. JAMES WALDRON
371 KINGS HIGHWAY WEST
HADDONFIELD, N.J.
PHONE 609-429-3742

SPACE VEHICLE GUIDANCE SYSTEM NETWORK

DATE 6-10-63   DWG NO. 107

PERT Schedule

Inertial Guidance System Project

(SLACK shows effect of T_S of 25 at Event # 310)

I	J	te	DESCRIPTION	EXPECTED FIN TIME	LATEST FIN TIME	SCHEDULED FIN TIME	SLACK
0	110	5.2	PRELIM DES VEH FRAME DIMEN	5.2	8.2	3.7	-1.5
0	115	0	DUMMY	0	16.1	11.9	11.9
0	120	7.2	SYSTEM DES	7.2	8.2	3.7	-3.5
0	130	8.2	PRELIM DES TURBO GEN	8.2	12.9	7.7	-0.5
0	135	8.2	PRELIM DES GUID MTR CONTROL UNIT	8.2	8.2	3.7	-4.5
0	140	1.8	INVESTIGATE AREAS WRITE EM ENVIR	1.8	11.9		10.1
120	125	0	DUMMY	7.2	9.4	5.2	-2.0
120	145	0	DUMMY	8.2	15.9	10.7	-3.5
120	150	0	DUMMY	8.2	8.2	3.7	-3.5
125	170	5.5	PRELIM DES SENSOR STABLE PLAT	12.7	14.9	10.7	-2.0
130	145	0	DUMMY	8.2	15.9	10.7	-0.5
130	210	10.3	FINAL DES TURBO GEN	18.5	23.2	18.0	-0.5
135	150	0	DUMMY	8.2	8.2	3.7	-4.5
135	240	10.7	FINAL DES GUID MTR CONTROL UNIT	18.9	21.2	16.7	-2.2
140	160	3.2	DESIGN WRITE EM	5.0	15.1		10.1
140	165	4.3	DESIGN ENVIR TEST FACILITY	3.8	25.2	15.0	21.4
145	195	9.2	PRELIM DES REC PWR SUPPLY	12.5	20.2		2.5
150	215	7.0	PRELIM DES GUID COMPUTER	17.4	17.4	12.9	-4.5
160	225	2.0	CONSTRUCT WRITE EM	12.0	22.1		10.1
165	230	5.8	CLIENT APPL	5.8	27.2		21.4
170	175	3.3	FAB TEST SENSOR PROTOTYPE	16.0	18.2	14.0	-2.0
175	180	2.0	CLIENT APPL	18.0	20.2	16.0	-2.0
180	190	4.2	FINAL DES SENSOR STBLE PLAT	22.2	24.4	16.2	-2.0
190	200	1.0	FAB EAG PWR SUPPLY	13.5	21.2	16.0	2.5
200	210	2.0	CLIENT APPL	15.5	23.2	18.0	2.5
210	250	2.2	FINAL DES PWR SUPPLY	20.7	25.4	20.2	-0.5
215	220	1.8	FAB GUID COMPUTER BREAD BD	19.2	19.2	14.7	-4.5
220	240	2.0	CLIENT APPL	21.2	21.2	16.7	-4.5
225	275	2.3	ACCEPTANCE TEST WRITE EM	14.3	24.4		10.1
230	735	3.0	CONSTRUCT ENVIR TEST FACILITY	9.1	30.5		21.4
230	330	3.0	ACCEPTANCE TEST ENVIR FAC	12.1	33.5		21.4
235	245	5.5	FINAL DES GUID COMPUTER	24.7	24.7	20.2	-4.5
240	255	0	DUMMY	24.7	25.4	20.2	-0.5
245	300	0	DUMMY	24.7	24.7		0.0
250	265	0	DUMMY	20.7	25.4	20.2	6.2
250	280	0	DUMMY	20.7	26.9		6.2
190	265	0	DUMMY	22.2	25.4	20.2	-2.0
255	270	0	DUMMY	22.2	23.4		2.2
255	310	0	DUMMY	11.4	30.2	25.0	6.8
260	265	0	DUMMY	8.3	24.4		13.6
260	270	0	DUMMY	8.3	24.4	20.2	11.9
260	300	0	DUMMY	8.3	24.9		16.1
265	285	2.8	FAB SYSTEM MOCK UP	27.5	28.2	23.0	-4.5
270	290	4.3	BUILD SENSOR STBLE PLAT	26.5	28.7		-2.2
275	270	0	DUMMY	14.3	24.4		10.1
275	280	0	DUMMY	14.3	26.9		12.6
275	300	0	DUMMY	14.3	24.7		10.4
280	295	1.8	FAB REC PWR SUPPLY	22.5	28.7		6.2
285	305	2.0	DEL TO VEHICL MFG, APPL	29.5	30.2	25.0	-4.5
290	320	0	DUMMY	26.5	28.7		2.2
295	320	0	DUMMY	22.5	28.7		6.2
300	315	4.0	FAB GUIDANCE COMPUTER	28.7	28.7		0.0
300	340	0	DUMMY	28.7	30.2	25.0	1.5
305	310	10.8	DUMMY	29.5	41.0		0.7
310	340	0	CONSTRUCT VEHICLE	40.3	28.7		0.0
315	320	3.0	ASSEMBLE GUIDANCE SYSTEM	31.7	31.7		0.0
325	325	1.8	PRELIMINARY TEST	33.5	33.5		0.0
330	335	5.7	ENVIR TEST AND ACCEPT	39.2	39.2		0.0
335	340	1.8	DELIVER TO VEHICLE MFG	41.0	41.0		0.0
115	255	6.2	FINAL VEHICLE FRAME DIMEN	11.4	25.4	20.2	8.8
110	260	8.3	SECURITY CLEARANCE	6.3	24.4	20.2	11.9
110	120	0	DUMMY	5.2	8.2	3.7	-1.5

Problem # 10.2

Probability of reaching Event # 310 (Start Vehicle Construction) by Week # 25 :

Longest Path Activity		Range Squared
0 — 135	$(12 - 5)^2$	49
135 — 150	0 (Dummy)	0
150 — 215	$(11 - 4)^2$	49
215 — 220	$(2 - 1)^2$	1
220 — 240	$(3 - 1)^2$	4
240 — 245	$(7 - 2)^2$	25
245 — 265	0 (Dummy)	
265 — 285	$(4 - 1)^2$	9
285 — 305	$(2 - 2)^2$	0
305 — 310	0 (Dummy)	0
		137

$$\sigma^2 \text{ (Variance)} = \frac{137}{36} = 3.81$$

$$\sigma \text{ (Deviation)} = \sqrt{3.81} = 1.95$$

$$Z = \frac{25 - 29.5}{1.95} = \frac{-4.5}{1.95} = -2.31$$

From Standard Deviation Function Table:

$$\text{Pr (Probability)} = -0.0154$$

Or one chance in a hundred of starting Vehicle Construction by Week # 25.

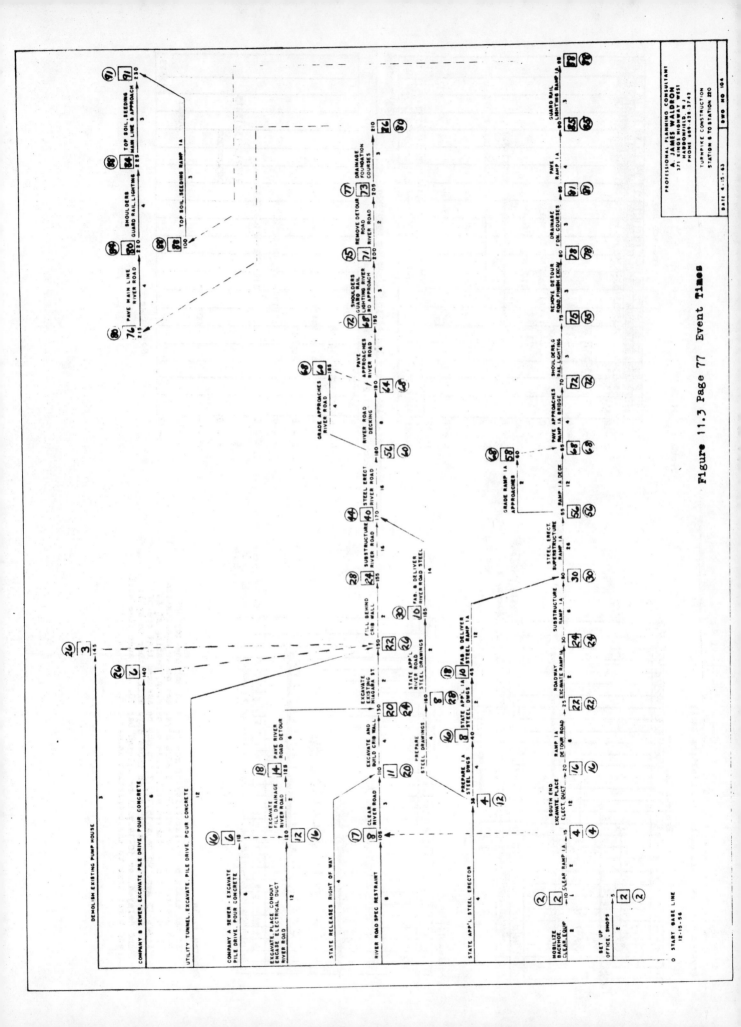

Figure 11.3 Page 77 Event Times

Problem # 11.2a

Activities are listed in "1" major sequence

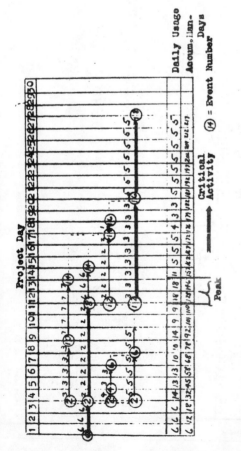

Problem # 11.1a : All Activities Starting Early

Project Day

Daily Usage
Accum. Man-
Days

⟶ Critical Activity

(H) = Event Number

Problem 11.1b

Daily Usage
Accum. Man
Days

Problem 11.1c

If the maximum availability of men is 11 on any one day, the project will be extended 3 more days. Any other combination of 10-14 and 8 - 14 above will cause an overlap, peaking at at least 13 men. On days 15,16,17, it may be assumed that one of the three activities scheduled those days will either be split and delayed, or 10-14 will be delayed Day 18. Any such case will add three days to the completion date.

No change

PROFESSIONAL PLANNING CONSULTANT
A. JAMES WALDRON
371 KINGS HIGHWAY WEST
HADDONFIELD, N.J.
PHONE 609-428-3742

TIME SCALE CHART

DWG NO. 103

DATE 4-6-63

Problem #1.3

* CRITICAL ACTIVITY
C = COMMON ACTIVITY FOR PLUMBERS AND LABORERS

PROBLEM #1.32
Peck Carpenter Requirements

CARPENTERS

PROBLEM #1.36
Use of threshold
and reduced crews

CARPENTERS

PROBLEM #1.35

LABORERS

PLUMBERS

Daily Requt.

Split

Notize split 74 & 79 of 25-75 for smoothing
of resource scheduling (plumbers)

None of the cases
affects Day #59
completion date

PROBLEM # 12.1

HOUSE WEIGHED SCHEDULE

PROBLEM # 12.3

CAR WASH WEIGHED SCHEDULE

LAG TIME

PROFESSIONAL PLANNING CONSULTANT
A. JAMES WALDRON
27 KINGS HIGHWAY WEST
HADDONFIELD, N.J.
PHONE 609-428-3748

PROBLEMS 12.1 & 12.3
TIME SCALE CHART

DATE 4-6-63   DWG. NO. 103

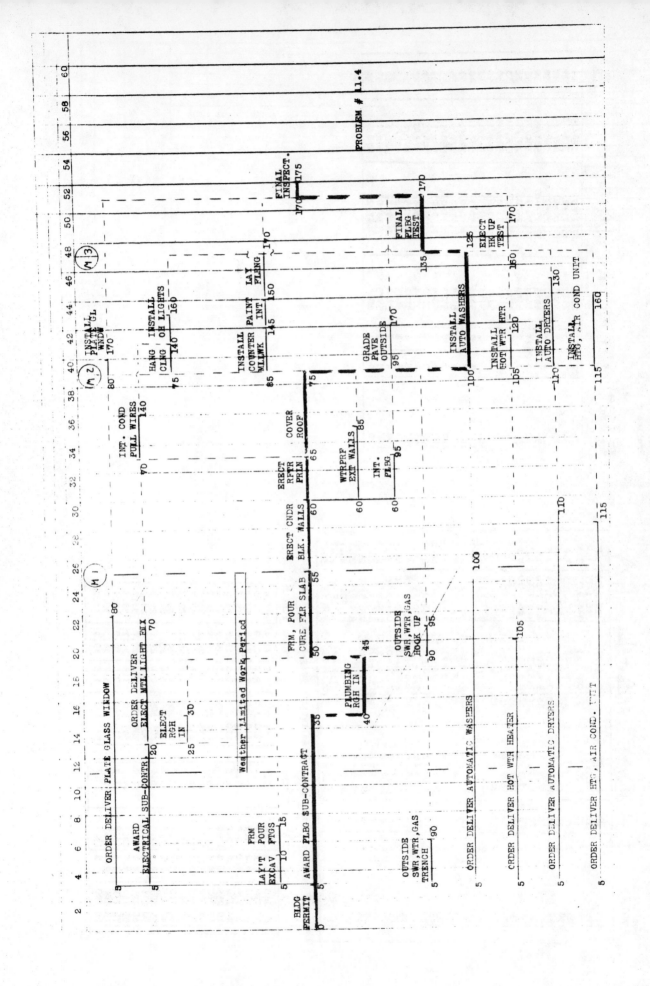

PROBLEM # 11.4

Problem # 12.2

Non Critical Event No.	X	Y	$(T_L)_j$ J x X_j	$Y_j[\Delta T_{L,1} + D]_{max}$	$\widehat{T_L}$
110	5	26	8.2 x 5	26(0 + 5.2)	176.2/31	5.7
115	1	16	16.1x1	16(0+0)	16.1/17	1.0
120	7	26	8.2 x 7	26(0+7.2)	244.9/33	7.4
125	7	21	9.4 x 7	26(7.4 + 0)	258.2/33	7.6
130	7	15	12.9 x 7	21(0+8.2)	362.5/26	9.4
140	7	13	11.9 x 1	13(0+1.8)	75.1/14	2.5
145	7	19	15.9 x 7	19(9.4 + 0)	369.9/26	11.1
160	3	10	15.1 x 3	10(2.5 + 3.2)	102.3/13	7.9
165	2	19	25.2 x 2	8(2.5+2.0)	86.4/10	8.6
170	7	15	14.9 x 7	19(7.6+5.5)	553.3/26	13.6
175	4	15	18.2 x 4	15(13.6+3.3)	526.3/19	17.2
180	2	13	20.2 x 2	13(17.2+2.0)	290/15	19.3
190	3	10	24.4 x 3	10(19.3+4.2)	508.2/13	23.7
200	0	15	0	15(16.4+1.0)	261/15	17.4
210	8	13	23.2 x 8	13(9.4+10.3)	441.7/21	21.0
225	4	6	22.1 x 4	6(7.9+7.0)	178/10	17.8
230	2	6	27.2 x 2	6(8.6+7.0)	118/8	14.3
235	4	2	30.5 x 4	2(14.3+3.3)	157.2/6	26.2
250	3	10	25.4 x 3	10(21.0+2.2)	308.2/13	23.7
255	5	10	25.4 x 5	10(5.7+6.2)	346/15	16.4
260	6	10	24.4 x 6	10(1.0+8.3)	239.4/16	15.0
265	2	4	26.4 x 2	4(23.7+0)	299.4/16	24.9
270	6	4	24.4 x 6	4(17.8+2.3)	141.2/10	24.1
280	6	1	26.9 x 6	1(23.7+0)	129.1/6	21.6
285	3	7	28.2 x 3	7(23.7+2.8)	185.1/7	26.5
290	7	0	28.7 x 7	0	276.5/10	27.9
295	1	0	28.7 x 1	0	14.5/4	28.7
305	0	7	0	7(27.9+2)	20.7	29.9
310	5	7	30.2 x 5	7(29.9+0)	560/12	30.0

Problem # 12.3

Non Critical Event No.	X	Y	$(T_L)_j$ J x X_j	$Y_j[\Delta T_{L,1} + D]_{max}$	$\widehat{T_L}$
15	2	17	29x2	17(17+3)	398/19	21
20	1	15	31x1	15(21+2)	376/16	24
25	1	14	35x1	14(24+4)	427/15	29
30	3	2	35x2	2(29+0)	128/4	32
35	3	17	43x3	17(17+10)	588/20	30
40	3	14	46x3	14(30+3)	590/17	35
50	2	9	51x2	9(21+0)	291/11	26
55	3	8	53x2	8(26+2)	350/10	33
65	4	13	48x1	13(35+2)	529/14	38
80	2	9	52x4	9(43+6)	649/13	50
90	1	12	51x1	12(38+3)	543/13	42
100	3	9	54x3	9(42+3)	567/12	48
105	5	9	54x5	9(50+0)	720/14	51
110	3	8	61x3	8(47+0)	277/5	55
120	4	8	58x3	8(47+0)	550/11	55
125	2	7	58x4	7(50+0)	582/11	53
135	2	0	58x2	0	104/2	58
145	4	4	57x4	4(23+4)	376/8	47
150	2	4	57x2	4(17+5)	202/6	34
155	5	4	57x2	4(51+6)	528/9	59
160	5	2	57x5	2(41+0)	322/6	54
169	5	7	58x5	7(56+0)	682/12	57
170	4	4	60x4	4(57+2)	476/8	60

"Ideal" Schedule (Time Consuming activities only. Dummies Eliminated)

i	j	t_e	Start	Finish
0	110	5.2	0.5	5.7
0	120	7.2	0.2	7.4
0	130	8.2	1.2	9.4
0	140	8.2	0.0	8.2
0	140	1.8	0.7	2.5
110	255	6.2	10.2	16.4
115	260	8.3	6.7	15.0
120	170	5.5	8.1	13.6
130	210	10.3	10.7	21.0
135	240	10.7	8.2	18.9
140	160	3.2	4.7	7.9
140	165	2.0	6.6	8.6
145	195	4.3	13.1	16.4
150	215	9.2	8.2	17.4
160	225	7.0	10.8	17.8
165	230	2.0	12.3	14.3
170	175	3.3	13.9	17.2
175	180	2.0	17.3	19.3
180	190	4.2	19.5	23.7

i	j	t_e	Start	Finish
195	200	1.0	16.4	17.4
200	210	2.0	19.0	21.0
210	250	2.2	21.5	23.7
215	220	1.8	17.4	19.2
220	240	2.3	19.3	21.6
225	275	3.3	22.9	26.2
230	235	3.0	26.2	29.2
235	330	5.0	21.6	24.7
240	245	3.5	21.2	24.7
265	265	2.8	25.1	27.9
270	290	4.3	24.1	28.4
285	295	1.8	26.5	28.3
295	305	2.0	27.9	29.9
305	315	4.0	24.7	28.7
320	325	1.8	28.7	31.7
325	330	10.8	31.7	33.5
310	340	30.0	30.0	40.8
330	335	5.7	33.5	39.2
335	340	1.8	39.2	41.0

Problem # 13.1

ACTIVITY COST INFORMATION

Activity i – j	Normal Time	Normal Cost $	Crash Time	Crash Cost $	Slope ($/Day)
5 – 10	3	700	3	700	0
5 – 70	4	950	2	1350	200
10 – 15	5	1320	2	1770	150
15 – 20	5	850	3	1580	365
15 – 25	3	470	3	470	0
15 – 45	4	2200	3	2450	250
20 – 30	2	430	1	605	175
20 – 35	6	1200	2	4000	700
25 – 75	10	350	5	1350	200
30 – 50	7	650	4	1250	200
35 – 45	8	1130	3	3630	500
40 – 50	3	490	2	665	175
45 – 55	7	2060	3	3060	250
50 – 60	9	480	5	1280	200
55 – 65	4	370	1	850	160
65 – 80	10	840	6	1560	180
70 – 90	12	1320	3	2220	100
75 – 85	3	490	2	840	350
80 – 95	5	730	4	950	220
80 – 100	4	490	2	770	140
85 – 100	7	1870	3	2830	240
90 – 100	10	750	6	1390	160
100 – 110	4	500	4	500	0

Minimum time on a crash program is 32 days for completion.

Minimum additional cost is $10,670 more than normal, based on the following times and costs

Crashed Activities

i	j	Additional Total Cost	i	j	Additional Total Cost
10	15	$450	80	95	$220
15	20	$730	35	45	$2500
20	30	$175	45	55	$1000
30	50	$600	55	65	$480
50	60	$800	40	50	$175
65	80	$720			

Expedited Activities

i	j	Additional Total Cost	Time (Days)
20	35	$2100	3
85	100	$720	3

All other activities remain normal.

Problem # 14.1

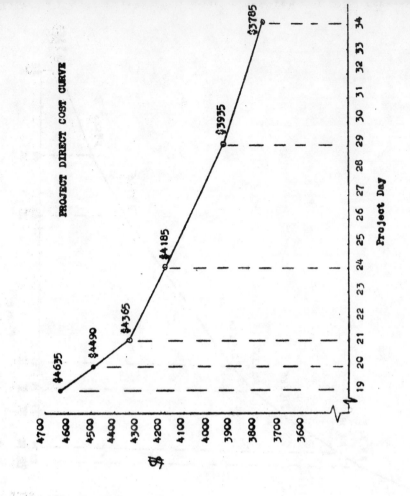

PROJECT DIRECT COST CURVE

$4635, $4490, $4365, $4185, $3935, $3785

(Y-axis $: 3600–4700; X-axis Project Day: 19–34)

Problem # 14.2   PHASE II SCHEDULE   Problem # 14.3

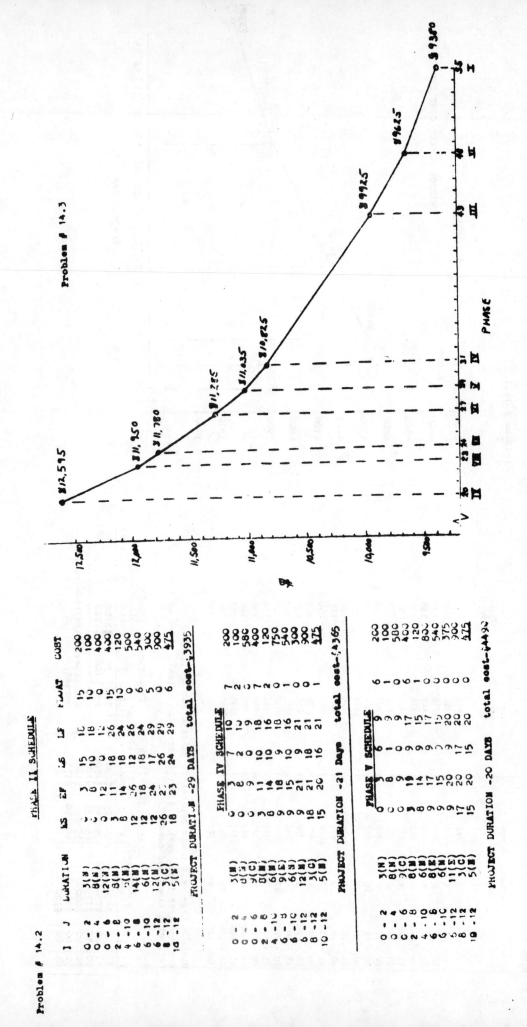

I J	DURATION	ES	EF	LS	LF	FLOAT	COST
0 - 2	3(N)	3	3	10	15	15	200
0 - 4	8(N)	3	8	18	18	12	100
0 - 6	12(N)	3	12	12	12	0	400
2 - 6	8(N)	8	11	18	26	15	400
4 - 10	3(N)	8	14	18	24	12	120
6 - 8	14(N)	12	26	12	26	0	500
6 - 10	6(N)	12	18	24	24	6	540
6 - 12	12(N)	12	24	17	29	5	300
8 - 12	3(C)	26	26	26	29	5	900
10 - 12	5(N)	18	23	24	29	6	475

PROJECT DURATION -29 DAYS   total cost--3935

FHASE IV SCHEDULE

	DURATION						COST
0 - 2	3(N)	3	3	7	10	7	200
0 - 4	8(N)	3	12	9	18	2	100
0 - 6	9(C)	0	9	0	9	0	580
2 - 6	8(N)	3	11	10	18	7	400
4 - 10	9(N)	9	14	10	16	2	120
6 - 8	9(E)	9	18	9	18	0	750
6 - 10	6(N)	9	15	10	16	1	540
6 - 12	12(N)	9	21	9	21	0	300
8 - 12	3(C)	18	21	18	21	0	900
10 - 12	5(N)	15	20	16	21	1	475

PROJECT DURATION -21 Days   total cost--4365

PHASE V SCHEDULE

	DURATION						COST
0 - 2	3(N)	0	3	6	9	6	200
0 - 4	8(N)	0	8	9	9	1	100
0 - 6	9(C)	0	9	17	17	0	580
2 - 6	8(N)	3	9	9	15	6	400
4 - 10	6(N)	8	14	9	17	1	120
6 - 6	8(E)	9	17	9	17	0	800
6 - 10	6(N)	9	15	12	20	0	540
6 - 12	11(C)	9	20	9	20	0	375
8 - 12	3(C)	17	20	17	20	0	900
10 - 12	5(N)	15	20	15	20	0	475

PROJECT DURATION -20 DAYS   total cost--4490

19

Notes on Problems # 14.3 and 14.4

THE KEY TO THESE PROBLEMS IS THE "PHASE" IN WHICH THE "FLOW CAPACITY" (SLOPE) IS REMOVED FROM ACTIVITY 18-21. In the forward pass of Phase III, we arrive at Event #6 with a $\frac{26}{15}$

Looking forward from Event #6, the parallel paths of 6-18,18-21 with 6-9, 9-21, both have the same total length of 4 days; but 6-9 and 9-21 are at their crash limit (2 days each). This is equivalent to looking at one arrow, 6-21, with the sum of the slopes of the parallel branches, or $50/Day (18-21) plus ∞ (both 6-9 and 9-21). $50/Day plus ∞ is the same as ∞. Thus, the backward pass of Phase III does not remove the capacity from 18-21. The same situation arises in Phase IV forward pass. We arrive at Event #6 with a $\frac{21}{5}$

At Event #18, two concurrent time paths arrive there at day 21. But only the least slope from 0-12, 12-15, 15-18, of $25/Day is carried thru 18-21. The backward pass of Phase IV removes only the $25/Day capacity from 18-21. Activity 18-21 is not reduced to its Crash limit until the backward pass of Phase VII.

Activities 0-7 and 7-9 always stay normal (The sum of their normal times, 5+7, equals the sum of crash durations into Event #9) and need not be considered in the cost curve generation.

PHASE III SCHEDULE — 43 Day Project Duration

I - J	Duration	ES	EF	LS	LF	TF	COST
0 - 3	12(N)	0	12	0	12	0	1800
0 - 7	5(N)	0	5	16	21	16	1000
0 - 12	8(N)	0	8	5	13	5	700
3 - 6	14(N)	12	26	12	26	0	645
6 - 18	2(C)	26	28	26	28	0	1150
6 - 18	0	26	26	26	26	0	--
7 - 9	7(N)	5	12	21	28	16	900
9 - 21	2(C)	28	30	28	30	0	620
12- 15	3(N)	8	11	13	16	5	550
15- 18	10(N)	11	21	16	26	5	1300
18- 21	4(N)	26	30	26	30	0	450
21- 24	13(N)	30	43	30	43	0	810

Total Cost: 9925

PHASE VII SCHEDULE — 24 Day Project Schedule

I - J	Duration	ES	EF	LS	LF	TF	COST
0 - 3	7(C)	0	7	0	7	0	2175
0 - 7	5(N)	0	5	4	9	4	1000
0 - 12	8(N)	0	8	0	8	0	700
3 - 6	7(E)	7	14	7	14	0	1325
6 - 18	0	14	16	14	16	0	1150
6 - 18	0	14	14	14	14	0	--
7 - 9	2(C)	5	7	12	16	4	900
9 - 21	2(C)	9	14	9	14	0	620
12- 15	1(C)	8	9	8	9	0	600
15- 18	5(C)	9	14	9	14	0	1525
18- 21	4(N)	14	18	14	18	0	450
21- 24	6(C)	18	24	18	24	0	1335

Total Cost: $11,780

CRASH SCHEDULE — 20 Day Project Schedule

I - J	Duration	ES	EF	LS	LF	TF	COST
0 - 3	7(C)	0	7	0	7	0	2175
0 - 7	5(N)	0	5	0	5	0	1000
0 - 12	5(C)	0	5	0	5	0	985
3 - 6	3(C)	7	10	7	10	0	1805
6 - 18	2(C)	10	12	10	12	0	1150
6 - 18	0	10	10	11	11	1	--
7 - 9	7(N)	5	12	5	12	0	900
9 - 21	2(C)	12	14	12	14	0	620
12- 15	1(C)	5	6	5	6	0	600
15- 18	5(C)	6	11	6	11	0	1525
18- 21	3(C)	11	14	11	14	0	500
21- 24	6(C)	14	20	14	20	0	1335

Total Cost: 12,595

Problem # 14.5

Phase I / Phase II / Phase III / Phase IV

I	J	M	C	Phase I					Phase II					Phase III					Phase IV				
				T_{Ej}	LSCF	NEW B	NEW	D	T_{Ej}	LSCF	NEW B	NEW	D	T_{Ej}	LSCF	NEW B	NEW	D	T_{Ej}	LSCF	NEW B	NEW	D
10	15	5	2	8	150	8	2	2	8	8	8	2	2	5	8	8	2	2	5	8	8	2	2
15	20	5	5	13	150	215	5	5	10	215	205	5	5	10	205	185	5	5	10	185	125	5	5
20	30	5	1	15	150	175	2	2	12	175	175	2	2	12	175	175	2	2	12	175	175	2	2
20	35	6	2	19	150	550	6	6	16	215	540	6	6	16	205	520	6	6	16	185	460	6	6
35	45	8	7	27	150	350	8	8	24	340	320	8	8	24	205	320	8	8	24	185	260	8	8
30	50	7	2	22	150	200	7	7	19	175	200	7	7	19	200	200	7	7	19	175	200	7	7
40	50	5	7	22	150	175	3	3	19	175	175	3	3	19	175	175	3	3	19	175	175	3	3
45	55	9	4	34	150	100	7	7	31	100	70	7	7	31	90	70	7	7	31	70	70	9	9
50	60	10	6	38	150	200	9	9	28	175	200	9	9	28	175	200	9	9	28	200	200	1	1
55	65	9	5	48	150	10	4	4	35	10	8	1	1	42	90	8	6	6	32	70	8	6	6
65	80	10	7	53	150	30	10	10	45	30	20	10	10	47	20	220	9	9	38	70	220	5	5
80	95	5	3	55	150	220	5	5	50	220	220	5	5	49	20	60	5	5	43	60	88	7	7
85	100	7	3	59	150	90	7	7	52	10	80	7	7	53	20	-	7	7	45	60	88	5	5
100	110	4	4		150	-	4	4	56	10	-	4	4						49	60	-	4	4

Indirect Cost Equals $30/Day insurance benefit for adults, plus $40/Day insurance benefit for children, plus $130/Day for Construction Supervisor, or $200/Day.

Project Duration	Direct Cost	Indirect Cost	Total Cost
59	$20,640	$11,800	$32,440
56	21,090	11,200	32,290
53	21,570	10,500	32,170
* 49	22,290	9,800	32,090 *
47	22,770	9,400	32,170
43	23,770	8,600	32,370
41	24,500	8,200	32,700
40	24,960	8,000	32,960
36	27,760	7,200	34,960
35	28,610	7,000	35,610
32	31,310	6,400	37,710

Least Total Cost Schedule to XYZ Insurance Co is the 49 day Project Schedule. Individual activity schedules are shown on left.

Benefit/Cost Ratio: Reduction in Indirect Costs / Additional Direct Cost

$$\frac{11,800 - 9,800}{22,290 - 20,640} = \frac{2,000}{1,650} = 1.21$$

Problem # 14.5

Least Cost Schedule, Building the House for XYZ Co.

49 Day Project Schedule

I	J	DUR	ES	EF	LS	LF	TF
5	10	3	0	3	0	3	0
5	70	4	0	4	19	23	19
10	15	2	3	5	3	5	0
10	70	3	3	6	20	23	17
15	20	5	5	10	5	10	0
15	25	3	5	8	22	25	17
15	45	4	5	9	20	24	15
20	30	2	10	12	14	16	4
20	35	6	10	16	10	16	0
25	75	10	8	18	25	35	17
30	40	7	12	19	16	23	4
30	50	0	12	12	20	20	8
35	45	8	16	24	16	24	0
35	50	3	16	19	20	23	4
40	50	5	19	24	23	28	4
45	55	7	24	31	24	31	0
50	60	9	19	28	23	32	4
55	65	1	31	32	31	32	0
60	65	0	28	28	32	32	7
60	75	6	28	38	32	38	0
65	75	0	32	4	35	35	31
70	90	12	4	16	23	35	19
75	85	0	28	31	38	38	7
80	85	3	38	38	38	38	0
80	95	5	38	43	40	45	2
80	100	4	38	42	41	45	3
85	100	7	38	45	38	45	0
90	100	10	16	26	35	45	19
95	100	0	43	43	45	45	2
100	110	4	45	49	45	49	0

Problem # 17.1

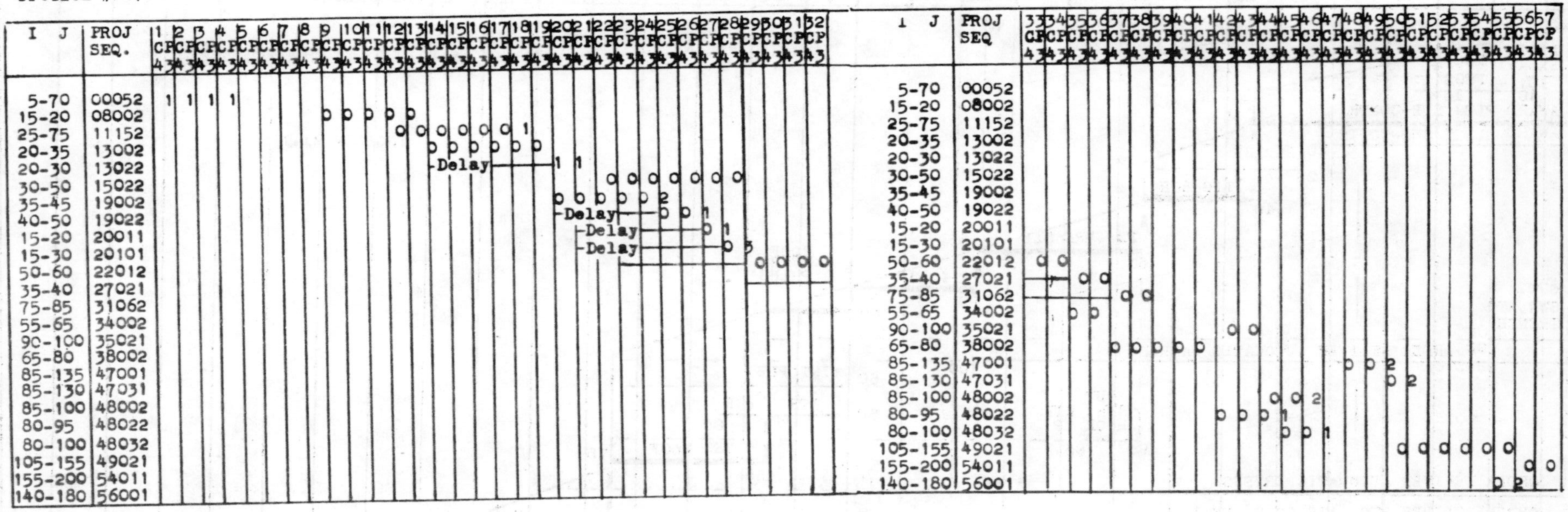

I J	PROJ SEQ.	1 CP 43	2 CP 43	3 CP 43	4 CP 43	5 CP 43	6 CP 43	7 CP 43	8 CP 43	9 CP 43	10 CP 43	11 CP 43	12 CP 43	13 CP 43	14 CP 43	15 CP 43	16 CP 43	17 CP 43	18 CP 43	19 CP 43	20 CP 43	21 CP 43	22 CP 43	23 CP 43	24 CP 43	25 CP 43	26 CP 43	27 CP 43	28 CP 43	29 CP 43	30 CP 43	31 CP 43	32 CP 43
5-70	00052	1	1	1	1																												
15-20	08002									0	0	0	0	0																			
25-75	11152												0	0	0	0	0	0	1														
20-35	13002														0	0	0	0	0	0													
20-30	13022														-Delay						1	1											
30-50	15022																						0	0	0	0	0	0	0				
35-45	19002																				0	0	0	0	0	2							
40-50	19022																				-Delay					0	0	1					
15-20	20011																					-Delay						0	1				
15-30	20101																					-Delay							0	3			
50-60	22012																													0	0	0	0
35-40	27021																																
75-85	31062																																
55-65	34002																																
90-100	35021																																
65-80	38002																																
85-135	47001																																
85-130	47031																																
85-100	48002																																
80-95	48022																																
80-100	48032																																
105-155	49021																																
155-200	54011																																
140-180	56001																																

I J	PROJ SEQ	33 CP 43	34 CP 43	35 CP 43	36 CP 43	37 CP 43	38 CP 43	39 CP 43	40 CP 43	41 CP 43	42 CP 43	43 CP 43	44 CP 43	45 CP 43	46 CP 43	47 CP 43	48 CP 43	49 CP 43	50 CP 43	51 CP 43	52 CP 43	53 CP 43	54 CP 43	55 CP 43	56 CP 43	57 CP 43
5-70	00052																									
15-20	08002																									
25-75	11152																									
20-35	13002																									
20-30	13022																									
30-50	15022																									
35-45	19002																									
40-50	19022																									
15-20	20011																									
15-30	20101																									
50-60	22012	0	0																							
35-40	27021			0	0																					
75-85	31062					0	0																			
55-65	34002			0	0																					
90-100	35021										0	0														
65-80	38002					0	0	0	0	0																
85-135	47001																0	0	2							
85-130	47031																		0	2						
85-100	48002												0	0	2											
80-95	48022										0	0	0	1												
80-100	48032													0	0	1										
105-155	49021																		0	0	0	0	0	0		
155-200	54011																								0	0
140-180	56001																							0	2	

Proj. Priority # 1 (Last Digit): Car Wash, which has 12 Day Lag in Start

Proj. Priority # 2 (Last Digit): House

Only activities involving carpenters and plumbers are scheduled. Remainders shown per day after assignment. Starting times are based on logical sequence. However, scheduled duration may differ from original estimate, since schedule is based on man-days. For example, 65-80 of the house project was originally estimated for 10 days with 2 carpenters. 4 carpenters were found, thus it was scheduled above in 5 days(both are 20 man day totals). There is an improvement in completion date with the availablity of 4 carpenters and 3 plumbers.

Problem 17.2

Even with the change in project priority, the sequence remains the same, thus the schedule found in 17.1 above will also pertain.

I J	PROJ. SEQ.	I J	PROJ SEQ	I J	PROJ SEQ.		
5-70	00051	35-45	19001	75-85	31061	85-100	48001
15-20	08001	40-50	19021	55-65	34001	80-95	48021
25-75	11151	15-20	20012	90-100	35022	80-100	48031
20-35	13001	15-30	20102	65-80	38001	105-155	49022
20-30	13021	50-60	22011	85-135	47002	155-200	54012
30-50	15021	35-40	27022	85-130	47032	140/180	56001

All predecessor activities not involving carpenters and plumbers are assumed to have been scheduled with their original normal duration.

The Project Sequence is based on: 1st Two Digits - Early Start Time
2nd Two Digits - Distribiuted Float
Last Digit - Project Priority

Problem # 16.1

"J" of Final Event: = 936
Network numbered in steps of 3's
No. of Events(approx): 936/3 = 312 Events
No. of arrows(approx): 312 x 1.5 = 470 arrows
K 936+470 = 1406
K of IBM #1620 computer = 1614, so it can be used to calculate the schedule of this project.

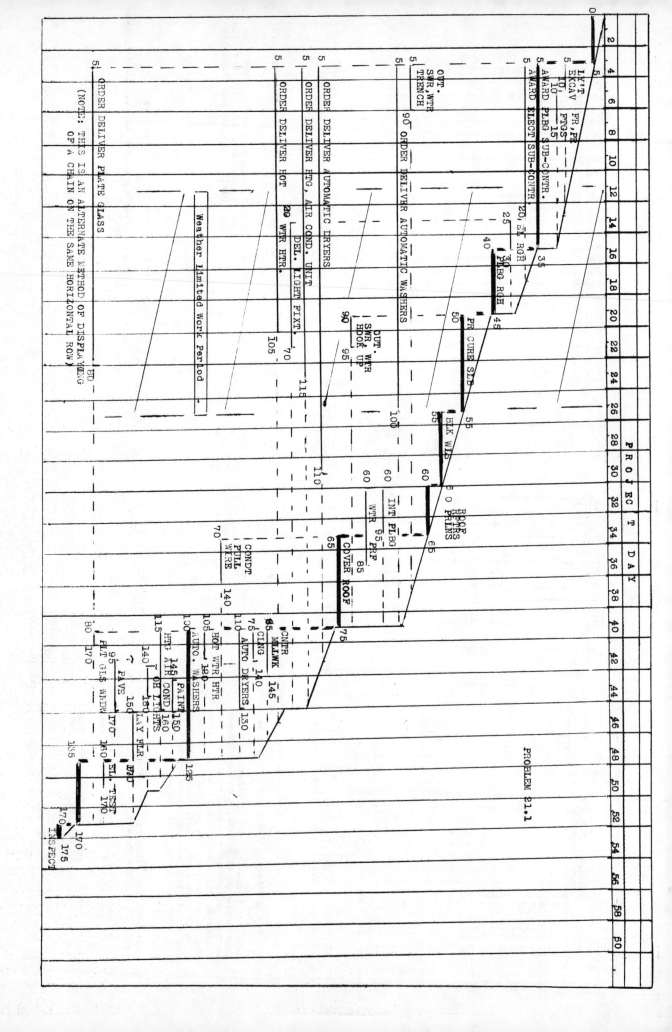

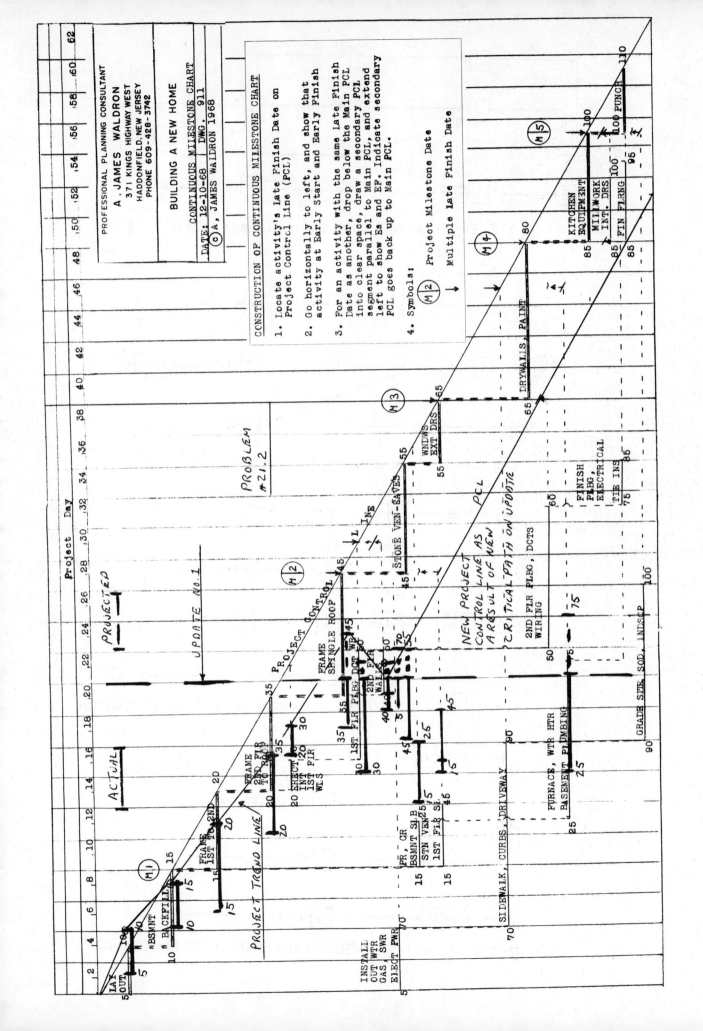

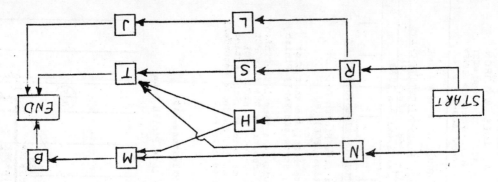

c. Precedence Diagram for Problem 2.4

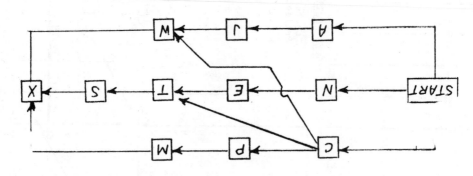

b. Precedence Diagram for Problem 2.3

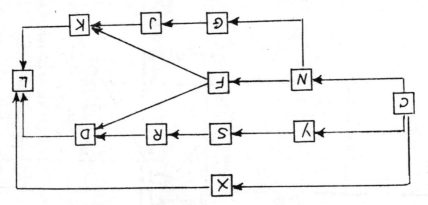

PROBLEM 20.2 a. Precedence Diagram for Problem 2.2

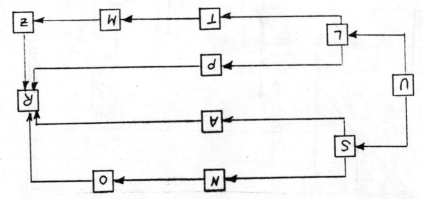

PROBLEM 20.1 (Precedence Diagram for Problem 2.1)